Lecture Notes in Physics

Bisher erschienen/Already published

Vol. 1: J. C. Erdmann. Wärmeleitung in Kristallen, theoretische Grundlagen und fortgeschrittene experimentelle Methoden. II, 283 Seiten. 1969.

Vol. 2: K. Hepp, Théorie de la renormalisation. III, 215 pages. 1969.

Vol. 3: A. Martin, Scattering Theory: Unitarity, Analyticity and Crossing. IV, 125 pages. 1969.

Vol. 4: G. Ludwig, Deutung des Begriffs „physikalische Theorie" und axiomatische Grundlegung der Hilbertraumstruktur der Quantenmechanik durch Hauptsätze des Messens. 1970. Vergriffen.

Vol. 5: Schaaf, The Reduction of the Product of Two Irreducible Unitary Representations of the Proper Orthochronous Quantummechanical Poincare Group. IV, 120 pages. 1970.

Vol. 6: Group Representations in Mathematics and Physics. Edited by V. Bargmann. V, 340 pages. 1970.

Vol. 7: R. Balescu, J. L. Lebowitz, I. Prigogine, P. Résibois, Z. W. Salsburg, Lectures in Statistical Physics. V, 181 pages. 1971.

Vol. 8: Proceedings of the Second International Conference on Numerical Methods in Fluid Dynamics. Edited by M. Holt. 1971. Out of print.

Vol. 9: D. W. Robinson, The Thermodynamic Pressure in Quantum Statistical Mechanics. V, 115 pages. 1971.

Vol. 10: J. M. Stewart, Non-Equilibrium-Relativistic Kinetic Theory. III, 113 pages. 1971.

Vol. 11: O. Steinmann, Pertubation Expansions in Axiomatic Field Theory. III. 126 pages. 1976.

Vol. 12: Statistical Models and Turbulence. Edited by C. Van Atta and M. Rosenblatt. Reprint of the First Edition. VIII, 492 pages. 1975.

Vol. 13: M. Ryan, Hamiltonian Cosmology. VII, 169 pages. 1972.

Vol. 14: Methods of Local and Global Differential Geometry in General Relativity. Edited by D. Farnsworth, J. Fink, J. Porter, and A. Thompson. V, 188 pages.

Vol. 15: M. Fierz, Vorlesungen zur Entwicklungsgeschichte der Mechanik. V, 97 Seiten. 1972.

Vol. 16: H.-O. Goergii, Phasenübergang 1. Art bei Gittergasmodellen. IX, 167 Seiten. 1972.

Vol. 17: Strong Interaction Physics. Edited by W. Rühl and A. Vancura. V, 405 pages. 1973.

Vol. 18: Proceedings of the Third International Conference on Numerical Methods in Fluid Mechanics, Vol. I. Edited by H. Cabannes and R. Temam. VII, 186 pages. 1973.

Vol. 19: Proceedings of the Third International Conference on Nemerical Methods in Fluid Mechanics, Vol. II. Edited by H. Cabannes and R. Temam. VII, 275 pages. 1973.

Vol. 20: Statistical Mechanics and Mathematical Problems. Edited by A. Lenard. VIII, 247 pages. 1973.

Vol. 21: Optimization and Stability Problems in Continuum Mechanics. Edited by P. K. C. Wang. V, 94 pages. 1973.

Vol. 22: Proceedings of the Europhysics Study Conference on Intermediate Processes in Nuclear Reactions. Edited by N. Cindro, P. Kulišic and Th. Mayer-Kuckuk. XIV, 329 pages. 1973.

Vol. 23: Nuclear Structure Physics. Proceedings 1973. Edited by U. Smilansky, I. Talmi, and H. A. Weidenmüller. XII, 296 pages. 1973.

Vol. 24: R. F. Snipes, Statistical Mechanical Theory of the Electrolytic Transport of Non-electrolytes. V, 210 pages. 1973.

Vol. 25: Constructive Quantum Field Theory. The 1973 "Ettore Majorana" International School of Mathematical Physics. Edited by G. Velo and A. Wightman. III. 331 pages. 1973.

Vol. 26: A. Hubert, Theorie der Domänenwände in geordneten Medien. XII, 377 Seiten. 1974.

Vol. 27: R. K. Zeytounian, Notes sur les Ecoulements Rotationnels de Fluides Parfaits. XIII, 407 pages. 1974.

Vol. 28: Lectures in Statistical Physics. Edited by W. C. Schieve and J. S. Turner. V. 342 pages. 1974.

Vol. 29: Foundations of Quantum Mechanics and Ordered Linear Spaces. Advanced Study Institute, Marburg 1973. Edited by A. Hartkämper and H. Neumann. VI, 355 pages. 1974.

Vol. 30: Polarization Nuclear Physics. Proceedings 1973. Edited by D. Fick. IX, 292 pages. 1974.

Vol. 31: Transport Phenomena. Sitges International Schools of Statistical Mechanics, June 1974. Edited by G. Kirczenow and J. Marro. XIV, 517 pages. 1974.

Lecture Notes in Physics

Edited by J. Ehlers, München, K. Hepp, Zürich,
H. A. Weidenmüller, Heidelberg, and J. Zittartz, Köln
Managing Editor: W. Beiglböck, Heidelberg

54

Critical Phenomena

Sitges International School on
Statistical Mechanics, June 1976
Sitges, Barcelona/Spain
Director: L. Garrido

Edited by J. Brey and R. B. Jones

Springer-Verlag
Berlin Heidelberg GmbH 1976

Director
Luis Garrido
Instituto de Fisica Teórica
Diagonal, 647
Barcelona–14/Spain

Editors
Javier Brey
Universitade Sevilla
Facultad de Ciencias
Sevilla/Spain

Robert B. Jones
Queen Mary College
University of London
Mile End Road
London/Great Britain

Library of Congress Cataloging in Publication Data

Sitges International School of Statical Mechanics, 1976.
Critical phenomena.

(Lecture notes in physics ; 54)
Includes bibliographical references.
1. Critical phenomena (Physics)
I. Brey, Javier, 1947- II. Jones, Robert B., 1940- III. Title. IV. Series.
QC173.4.C74S57 1976 536'.401 76-29038

ISBN 978-3-540-07862-3 ISBN 978-3-540-38106-8 (eBook)
DOI 10.1007/978-3-540-38106-8

Originally published by Springer-Verlag Berlin Heidelberg New York in 1976

ACKNOWLEDGEMENTS

The Sitges International School of Statistical Mechanics was honored by the patronage of His Majesty the King, Juan Carlos I.

Acknowledgement must be made to those institutions without whose ready cooperation and financial assistance the School would not have been possible. I wish to express my particular gratitude to:

Ministerio de Educación y Ciencia
Ayuntamiento de Barcelona
Diputación de Barcelona
Rectorado de la Universidad de Barcelona
Ayuntamiento de Sitges

Finally I must not fail to thank all those who constantly encouraged and helped me and so ensured the success of this venture. I especially would mention Prof. de la Rubia, Dr. Martínez Sardá, Dr. Seglar and, in particular, my wife.

L. Garrido

CONTENTS

LECTURERS

Prof. R. BROUT, Brussels, Belgium
Prof. Ch. ENZ, Geneva, Switzerland
Prof. G. GALLAVOTTI, Rome, Italy
Prof. M.S. GREEN, Philadelphia, USA
Prof. R. HAAG, Hamburg, Germany
Prof. L.P. KADANOFF, Providence, USA
Prof. J.L. LEBOWITZ, New York, USA
Prof. S.K. MA, Saclay, France
Prof. S. MIRACLE-SOLE, Zaragoza, Spain
Prof. P. SZEPFALUSY, Budapest, Hungary
Prof. F. WEGNER, Heidelberg, Germany

PARTICIPANTS

Mr. K. AMSLER, Bern, Switzerland
Miss M. AUBERSON, Neuchâtel, Switzerland
Dr. M. AUSLOOS, Liège, Belgium
Dr. P. AUSLOOS-CLIPPE, Liège, Belgium
Dr. G. BASKARAN, Bangalore, India
Dr. J. BASTEN, Petten, Holland
Prof. A. BERNALTE, Bilbao, Spain
Dr. J. BIEL, Barcelona, Spain
Prof. L.J. BOYA, Zaragoza, Spain
Mr. J. BRICMONT, Louvain-le-Neuve, Belgium
Dr. A. BUNDE, Antwerp, Belgium
Mr. C. CAMAROTTA, Naples, Italy
Mr. V. CANIVELL, Barcelona, Spain
Mr. H. CAPES, Fontenay-aux-Roses, France
Dr. P. CHANDRA SEKHAR, Bangalore, India
Dr. N. CLAVAGUERA, Barcelona, Spain
Dr. A. CORDOBA, Sevilla, Spain
Dr. P. DAMAY, Lille, France
Dr. F. DEBACKER-MATHOT, Louvain-le-Neuve, Belgium

Mr. B. DEMOEN, Heverlee, Belgium
Mr. E. ELIZALDE, Barcelona, Spain
Prof. R. FERNÁNDEZ ALVAREZ-ESTRADA, Madrid, Spain
Dr. R. FIGARI, Naples, Italy
Miss M. FILIPE, Oxford, England
Dr. R. FOLK, Linz, Austria
Dr. B. GIDAS, Seattle, USA
Mr. E. GLÖTZL, Linz, Austria
Dr. H. GRABERT, Stuttgart, Germany
Dr. J. GOMIS, Barcelona, Spain
Prof. J.D. GUNTON, Philadelphia, USA
Prof. H. HAHN, Braunschweig, Germany
Mr. D. HELLWIG, Essen, Germany
Mrs. D. HERRMAN-RONZAUD, Grenoble, France
Dr. A. HUBER, Kiel, Germany
Mr. D. JOU MIRABENT, Barcelona, Spain
Mr. H. KING, Essen, Germany
Prof. N. KUMAR, Trieste, Italy
Dr. M. LEAL, Valladolid, Spain
Dr. K. LENDI, Zürich, Switzerland
Prof. F. di LIBERTO, Naples, Italy
Mr. A. MAC LEOD, St. Andrews, Scotland
Mr. J.M. MARCAIDE, Zaragoza, Spain
Miss M. MARQUES, Porto, Portugal
Dr. J. MARRO, Barcelona, Spain
Mr. J. MATEOS, Zaragoza, Spain
Dr. D. MAYER, Bures-sur-Yvette, France
Dr. N. MENYHARD-SZEPFALUSY, Budapest, Hungary
Dr. A. MESSAGER, Marseille, France
Dr. G. MONROY, Naples, Italy
Prof. M.T. MORA, Valladolid, Spain
Dr. R. MORF, Zürich, Switzerland
Dr. M. MURMANN, Heidelberg, Germany
Dr. Ch. NAPPI, Naples, Italy
Dr. J. NAUDTS, Antwerp, Belgium
Dr. M. PARAMIO, Zaragoza, Spain
Mr. J.M. PARRA, Barcelona, Spain
Prof. A. PAVLOVIC, Grenoble, France
Mr. L. PESQUERA, Valladolid, Spain
Dr. E.B. POHLMEYER, Hamburg, Germany
Dr. M. QUIROS, Madrid, Spain

Dr. M. RAMON, Madrid, Spain
Dr. D. REISETBAUER, Linz, Austria
Dr. G. REITER, Göteborg, Sweden
Mrs. M. REYNAERT, Heverlee, Belgium
Dr. M. RIVAS, Bilbao, Spain
Prof. E. ROJAS, Sevilla, Spain
Dr. V. RUIZ, Valencia, Spain
Mr. J.M. SANCHO, Barcelona, Spain
Mr. M. SAN MIGUEL, Barcelona, Spain
Prof. E. SANTOS, Valladolid, Spain
Miss M. SANTOS, Birmingham, England
Dr. J. SARBACH, Zürich, Switzerland
Dr. J. SCHOUTEN, Amsterdam, Holland
Dr. P. SEGLAR, Barcelona, Spain
Dr. P. SHEPHERD, Exeter, England
Dr. P. SHUKLA, Oxford, England
Mr. L. SNEDDON, Oxford, England
Prof. P. TOMBESI, Rome, Italy
Miss K. UZELAC, Zagreb, Yugoslavia
Dr. S. VELASCO, Salamanca, Spain
Prof. M. ZANNETTI, Salerno, Italy
Mr. T. ZIMAN, Oxford, England

DIRECTOR

Prof. L. GARRIDO, Barcelona, Spain

EDITORS

Prof. J. BREY, Sevilla, Spain
Dr. R.B. JONES, London, England

SECRETARIES

Mrs. Ch. BALY, Geneva, Switzerland
Mrs. R. CHESTER, Edinburgh, Scotland
Miss P. UDINA, Barcelona, Spain

CRITICAL PHENOMENA AND SCALE INVARIANCE

F.J. WEGNER

Institut für Theoretische Physik
Universität Heidelberg

Critical Phenomena and Scale Invariance

F.J. WEGNER

Institut für Theoretische Physik
Universität Heidelberg

The concept of scale invariance has turned out to be a very fruitful idea to explain critical phenomena. This idea gives a very intuitive picture of the behaviour in the critical region. It is based on the idea of a fixed point hamiltonian which is invariant under change of the length scale. This theory has confirmed most of the phenomenological assumptions and heuristic observations on critical systems, and has reproduced the features of exact model solutions. Moreover, the theory gave a deeper insight into the complicated non-analytic behaviour at the critical point.

In view of the numerous papers on this subject the reader is in many cases referred to the review articles [1-10] and the extensive literature cited therein. Whereas the references [1-3] appeared before Wilson's formulation of the renormalization group the references [4-10] report on this concept and its consequences.

I. Scaling : Phenomenological Description [1-3], [9], [11]

1.1. Critical Point

Let us consider a system like a ferromagnet or a liquid-vapour system at the critical point. Through a microscope we will observe domains (or droplets) of various size. There will be droplets in bubbles and bubbles in droplets since the critical point is characterized by strong fluctuations. Now let us switch to a lower magnification and reduce the contrast appropriately. Then we will observe essentially the same picture, that is the same distribution of domain (or droplets) sizes as before. Our system is scale invariant. Suppose the original magnification yielded a correlation of the order parameter S(r)

$$\langle S(0)S(r)\rangle_{crit.} = g(r) . \tag{1.1}$$

Then by changing the magnification we change the length scale by a factor b

$$r = br' \tag{1.2}$$

and the density of the order parameter by

$$S(r)d^d r = \tilde{S}(r')d^d r' = b^{-d}\tilde{S}(r')d^d r$$

where d is the dimensionality of the system. The change in the contrast adds another factor we call b^{y_h}, thus

$$S(r) = b^{y_h - d} S'(r') \tag{1.3}$$

and

$$\langle S(0)S(r)\rangle = b^{2y_h - 2d} \langle S'(0)S'(r')\rangle . \tag{1.4}$$

Since, however, the observed picture did not change we have

$$\langle S'(0)S'(r')\rangle_{crit.} = g(r') \tag{1.5}$$

which yields

$$g(r) = b^{2(y_h - d)} g(r/b) . \tag{1.6}$$

This equation holds for a change of the length scale (magnification) by a factor b. Obviously we may apply it p times

$$g(r) = b^{2(y_h-d)p} g(\frac{r}{b^p}) \tag{1.7}$$

and we may choose b close to unity. Then with

$$b^p = L \tag{1.8}$$

one obtains for arbitrary change of the length scale L

$$g(r) = L^{2(y_h-d)} g(\frac{r}{L}) \ . \tag{1.9}$$

Choosing finally L by L=r one obtains the power law for the order parameter correlation at criticality

$$\langle S(0)S(r)\rangle_{crit.} = \frac{g(1)}{r^{d-2+\eta}} \tag{1.10}$$

with the critical exponent

$$\eta = d + 2 - 2y_h \ . \tag{1.11}$$

1.2. Critical Region

In the critical region our system is characterized by the temperature difference $\tau = (T-T_c)/T_c$ and by a magnetic field h. Again observe the sample through the microscope. Switching the magnification and the contrast as before we will observe a different picture : The system is no longer scale invariant. However, the lower magnification will show the same picture as a sample determined by τ' and h' in the original magnification. Thus the change of the length scale and the order parameter is equivalent to a change of τ and h to τ' and h'. In particular one obtains for the correlation function

$$\begin{aligned} G(r,\tau,h) &= \langle S(0)S(r)\rangle_{\tau,h} \\ &= b^{2(y_h-d)} \langle S'(0)S'(r')\rangle_{\tau',h'} = b^{2(y_h-d)} G(\frac{r}{b},\tau',h') \ . \end{aligned} \tag{1.12}$$

Away from criticality correlation functions decay at large distances with an exponential

$$G(r,\tau,h) \sim \exp(-r/\xi(\tau,h)) \ . \tag{1.13}$$

From equation (1.12) it follows that the correlation length ξ obeys

$$\xi(\tau,h) = b\,\xi(\tau',h') \ . \tag{1.14}$$

By the same token one observes that the mean magnetization obeys

$$m(\tau,h) = \langle S(r)\rangle_{\tau,h} = b^{y_h-d} m(\tau',h') \tag{1.15}$$

and a similar result is obtained for the susceptibility

$$\chi(\tau,h) = \int \langle (S(0)-m)\ (S(r)-m)\rangle_{\tau,h} d^d r$$

$$= b^{2y_n-d} \int \langle (S'(0)-m')\ (S'(r')-m')\rangle_{\tau',h'} d^d r' \qquad (1.16)$$

$$= b^{2y_n-d}\ \chi(\tau',h') \ .$$

It will turn out later that τ' and h' are non singular functions of τ and h. Since $\tau=0$, h=0 implies $\tau'=0$, $h'=0$ we may write in lowest order

$$\tau' = \lambda_E\ \tau + \lambda' h \qquad (1.17a)$$

$$h' = \lambda''\ \tau + \lambda_h\ h \quad . \qquad (1.17b)$$

A change of sign in the magnetic field h will change sign of h' but will leave τ' invariant. Thus, $\lambda' = \lambda'' = 0$. Consider the case $\tau=0$.

Then apparently

$$m(0,h) = b^{y_h-d} m(0, \lambda_h h) \qquad (1.18)$$

$$\chi(0,h) = b^{2y_h-d}\ \chi(0, \lambda_h h) \quad . \qquad (1.19)$$

Differentiate (1.18) with respect to h

$$\chi(0,h) = b^{y_h-d}\ \lambda_h\ \chi(0, \lambda_h h) \quad . \qquad (1.20)$$

Comparison of eqs. (1.19) and (1.20) yields

$$\lambda_h = b^{y_h} \quad . \qquad (1.21)$$

Let us write in a similar way

$$\lambda_E = b^{y_E} \quad . \qquad (1.22)$$

Choosing $b^p = L$ as before one obtains the homogeneity laws

$$G(r, \tau,h) = L^{2(y_h-d)}\ G(\tfrac{r}{L}, \tau L^{y_E}, hL^{y_h}) \qquad (1.23)$$

$$\xi(\tau,h) = L\ \xi(\tau L^{y_E}, hL^{y_h}) \qquad (1.24)$$

$$m(\tau,h) = L^{y_h-d} m(\tau L^{y_E}, hL^{y_h}) \tag{1.25}$$

$$\chi(\tau,h) = L^{2y_h-d} \chi(\tau L^{y_E}, hL^{y_h}) \quad . \tag{1.26}$$

Equation (1.25) may be integrated with respect to h yielding the density of the free energy

$$F_{sing}(\tau,h) = L^{-d} F_{sing}(\tau L^{y_E}, hL^{y_h}) \quad . \tag{1.27}$$

It reflects that the density of the free energy changes by the factor L^d of the change of the volume, thus conserving the total free energy.

To be precise F is defined as the free energy $\mathcal{F}$ divided by V and kT

$$F = \mathcal{F}/(kTV) \quad . \tag{1.28}$$

The function F_{sing} in equation (1.27) is not the complete density of the free energy. There is a regular part to F which does not contribute to the critical behaviour.

$$F = F_{sing}(\tau,h) + F_{reg}(\tau,h) \quad . \tag{1.29}$$

Differentiating (1.27) twice with respect to τ one obtains the singular part of the specific heat

$$C_{sing}(\tau,h) = L^{2y_E-d} C_{sing}(\tau L^{y_E}, hL^{y_h}) \quad . \tag{1.30}$$

One usually introduces a number of critical exponents to describe the critical behaviour as a function of τ for vanishing magnetic field. The choice $|\tau| L^{y_E}=1$ yields the power laws

$$\xi(\tau,0) = A_{\xi\pm} |\tau|^{-\nu} \tag{1.31a}$$

$$m(\tau,0) = A_m |\tau|^{\beta} \tag{1.31b}$$

$$\chi(\tau,0) = A_{\chi\pm} |\tau|^{-\gamma} \tag{1.31c}$$

$$C_{sing}(\tau,0) = A_{C\pm} |\tau|^{-\alpha} \tag{1.31d}$$

with

$$\nu = 1/y_E \tag{1.32a}$$

$$\beta = (d-y_h)/y_E \tag{1.32b}$$

$$\gamma = (2y_h-d)/y_E \tag{1.32c}$$

$$\alpha = (2y_E - d)/y_E \quad . \tag{1.32d}$$

Since only two scaling exponents y_h and y_E determine $\alpha, \beta, \gamma, \eta$ and ν three independent relations hold

$$\alpha + 2\beta + \gamma = 2 \tag{1.33a}$$

$$2 - \alpha = d\nu \tag{1.33b}$$

$$\gamma = \nu(2 - \eta) \quad . \tag{1.33c}$$

These relations are called scaling laws.

II. Renormalization Group Equation

2.1. Basic Idea

In the preceding section we developed an intuitive picture for the homogeneity laws in critical phenomena. In this section we will give a justification of this picture on the basis of renormalization group ideas (Kadanoff [11] , Wilson [12] , Wegner [13]). The observation that at criticality the correlation function is scale invariant suggests that the effective interaction at criticality is invariant with respect to the change of the length scale, too. We call the procedure which changes the scale of the hamiltonian (effective interaction) renormalization group (RG) procedure and the corresponding transformation is called RG transformation.

We denote the hamiltonian function $\mathcal{H}$ and introduce

$$H = \mathcal{H}/(k_B T) \; . \tag{2.1}$$

For simplicity's sake we call H hamiltonian (or effective interaction). From the definition of F, eq. (1.28), we have

$$- F = \frac{1}{V} \ln \text{trace} \exp (-H) \; . \tag{2.2}$$

The RG transformation consists of two parts

(i) a change of the length scale by a factor $b = e^{\ell}$ in all linear dimensions (we leave the partition function $Z = \text{trace} \exp(-H)$ invariant). Since the volume shrinks by a factor $e^{-d\ell}$ we obtain

$$F_0 = e^{-d\ell} F_\ell \; . \tag{2.3}$$

(ii) an elimination and/or transformation of the degrees of freedom S which leaves the free energy invariant.

This second step is necessary to obtain a hamiltonian which is comparable with the original hamiltonian. The density of degrees of freedom S has to be reduced to the original one by thinning out degrees of freedom which describe short range fluctuations (Fourier components with large wave vector). On the other hand, the system has to be extended to the volume of the original system. Then a one-to-one correspondence of the original and the new variables S is possible and the hamiltonians are comparable. Considering our sample through the microscope we observe in lower magnification less details which corresponds to the elimination process. On the other hand, we observe a large part of the sample which corresponds to the extension of the volume.

The total RG transformation transforms H_o into H_ℓ

$$H_\ell = R_\ell (H_o) \quad . \tag{2.4}$$

$$F(H_o) = e^{-d\ell} F(H_\ell) \tag{2.5}$$

In many cases it is possible to formulate the RG procedure for infinitesimal ℓ. Then equation (2.4) can be cast in differential form :

$$\frac{dH}{d\ell} = \mathcal{G} H \tag{2.6}$$

where $\mathcal{G}$ is the generator of the RG. Generally the generator $\mathcal{G}$ acts in a nonlinear way on H.

2.2. Fixed Point, Classification of Operators

Now two assumptions are made :

(i) It is assumed that a fixed point hamiltonian H^* exists

$$\mathcal{G} H^* = 0 \quad . \tag{2.7}$$

This is a hamiltonian which maps into itself.

(ii) It is assumed for a critical hamiltonian

$$\lim_{\ell \to \infty} H_\ell = H^* \quad . \tag{2.8}$$

This is a much weaker condition than to demand that at criticality the system is described by H^*. Condition (2.8) demands only that the effective hamiltonian approaches asymptotically a fixed point upon application of the RG.

To discuss the behaviour around criticality we split the generator in a linear part $\mathcal{L}$ and contributions of higher order Q .

$$\mathcal{G}(H^* + \Delta H) = \mathcal{L} \Delta H + Q(\Delta H) \tag{2.9}$$

where

$$Q(\Delta H) = O(\Delta H)^2 \quad . \tag{2.10}$$

We define eigenoperators O_i by the eigenvalue equation

$$\mathcal{L} O_i = y_i \, O_i \tag{2.11}$$

and assume in the following that the eigenoperators form a complete set of operators so that any hamiltonian H_o can be expanded

$$H_o = H^* + \sum_i \mu_i O_i \quad . \tag{2.12}$$

Then we obtain in linear order in μ

$$H_\ell = H^* + \sum_i \mu_i e^{y_i \ell} O_i \quad . \tag{2.13}$$

Corresponding to the eigenvalues y one distinguishes

$$\begin{aligned} y &> 0 \quad \text{relevant operators,} \\ y &= 0 \quad \text{marginal operators,} \\ y &< 0 \quad \text{irrelevant operators.} \end{aligned} \tag{2.14}$$

From equation (2.13) one finds immediately that at the critical point the fields (in high energy physics sources) μ_i of all relevant operators have to vanish.

Marginal operators may act as relevant or irrelevant operators depending on whether the nonlinear terms will drive μ_i to zero or away from zero. There is also the rare case that a marginal operator yields a line of fixed points as apparently in the eight-vertex-model.

There is a special operator, the constant (that is independent of S) contribution $\mu_o V$, which formally has

$$O_o = V \, , \; y_o = d \tag{2.15}$$

since V shrinks by a factor $e^{-d\ell}$ which is compensated by a growth of μ_o by a factor $e^{d\ell}$. However, the addition of a constant to the hamiltonian does not change its critical behaviour. Therefore $\mu_o=0$ is not necessary for criticality. This is the origin of the regular part of the free energy.

The type of the critical behaviour depends on the number of symmetry conserving relevant operators. (Symmetry conserving means that the symmetry of the hamiltonian is conserved, it does not exclude a spontaneously broken symmetry of the system). Let us expand

$$\mathcal{H} = \sum_i \mu_i^o O_i \quad , \tag{2.16}$$

$$H^* = \sum \mu_i^* O_i \quad , \tag{2.17}$$

then we obtain

$$\mu_i = \beta \mu_i^o - \mu_i^* \, . \tag{2.18}$$

For a normal critical point one has one relevant symmetry conserving operator (apart from O_o) O_E which determines the critical temperature

$$\mu_E \equiv \tau = \beta \mu_E^o - \mu_E^* = (\beta - \beta_c) \mu_E^o \; ; \; \beta_c = \frac{\mu_E^*}{\mu_E^o} \tag{2.19}$$

Crudely speaking O_E is proportional to the hamiltonian minus its expectation value at the critical point. At a tricritical point one has two relevant symmetry conserving operators (apart from O_o) and consequently two conditions for criticality.

The RG procedure is not uniquely defined. This has the following consequences (Wegner [14]) : H^* depends on the RG transformation. Varying this transformation infinitesimally changes H^* by operators which we call redundant. The formal critical exponents of redundant operators depend on the choice of the RG, but the redundant operators do not contribute to the free energy. Thus, the exponents of the redundant operators do not have any physical meaning. The exponents of the other operators are invariant with respect to an infinitesimal variation of the RG equation.

2.3. Scaling of the Free Energy

Suppose it is sufficient to consider only two operators O_E and the magnetization O_h

$$H_o = H^* + \tau O_E + hO_h \tag{2.20}$$

which yields

$$H_\ell = H^* + \tau e^{y_E \ell} O_E + h e^{y_h \ell} O_h . \tag{2.21}$$

Then we obtain from eq. (2.5)

$$F(\tau,h) = e^{-d\ell} F(\tau e^{y_E \ell}, h e^{y_h \ell}) \tag{2.22}$$

which agrees for $L=e^\ell$ precisely with eq. (1.27) of our intuitive picture.

Normally one has an infinite number of perturbations in equation (2.20). To study their effect on the homogeneity law we add at least one further operator O_i

$$H_o = H^* + \tau O_E + hO_h + \mu_i O_i \tag{2.23}$$

and obtain

$$H_\ell = H^* + \tau e^{y_E \ell} O_E + h e^{y_h \ell} O_h + \mu_i e^{y_i \ell} O_i \tag{2.24}$$

$$F(\tau,h,\mu_i) = |\tau|^{d/y_E} F(\pm 1, \frac{h}{|\tau|^{y_h/y_E}}, \frac{\mu_i}{|\tau|^{y_i/y_E}}) . \tag{2.25}$$

We are interested in the critical behaviour, that is in the limit $\tau \to 0$

$$\lim_{\tau \to 0} \frac{\mu_i}{|\tau|^{y_i/y_E}} = \begin{cases} 0 & y_i < 0 \text{ or } \mu_i = 0 \\ \pm\infty & y_i > 0 \text{ and } \mu_i \neq 0 \end{cases} \tag{2.26}$$

If O_i is relevant ($y_i > 0$) then μ_i has explicitly to be taken into account. For irrelevant operators the term $\mu_i / |\tau|^{y_i/y_E}$ can be neglected if F can be expanded in powers of μ_i. Note that the right hand side of eq. (2.26) contains the free energy well apart from criticality. The irrelevant operator yields a correction to scaling

$$\begin{aligned} F &= |\tau|^{d/y_E} F(\pm 1, \frac{h}{|\tau|^{y_h/y_E}}, 0) \\ &+ \mu_i \; |\tau|^{(d-y_i)/y_E} F'(\pm 1, \frac{h}{|\tau|^{y_h/y_E}}, 0) + \ldots \end{aligned} \tag{2.27}$$

as observed in superfluid He (Ahlers 15). If F cannot be expanded in powers of μ_i, then Fisher's idea of the anomalous dimension of the vacuum might apply [16].

2.4. Correlations

Until now we considered only translational invariant perturbations. We may also consider the eigenvalue equation for an operator O_i acting only close to the origin of coordinate space

$$\mathcal{L}\tilde{O}_i = - x_i \tilde{O}_i \tag{2.28}$$

Then in linear approximation one obtains from

$$H_o = H^* + \lambda_i \tilde{O}_i \tag{2.29}$$

the hamiltonian

$$H_\ell = H^* + \lambda_i \, e^{-x_i \ell} \, \tilde{O}_i \; . \tag{2.30}$$

Instead of $\tilde{O}_i$ which is a function of the $S(r')$ we may consider $\tilde{O}_i(r)$ which is obtained by replacing $S(r')$ by $S(r'+r)$. This operator $\tilde{O}_i(r)$ acts in the vicinity of r. From

$$H_o = H^* + \lambda_i \tilde{O}_i(r) \tag{2.31}$$

one obtains

$$H_\ell = H^* + \lambda_i \, e^{-x_i \ell} \, \tilde{O}_i(re^{-\ell}) \tag{2.32}$$

since the length scale shrinks by a factor $e^{-\ell}$ under the RG transformation.

We may now connect the exponents x_i and y_i by introducing the translational invariant operator

$$O_i = \int d^d r \; \tilde{O}_i(r) \tag{2.33}$$

Using eqs. (2.31) and (2.32) we obtain from

$$H_o = H^* + \mu_i O_i = H^* + \mu_i \int d^d r \, \tilde{O}_i(r) \tag{2.34}$$

the hamiltonian

$$\begin{aligned} H_\ell &= H^* + \mu_i e^{-x_i \ell} \int d^d r \, \tilde{O}_i(re^{-\ell}) \\ &= H^* + \mu_i e^{(d-x_i)\ell} \int d^d(re^{-\ell}) \, \tilde{O}_i(re^{-\ell}) \\ &= H^* + \mu_i e^{(d-x_i)\ell} O_i \, . \end{aligned} \tag{2.35}$$

Comparison with (2.13) yields

$$y_i = d - x_i \tag{2.36}$$

(unless $\int d^d r \, \tilde{O}_i(r)$ vanishes).

Then one obtains from

$$H_o = H^* + \tau O_E + hO_h + \lambda_1 \tilde{O}_h(0) + \lambda_2 \tilde{O}_h(r) \tag{2.37}$$

the hamiltonian

$$\begin{aligned} H_\ell = H^* &+ \tau e^{y_E \ell} O_E + h e^{y_h \ell} O_h + \lambda_1 e^{(y_h-d)\ell} \tilde{O}_h(0) \\ &+ \lambda_2 e^{(y_h-d)\ell} \tilde{O}_h(re^{-\ell}) \, . \end{aligned} \tag{2.38}$$

Differentiating the free energy of the hamiltonian (2.37) yields the correlation function

$$G(r, \tau, h) = - \frac{\partial^2}{\partial\lambda_1 \, \partial\lambda_2} (V_o F(H_o)) \Big|_{\lambda_1 = \lambda_2 = 0} + m^2(\tau, h). \tag{2.39}$$

Since $V_o F(H_o) = V_\ell \, F(H_\ell)$ one obtains

$$\begin{aligned} G(r,\tau,h) &= - \frac{\partial^2}{\partial\lambda_1 \partial\lambda_2} (V_\ell \, F(H_\ell)) \\ &= e^{2(y_h-d)\ell} G(re^{-\ell}, \tau e^{y_E \ell}, he^{y_h \ell}) + e^{2(y_n-d)\ell} m^2(\tau', h') \end{aligned} \tag{2.40}$$

which proves eq. (1.23).

III. An Example

There is a large variety of RG procedures. To demonstrate the ideas of the preceding sections a RG equation with smooth momentum cut-off first presented by Wilson [4] which can be generalized (Wegner [14], compare [8] and [17]), is most appropriate. We will give a less ambitious example here starting out from a formulation by Landau.

3.1. Landau Model

Landau [18] describes the effective interaction for the behaviour in the vicinity of the critical point by

$$H_o = \int d^d r \left\{ \frac{1}{2} (\nabla \vec{S}(r))^2 + \frac{1}{2} a_1 S^2(r) - \vec{h}\vec{S}(r) + \frac{1}{4} a_2 (S^2(r))^2 \right\}. \quad (3.1)$$

Here $\vec{S}(r)$ is an n-dimensional vector. A model of n-dimensional vectors with isotropic interaction is called isotropic n-vector model. It corresponds for

$n = 1$ to the Ising model,
$n = 2$ to the XY-model,
$n = 3$ to the isotropic Heisenberg model, and in the limit
$n = 0$ it yields the critical behaviour of polymers.

A rough argument for the interaction (3.1) runs as follows : Consider the system on a scale which does not allow one to detect single magnetic moments. Thus $S(r)$ is a magnetization density averaged over a region of linear extension Λ^{-1}. Then $S(r)$ contains only Fourier components up to $q=\Lambda$. Thus a given function $S(r)$ describes a large number of microscopic configurations. A calculation of the free energy from H_o should yield the free energy of the microscopic system. Thus H_o contains the entropy due to fluctuations on a length scale less than Λ^{-1}. Therefore H_o contains (i) the energy due to the configuration $S(r)$ which for a model of original interaction

$$\frac{1}{2} \Sigma I_{ij} s_i s_j - \hat{h} \Sigma s_i$$

yields terms proportional to $S(r)$, $S^2(r)$, and $(\nabla S(r))^2$. The last term acts to keep the magnetic moments at various points in space parallel.

(ii) H_0 contains the entropy which is an even function of S(r) and can be expanded in powers of $S^2(r)$, thus contributing to $a_1S^2(r)$ and $a_2(S^2(r))^2$. An appropriate choice of scale for S(r) yields eq. (3.1).

Landau simply minimized H_0, eq. (3.1), to obtain the free energy and the magnetization. For vanishing magnetic field this procedure yields

$$S(r) = \begin{cases} 0 & a_1 > 0 \\ \pm\sqrt{-\frac{a_1}{a_2}} & a_1 < 0 \end{cases} \quad . \tag{3.2}$$

Thus the critical point is determined by $a_1=0$. Energy and entropy contribute a negative and a positive term to a_1. At high temperatures the entropy wins with $a_1 > 0$ and at low temperatures the energy predominates yielding $a_1 < 0$

$$a_1 \sim 1 - \frac{T_c}{T} = \tau \ . \tag{3.3}$$

From eqs. (3.2) and (3.3) one obtains the critical exponent

$$\beta = \tfrac{1}{2} \ .$$

Variation of the magnetic field h yields

$$\frac{\partial S(r)}{\partial h} = \begin{cases} (2a_1)^{-1} & a_1 > 0 \\ -(4a_1)^{-1} & a_1 < 0 \end{cases} \tag{3.4}$$

with

$$\gamma = 1 \ . \tag{3.5}$$

Although this molecular field approximation gives a qualitative description of critical phenomena it does not yield correct critical exponents in general.

3.2. Renormalization of the Landau Interaction

It is now easy to perform the scale transformation (1.2), (1.3) with the choice $\eta=0$. Then the new interaction reads

$$H_1' = \int d^dr' \Big\{ \tfrac{1}{2}(\nabla S'(r'))^2 + \tfrac{1}{2} b^2 a_1 S'^2(r') - b^{d/2+1} hS'(r') + \tfrac{1}{4} b^{4-d} a_2 (S'^2(r'))^2 \Big\} \ . \tag{3.6}$$

This transformation contains the step (i) and the transformation of step (ii) of section 2.1.

We have to eliminate the Fourier components with $q > 1$. For this purpose we expand

$$S'(r') = \frac{1}{V} \sum_q S_q' \, e^{iqr} \tag{3.7}$$

which yields

$$H_1' = \frac{1}{V} \sum_q (q^2 + b^2 a_1) \vec{S}'_q \vec{S}'_{-q} - \vec{h} b^{d/2+1} \vec{S}'_o$$
$$+ \frac{a_2 b^{4-d}}{4 V^3} \sum (\vec{S}'_{q_1} \vec{S}'_{q_2}) (\vec{S}'_{q_3} \vec{S}'_{-q_1-q_2-q_3}) \,. \tag{3.8}$$

To obtain H_1 we have to perform the integrals

$$e^{-H_1} = \prod_{q>1} \left(\int dS_q \right) e^{-H_1'} \,. \tag{3.9}$$

We evaluate this integral only approximatively by asking : How do the coefficients a_1 and a_2 in front of $\vec{S}_o'^2$ and $(\vec{S}_o'^2)^2$ change? These changes come about from

$$\Delta H_1' = \frac{1}{2V} \sum_{q>\Lambda} (q^2 + b^2 a_1) \vec{S}'_q \vec{S}'_{-q}$$
$$+ \frac{a_2 b^{4-d}}{2V^3} \sum_{q>\Lambda} (\vec{S}'_q \vec{S}'_{-q}) (\vec{S}'_o)^2 \tag{3.10}$$
$$+ \frac{a_2 b^{4-d}}{V^3} \sum_{q>\Lambda} (\vec{S}'_q \vec{S}'_o) (\vec{S}'_{-q} \vec{S}'_o) \,.$$

For the choice $b=e^{\ell}$, ℓ infinitesimal, all other contributions are negligible.

We rewrite $\Delta H_1'$ in the form

$$\Delta H_1' = \frac{1}{2V} \sum_{\substack{q>\Lambda \\ \alpha\beta}} (q^2+b^2 a_1) S'_{q\alpha} S'_{-q\beta} (\delta_{\alpha\beta} + f_{\alpha\beta}) \tag{3.11}$$

where

$$f_{\alpha\beta} = c' (\delta_{\alpha\beta} \vec{S}_o'^2 + 2 S'_{o\alpha} S'_{o\beta}) \tag{3.12}$$

$$c' = \frac{a_2 b^{4-d}}{V^2 (q^2+b^2 a_1)} \,. \tag{3.13}$$

The subscripts α and β denote the n components of the vectors $\vec{S}$, $\alpha = 1 \ldots n$. Now

$$\int dS'_{q1} \ldots dS'_{qn} \, dS'_{-q1} \ldots dS'_{-qn}$$
$$\exp \left(- \frac{1}{V} \sum_{\alpha\beta} (q^2+b^2 a_1) S'_{q\alpha} S'_{-q\beta} (\delta_{\alpha\beta} + f_{\alpha\beta}) \right)$$
$$= \frac{\text{const.}}{(q^2+b^2 a_1)^n \det (1+f)} \tag{3.14}$$

which yields

$$\prod_q \int dS_q \exp(-\Delta H_1') = \text{const.} \exp\left(-\frac{1}{2} \sum_{q>\Lambda} \ln \det(1+f)\right). \quad (3.15)$$

We evaluate

$$\begin{aligned} &\frac{1}{2} \sum_{q>\Lambda} \ln \det(1+f) \\ &= \frac{1}{2} \sum_{q>\Lambda} \ln\left(1+\sum_\alpha f_{\alpha\alpha} + \frac{1}{2}\sum_{\alpha\beta}(f_{\alpha\alpha} f_{\beta\beta} - f_{\alpha\beta} f_{\beta\alpha}) + \ldots\right) \\ &= \frac{1}{2} \sum_{q>\Lambda} \left(\sum_\alpha f_{\alpha\alpha} - \frac{1}{2}\sum_{\alpha\beta} f_{\alpha\beta} f_{\beta\alpha} + \ldots\right). \end{aligned} \quad (3.16)$$

For infinitesimal l the sum yields

$$\sum_{q>\Lambda} g(q) = V\, l\, c\, g(\Lambda) \quad (3.17)$$

where c is some constant and c' reduces to

$$c' = \frac{a_2}{V^2(a_1+\Lambda^2)} \quad . \quad (3.18)$$

Thus we obtain

$$\sum_\alpha f_{\alpha\alpha} = (n+2) \frac{a_2}{V^2(a_1+\Lambda^2)} \vec{S}_0'^2 \quad (3.19)$$

$$\sum_{\alpha\beta} f_{\alpha\beta} f_{\beta\alpha} = (n+8) \frac{a_2^2}{V^4(a_1+\Lambda^2)^2} (\vec{S}_0'^2)^2 \quad (3.20)$$

and finally the contribution from the elimination procedure

$$\begin{aligned} &\frac{1}{2} \sum_{q>\Lambda} \ln \det(1+f) \\ &= \frac{1}{2}(n+2) \frac{l c a_2}{V(a_1+\Lambda^2)} \vec{S}_0'^2 - \frac{1}{4}(n+8) \frac{l c a_2^2}{V^3(a_1+\Lambda^2)^2} (\vec{S}_0'^2)^2 \quad . \end{aligned} \quad (3.21)$$

Taking into account the change of a_1, a_2 and h according to eqs. (3.6) and (3.21) one obtains the differential equations

$$\frac{da_1}{dl} = 2a_1 + \frac{a_2 c(n+2)}{a_1+\Lambda^2} \quad (3.22)$$

$$\frac{da_2}{dl} = (4-d)a_2 - \frac{a_2^2 c(n+8)}{(a_1+\Lambda^2)^2} \quad (3.23)$$

$$\frac{dh}{dl} = \left(\frac{d}{2}+1\right) h \quad . \quad (3.24)$$

3.3. ε-expansion

The eqs. (3.22) to (3.24) allow the calculation of the effective hamiltonian H_ℓ parametrized by $a_1(\ell)$, $a_2(\ell)$, $h(\ell)$. For a fixed point a_1^*, a_2^*, h^*

$$\frac{da_1}{d\ell} = \frac{da_2}{d\ell} = \frac{dh}{d\ell} = 0 \tag{3.25}$$

holds. The calculation of the preceding section is exact for $a_2=0$. We expect it to be a reasonable approximation for small a_2. From the Landau theory we expect a_1^* to be small. Therefore we expand eqs. (3.12) and (3.23) up to quadratic terms in a

$$\frac{da_1}{d\ell} = 2a_1 + \frac{c}{\Lambda^2}(n+2)a_2 - \frac{c}{\Lambda^4}(n+2)a_1a_2 \tag{3.26}$$

$$\frac{da_2}{d\ell} = \varepsilon a_2 - \frac{c}{\Lambda^4}(n+8)a_2^2 \tag{3.27}$$

where we denote

$$\varepsilon = 4-d \ . \tag{3.28}$$

Then we obtain the
(i) trivial fixed point

$$a_1^* = 0 \ , \quad a_2^* = 0 \ , \quad h^* = 0 \tag{3.29}$$

and
(ii) the non trivial fixed point

$$a_1^* = -\frac{2\,\varepsilon(n+2)\,\Lambda^2}{(n+8)} \ , \quad a_2^* = \frac{\varepsilon\Lambda^4}{c(n+8)} \ , \ h^* = 0 \ . \tag{3.30}$$

Linearization around the fixed point yields

$$\frac{da_1}{d\ell} = (2 - \frac{c}{\Lambda^4}(n+2)a_2^*)\,(a_1-a_1^*) + \ldots \quad (a_2-a_2^*) \tag{3.31}$$

$$\frac{da_2}{d\ell} = (\varepsilon - \frac{2c}{\Lambda^4}(n+8)a_2^*)\,(a_2-a_2^*) \tag{3.32}$$

and the scaling exponents for the trivial fixed point

$$y_E = 2 \ , \quad y_2 = \varepsilon \ , \quad y_h = 3 - \frac{\varepsilon}{2} \tag{3.33}$$

and for the nontrivial fixed point

$$y_E = 2 - \frac{n+2}{n+8}\,\varepsilon \ , \ y_2 = -\varepsilon \ , \ y_h = 3 - \frac{\varepsilon}{2} \ . \tag{3.34}$$

For $d < 4$ the nontrivial fixed point has one, the trivial fixed point two relevant symmetry conserving operators. Thus for $d < 4$ one expects the nontrivial fixed point to describe the behaviour near a normal critical point.

The trivial fixed point describes a tricritical behaviour in three dimensions (Riedel and Wegner [19])

$$\alpha = \tfrac{1}{2}\,, \quad \beta = \tfrac{1}{4}\,, \quad \gamma = 1\,. \tag{3.35}$$

IV. Results on the Isotropic n-Vector Model

The isotropic n-vector model has been subject to many investigations. Exact results have been obtained in a number of limits by various authors (for references see M. Fisher [6]) :

(i) The critical exponents can be expanded around dimensionality 4 as sketched in section III

$$\alpha = \frac{(4-n)}{2(n+8)} \varepsilon + O(\varepsilon^2) \tag{4.1}$$

$$\gamma = 1 + \frac{(n+2)}{2(n+8)} \varepsilon + O(\varepsilon^2) \ . \tag{4.2}$$

(ii) According to Stanley the critical behaviour for $n=\infty$ coincides with that of the spherical model

$$\alpha = \frac{d-4}{d-2} , \quad \gamma = \frac{2}{d-2} , \quad 2 < d \leq 4 \ . \tag{4.3}$$

An expansion around this limit yields for d=3

$$\alpha = -1 + \frac{32}{\pi^2 n} + O\left(\frac{1}{n^2}\right) , \quad \gamma = 2 - \frac{24}{\pi^2 n} + O\left(\frac{1}{n^2}\right) \ . \tag{4.4}$$

(iii) Very recently Brezin and Zinn-Justin [20] gave an expansion around dimensionality 2. For $d=2+\varepsilon'$ and $n>2$ they obtain

$$\alpha = -\frac{2}{\varepsilon'} + \frac{n}{n-2} + O(\varepsilon') \tag{4.5}$$

$$\gamma = \frac{2}{\varepsilon'} - \frac{3}{n-2} + O(\varepsilon') \ . \tag{4.6}$$

Their result suggests that in the limit d=2, n=2 both T_c and the exponent η depend crucially on the ratio (d-2)/(n-2). This agrees with Zittartz' result [21] that the two-dimensional XY-model exhibits at low temperatures a critical line with a continuously varying exponent η as a function of temperature. Apparently the point d=2, n=2 is a point of confluence for the contours of constant critical exponents.

(iv) For n=-2 one obtains formally the scaling exponents y_E and y_h of the Gaussian fixed point yielding

$$\alpha = \frac{1}{2}(4-d) , \quad \gamma = 1 \ . \tag{4.7}$$

(v) For d=1, $n < 1$ formal calculations yield

$$\alpha = \gamma = 1 . \quad (4.8)$$

(vi) As an isolated point the critical exponents of the two-dimensional Ising model

$$\alpha = 0 , \quad \gamma = 7/4 \quad (4.9)$$

are known. In figure 1 the limit values of these exponents are summarized. Expansions are available in the shaded regions. The contours for constant α and γ are sketched in figures 2 and 3. (Compare Fisher [6] and Wegner [9]).

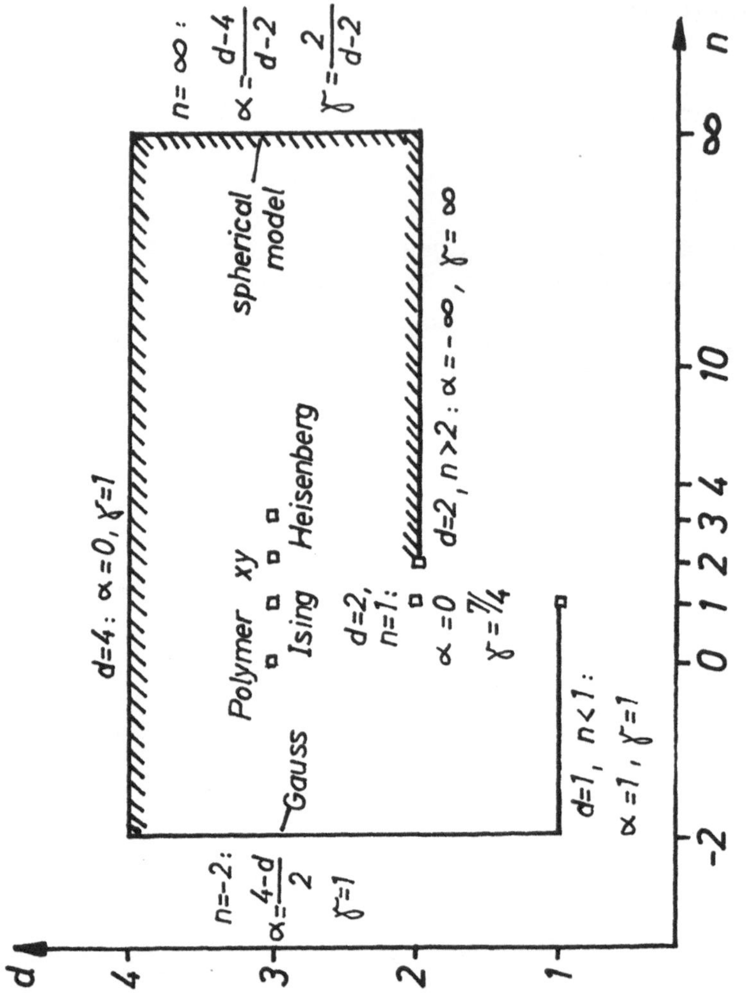

Fig. 1 The (n,d) plane for the isotropic n vector model : Exactly known critical exponents and (shaded) regions where expansions are available (after ref. 9).

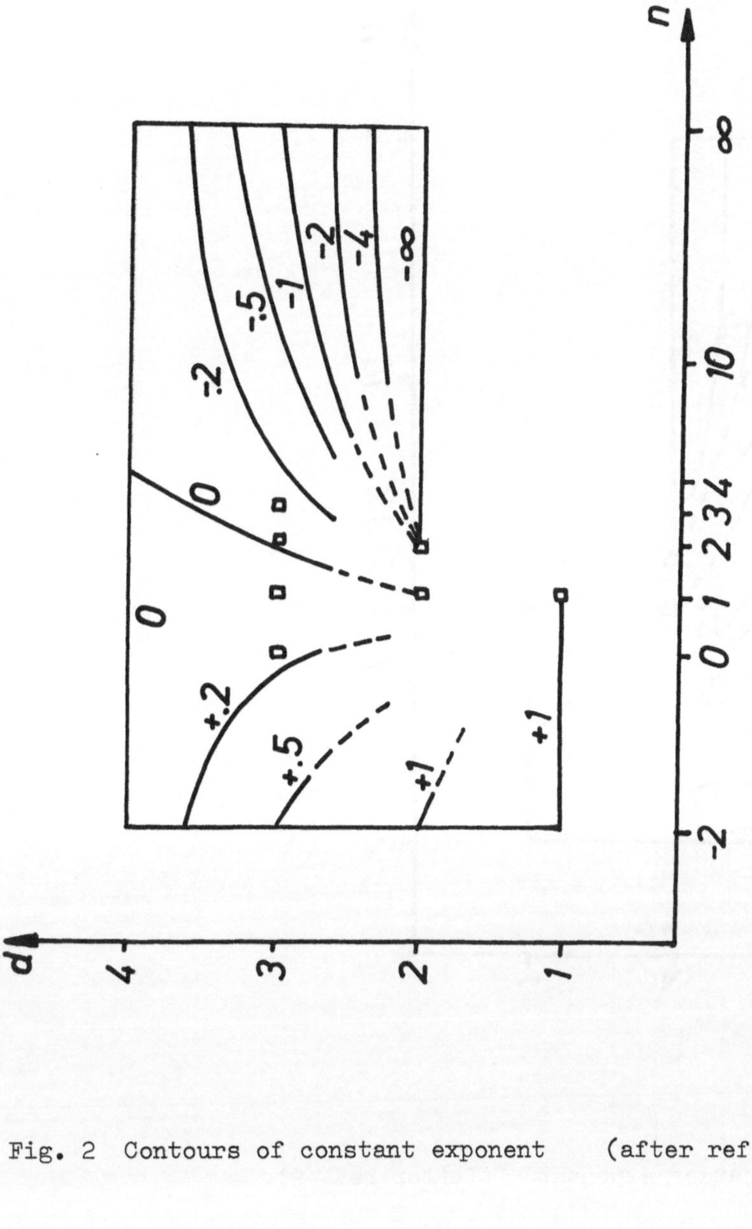

Fig. 2 Contours of constant exponent (after ref. 9).

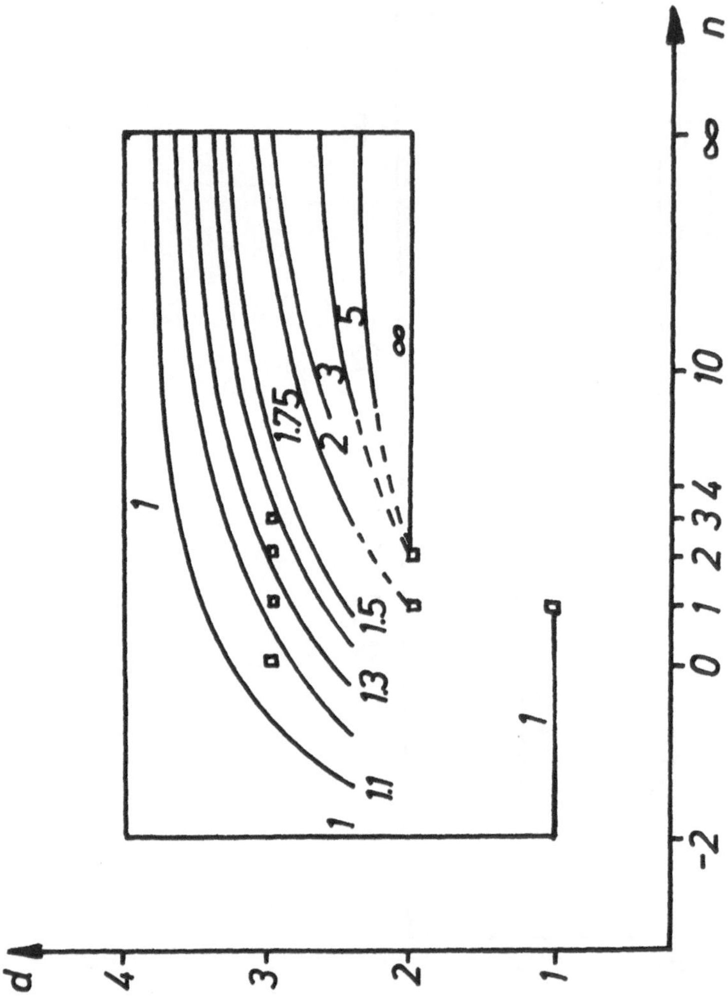

Fig. 3 Contours of constant exponent (after ref. 9).

V. Scaling Fields and Universality

Apart from certain exceptions the nonlinearities of the RG equation can be absorbed in scaling fields g_i so that

$$F\{g_i\} = e^{-d\ell} F\{g_i\, e^{y_i \ell}\} \tag{5.1}$$

holds exactly (Wegner [13]). The fields μ_i and the scaling fields are related by a Taylor expansion

$$\mu_i = g_i + \frac{1}{2} \sum b_{ijk}\, g_j g_k + O(g^3) \ . \tag{5.2}$$

The free energy does not depend on the scaling fields of the redundant operators.

Scaling fields can normally not be constructed if

$$y_i = y_{j_1} + y_{j_2} + \ldots + y_{j_n} \qquad n \geqslant 2 \ . \tag{5.3}$$

In such a case logarithmic factors arise. We refer the reader to reference [9] which provides a short introduction to these exceptions.

In terms of the scaling fields g the hamiltonian

$$H_0 = H\{g_i\} \tag{5.4}$$

transforms into

$$H_\ell = H\{g_i\, e^{y_i \ell}\} \ . \tag{5.5}$$

Thus the condition for criticality is that all relevant scaling fields g_i vanish.

From similar arguments as for equation (2.27) we expect that F can be expanded (for two relevant operators)

$$F\{g_i\} = g_0 + f(g_E, g_h) + \sum_i{}' g_i\, f_i(g_E, g_h) + \ldots \tag{5.6}$$

where the sum runs over all scaling fields g_i of irrelevant operators. g_0 is the regular part and $f(g_E, g_h)$ the leading singular part of F. According to eq. (5.1) one has

$$f(g_E\, e^{y_E \ell}, g_h\, e^{y_h \ell}) = e^{d\ell} f(g_E, g_h) \ , \tag{5.7}$$

$$f_i(g_E\, e^{y_E \ell},\ g_h\, e^{y_h \ell}) = e^{(d-y_i)\ell}\, f_i(g_E, g_h)\ . \tag{5.8}$$

5.1. Correlations

(An explicit discussion in q-space is given in Ref. [17]) .

The definition of the operators $\tilde{O}_i(r)$ can be modified in such a way that

$$\langle \tilde{O}_i(0)\tilde{O}_j(r)\rangle_{H_0} = e^{-(x_i + x_j)\ell} \langle \tilde{O}_i(0)\tilde{O}_j(re^{-\ell})\rangle_{H_\ell} \tag{5.9}$$

holds as long as r and $re^{-\ell}$ is large in comparison to the inverse momentum cut-off. As soon as the operators come to a distance of order Λ^{-1} the nonlinear part Q of the RG generator (2.9) will destroy this property. But then the product $\tilde{O}_i(0)\tilde{O}_j(\Lambda^{-1})$ can be expanded in single operators $\tilde{O}_k(0)$.

$$\tilde{O}_i(0)\tilde{O}_j\,(\Lambda^{-1}) = \sum_k C_{ijk}\, \tilde{O}_k(0)\ . \tag{5.10}$$

Then apparently

$$\tilde{O}_i(0)\tilde{O}_j(r) = \sum_k C_{ijk}\, \left(\frac{r}{\Lambda}\right)^{-x_i-x_j+x_k} \tilde{O}_k(0) \tag{5.11}$$

holds. Taking into account all nonlinear terms of the generator $\mathcal{G}$ one may define the operators $\tilde{O}_i$ as functions of g, so that

$$\int \tilde{O}_i(r)\, d^d r = \frac{\partial H}{\partial g_i} \tag{5.12}$$

and equation (5.11) reads

$$\tilde{O}_i(0)\tilde{O}_j(r) = \sum_k M_{ijk}\,(\{g\}, \vec{r})\, \tilde{O}_k(0) \tag{5.13}$$

where M splits into a scaling part S and a short-range part R

$$M_{ijk} = S_{ijk} + R_{ijk} \tag{5.14}$$

with

$$S_{ijk}\,(\{g\}, \vec{r}) = r^{-x_i-x_j+x_k}\, S_{ijk}\,(\{g r^y\}, \frac{\vec{r}}{r}) \tag{5.15}$$

The scaling part S determines the long-range behaviour of the correlation function. The short-range part R leads to finite correlations in the limit r=0. Thus at large distance one obtains the formal expansion

$$\langle \tilde{O}_i(0)\tilde{O}_j(r)\rangle = \sum_k r^{-x_i-x_j+x_k}\, S_{ijk}\,(\{g_i r^{y_i}\}, \frac{\vec{r}}{r})\, \frac{1}{V} \langle \frac{\partial H}{\partial g_i}\rangle \tag{5.16}$$

The coefficients S can be expanded in powers of the scaling fields g_i, but the expectation values $\langle \partial H/\partial g_i \rangle /V$ lead to nonanalyticities at the critical point for fixed r. Eq. (5.16) implies the homogeneity relation for any correlation length

$$\xi(\{g_i\}) = e^{\ell}\, \xi(\{g_i\, e^{y_i \ell}\}) \tag{5.17}$$

5.2. Universality

(Compare : References [22], [17] and the recent papers by Aharony, Hohenberg et al. [23], [24]).

Different phase transitions which are described by the same fixed point hamiltonian are said to belong to the same "critical class". These phase transitions have the same critical exponents. Moreover, it can be shown that they are also governed by the same scaling functions f, f_i,... for the free energy, eq. (5.6), and scaling functions S for the correlations, eq. (5.15). The property that a large number of quantities and in particular those describing the nonanalyticities agree for different phase transitions of the same critical class is called universality. It makes comparison of various systems and model calculations meaningful.

The functions f, f_i... and S are not uniquely defined. One is still free to choose the scales for g. Thus two choices for the scale of g will yield

$$\hat{f}(g_E,g_h) = f(\lambda_E\, g_E,\ \lambda_h\, g_h) \tag{5.18}$$

$$\hat{f}_i(g_E,g_h) = \lambda_i\, f(\lambda_E\, g_E,\ \lambda_h g_h) \tag{5.19}$$

$$\hat{S}_{ijk}(\{g_\ell\},\vec{r}) = \lambda_i^{-1}\, \lambda_j\, \lambda_k\, S_{ijk}(\{\lambda_\ell g_\ell\},\vec{r}) \tag{5.20}$$

$$\hat{\xi}(\{g_i\}) = \xi(\{\lambda_i g_i\}) \tag{5.21}$$

Any quantity constructed from these functions which is independent of the choice of the scales for g is universal. In particular the ratios $A_{\xi+}/A_{\xi-}$, $A_{\chi+}/A_{\chi-}$, A_{c+}/A_{c-} of the amplitudes of correlation length, susceptibility and specific heat are universal. Similarly $A_m^2/(A_{\chi-}F_-)$, $A^d_{\xi\pm}\, F_\pm$ are universal with

$$F_{sing}(\tau,0) = F_\pm\, |\tau|^{2-\alpha} \tag{5.22}$$

The fact that the length scale does not introduce a new scale factor has become known as two-scale-universality. If only the relevant variables g_E and g_h are considered, two scales have to be fixed (Stauffer, Ferer, and Wortis [25]).

We have already seen that the dimensionality d and the number n of critical components of a vector are crucial in determining the critical class. But even for given d and n there may be several critical classes. For a discussion of a number of fixed points see the review article by M. Fisher [6] and reference [10].

REFERENCES

1) M.E. FISHER, Rept. Prog. Phys. 30, 615 (1967).
2) L.P. KADANOFF et al., Rev. Mod. Phys. 39, 395 (1967).
3) M.S. GREEN (ed.), Proceedings of the International School of Physics "Enrico Fermi", Varenna, 51.
4) K.G. WILSON and J. KOGUT, Phys. Rept. 12C, 75 (1974).
5) S. MA, Rev. Mod. Phys. 45, 589 (1973).
6) M.E. FISHER, Rev. Mod. Phys. 46, 597 (1974).
7) VAN DER WAALS Centennial Conference on Statistical Mechanics, Physica 73 (1974).
8) F.J. WEGNER, Lecture Notes in Phys. 37, 171 (1975).
9) F.J. WEGNER, Festkörperprobleme XVI, Vieweg, Wiesbaden (1976).
10) C. DOMB and M.S. GREEN (eds.), Phase Transitions and Critical Phenomena, Academic Press, Vol. 6 (to appear).
11) L.P. KADANOFF, Physics 2, 263 (1966).
12) K.G. WILSON, Phys. Rev. B4, 3174, 3184 (1971).
13) F.J. WEGNER, Phys. Rev. B5, 4529 (1972).
14) F.J. WEGNER, J. Physics C7, 2098 (1974).
15) G. AHLERS, Phys. Rev. A8, 530 (1973).
16) M.E. FISHER, Nobel Symposium 24, 16 (1973).
17) F.J. WEGNER, in Ref. 10.
18) L.D. LANDAU, Phys. Z. Sowjetunion 11, 26 (1937).
19) E.K. RIEDEL and F.J. WEGNER, Phys. Rev. Lett. 29, 349 (1972).
20) E. BREZIN and J. ZINN-JUSTIN, Phys. Rev. Lett. 36, 691 (1976).
21) H. ZITTARTZ, Z. Phys. B23, 55 (1976).
22) L.P. KADANOFF, in C. Domb and M.S. Green (eds.), Phase Transitions and Critical Phenomena, Academic Press, Vol. 5A .
23) A. AHARONY, P.C. HOHENBERG, preprint.
24) P.C. HOHENBERG, A. AHARONY, B.I. HALPERIN and E.D. SIGGIA, preprint.
25) D. STAUFFER, M. FERER and M. WORTIS, Phys. Rev. B6, 3426 (1972).

INVARIANCE PROPERTIES OF THE RENORMALIZATION GROUP

M.S. GREEN

Temple University
Philadelphia, PA, 19122 (USA)

Invariance Properties of the Renormalization Group+

M.S. GREEN

Temple University, Philadelphia, PA, 19122 (USA)

I. Introduction

One of the features of the renormalization group approach to problems in many body physics is the apparent arbitrariness in the construction of a renormalization transformation. Since the physical results of the approach, critical exponents, scaling functions and the like are unique the procedure must have invariance properties which insure this uniqueness. Invariance of critical exponents has been demonstrated [1] in the context of the ϵ-expansion for a class of renormalization groups which has been called the Gaussian renormalization groups [2]. Wegner [3] has proved that critical exponents are unchanged under two types of perturbation of the renormalization group. These two theorems will lie at the basis of our discussion. We will not be able to discuss this question in its full generality but will limit ourselves to a general class of renormalization transformations proposed by Wegner. This class does not include renormalization transformations of the discrete type in which the Hamiltonian is a function of variables which may take on only a discrete set of values. It does include Wilson's recursion relation and it is nevertheless quite possible that it is representatives or the most general situation [2].

Our renormalization transformations are specified by a quantity $\psi_k(\sigma, \mathcal{H}(\sigma), \alpha)$ to be defined below, which, besides variables of physical significance, contains a parameter α which changes the renormalization transformation. We suppose that for α equal zero the renormalization group has a fixed point and consider what happens when α is changed infinitesimally. The results can be analysed in terms of two invariance theorems, the first due to Wegner [3] and the second to Wegner [3] and Jona-Lasinio [4]. One of the possible results of changing the parameter is that the new renormalization transformation has no fixed point near that of the original transformation. If, however, the change in α is such as to induce a new transformation which has a fixed point near that of the original transformation, the critical exponents are unchanged unless the original group had a so-called marginal operator of a special type. The exceptional case is the case in which critical exponents may depend on parameters in the Hamiltonian [3] .

II. Differential Renormalization Transformation in Wegner Form [3]

$$\mathcal{H}(\sigma) = \mu_0 + \frac{1}{2}\int_k \mu_2(k)\sigma_k\sigma_{-k} + \frac{1}{4!}\int \mu_4(k,k_1k_2)x$$
$$\sigma_k\sigma_{k_1}\sigma_{k_2}\sigma_{-k-k_1-k_2}$$

σ_k is k th Fourier component of a one-component field $\sigma(\vec{x})$.
$\mathcal{H} = \beta H$, H is the usual Hamiltonian.

$$\frac{\partial\mathcal{H}}{\partial\ell} = \int_k (\frac{d}{2}\sigma_k + k\cdot\nabla_k\sigma_k)\frac{\partial\mathcal{H}}{\partial\sigma_k} + G(\psi)\mathcal{H},$$

$$G(\psi)\mathcal{H} = \int_k \psi_k \frac{\partial\mathcal{H}}{\partial\sigma_k} - \frac{\partial\psi_k}{\partial\sigma_k} .$$

$\psi_k(\sigma, \mathcal{H}(\sigma))$ characterizes renormalization transformation.
First term on right in $\frac{\partial\mathcal{H}}{\partial\ell}$ represents scale change — changes volume of system but not thermodynamic potential. Second term keeps thermodynamic potential and volume fixed.

$$\exp -\mathcal{H}\, G(\psi)\mathcal{H} = \int_k \frac{\partial}{\partial\sigma_k} \psi_k \exp -\mathcal{H} ,$$

$$\frac{d}{d\ell}\int_\sigma \exp -\mathcal{H}(\sigma) = \int_\sigma \int_k \frac{\partial}{\partial\sigma_k}\psi_k \exp -\mathcal{H} = 0 .$$

ψ_k may be called flow vector in space of variables σ_k . $G(\psi)\mathcal{H}$ may be considered to be generated by an infinitesimal change of variables or flow :

$$\sigma_k = \sigma_k + d\ell\,\psi_k .$$

III. RG Linearized Around Fixed Point

$$\mathcal{H} = \mathcal{H}^* + \Delta\mathcal{H},$$

$$\psi_k = \psi_k^* + \mathcal{L}\Delta\mathcal{H},$$

$$\psi_k^*(\sigma) = \psi_k(\sigma, \mathcal{H}^*) , \quad \mathcal{L} = \left.\frac{\delta\psi_k}{\delta\mathcal{H}}\right|_{\mathcal{H}=\mathcal{H}^*} .$$

$$\frac{\partial\Delta\mathcal{H}}{\partial\ell} = L\,\Delta\mathcal{H} ,$$

$$L\Delta\mathcal{H} = G'(k\cdot\nabla_k\sigma_k + \frac{d\sigma}{2} + \psi_k^*)\Delta\mathcal{H} + G(\mathcal{L}\Delta\mathcal{H})\mathcal{H}^* .$$

The second term in the right hand side is linear in $\Delta\mathcal{H}$ because $G(\psi)\mathcal{H}$ is linear in ψ.

$$G'(\varphi_k) = \sum_k \varphi_k \frac{\partial}{\partial\sigma_k} .$$

Critical exponents are determined by eigenvalues of L.

$$LO_i = y_i O_i$$

IV. Redundant Perturbations to $\mathcal{H}^*$ [3]

Definition :

If $\Delta\mathcal{H} = G(\varphi_k)\mathcal{H}^*$ for some flow vector $\varphi_k(\sigma)$, $\Delta\mathcal{H}$ is called redundant.

1) $\mathcal{H}^* + \Delta\mathcal{H}$ is obtained from $\mathcal{H}^*$ by an infinitesimal change of coordinates

$$\sigma_k' = \sigma_k + \varphi_k(\sigma_k) \qquad (\varphi_k \text{ infinitesimal})$$

2) Space of all redundant perturbations is linear subspace.

3) $\mathcal{H}^* + \Delta\mathcal{H}$ has same thermodynamic potential as $\mathcal{H}$

Commutator identity

$$G'(\varphi)\; G(\psi)\mathcal{H} - G'(\varphi)\; G(\psi)\mathcal{H} = G(K_k(\varphi_k, \psi_k))\mathcal{H}$$

$$K_k(\varphi_k, \psi_k) = \sum_k \varphi_{k'} \frac{\partial \psi_k}{\partial \sigma_{k'}} - \psi_{k'} \frac{\partial \varphi_k}{\partial \sigma_{k'}}$$

$$L\, G(\varphi)\mathcal{H}^* = G'(D_k + \psi_k^*)\; G(\varphi)\mathcal{H}^* + G(\mathcal{L}\, G(\varphi)\,\mathcal{H}^*)\,\mathcal{H}^*$$

$$= G'(D_k + \psi_k)\;\; G(\varphi_k)\mathcal{H}^* - G'(\varphi_k)\; G(D_k + \psi_k)$$

$$+ \; G(\mathcal{L}\, G(\varphi)\,\mathcal{H}^*)\,\mathcal{H}^* = 0$$

$$= G(K_k(D_k + \psi_k, \varphi_k) + \mathcal{L}\, G(\varphi_k)\mathcal{H}^*)\,\mathcal{H}^*$$

1) A flow vector like D_k which contains $k \cdot \nabla_k \sigma_k$ changes scale and does not conserve thermodynamic potential/volume. $K_k(D_k, \varphi_k)$ contains no term $k \cdot \nabla_k \sigma_k$ and conserves thermodynamic potential/volume.

2) L operating on a redundant perturbation produces a redundant perturbation.

3) Linear subspace of redundant perturbations is an invariant subspace.

4) If eigenvectors, O_i , form a complete set they can be divided into two disjoint classes : a) non-redundant $\equiv$ scaling ;
b) redundant.

V. Flow Vector Containing a Parameter

$\psi_k(\sigma_k, \mathcal{H}(\sigma), \alpha)$

Suppose $\exists$ fixed point for $\alpha = 0$. What happens to F.P. and eigenvalues if α is changed infinitesimally.

Differentiate fixpoint eq. w.r.t. α , and set $\alpha = 0$,

$$L\,\mathcal{H}'^{*} + G(\psi'^{*})\mathcal{H}^{*} = 0 ,$$

$$\mathcal{H}'^{*} = \left.\frac{\partial \mathcal{H}^{*}}{\partial \alpha}\right|_{\alpha = 0} , \quad \psi'^{*} = \left.\frac{\partial \psi}{\partial \alpha}\right|_{\substack{\alpha = 0 \\ \mathcal{H} = \mathcal{H}^{*}}} .$$

Linear inhomogeneous eq. for $\mathcal{H}'^{*}$.

Solution or lack of solution to this eq. will tell us what happens to fixed point .

VI. What Happens to Fixed Point

Case a) L has no null vector = marginal O_i ($L\,\Delta\mathcal{H} \neq 0$).
Always $\exists$ a redundant solution because inhomogeneity redundant.

$$\mathcal{H}' = G(\chi)\mathcal{H}_0^* ,$$
$$\mathcal{H}_\alpha^* = \mathcal{H}_0^* + \alpha G(\chi)\mathcal{H}_0^*$$

where χ satisfies the equation ,

$$K(D_k + \Psi_k, \chi_k) + \mathcal{L}\, G(\chi)\mathcal{H}^* + \Psi' = 0$$

Case b) L has one (or more) redundant null-vectors.
$\exists$ a solution only if Ψ' satisfies one (or more) conditions.
If $\exists$ a solution there will be a one (or more) dimensional manifold of redundant solutions

Case c) L has a non-redundant null-vector = marginal <u>scaling</u> operator.

VII. What Happens to Eigenvalues

A) Perturbation to L .

Compute $L' = \left.\frac{\partial L}{\partial \alpha}\right|_{\alpha = 0}$.

Noting that fixed point depends on α

$$L' \Delta H = G'(\psi_k' + \mathcal{L}_k H'^*)\Delta H + G(\mathcal{L}_k' \Delta H) H^*$$

$$\mathcal{L}_k' = \left.\frac{d\mathcal{L}}{d\alpha}\right|_{\alpha = 0} \qquad \text{(noting F.P. depends on } \alpha \text{)} .$$

In case a) and b) $H'^* = G(\chi) H^*$ where χ satisfies same equation as above, we may manipulate L' into form

$$L' \Delta H = (G'(\chi)\, L - L G'(\varphi))\Delta H$$
$$+ \; G(K_k(\mathcal{L}\Delta H, \chi) \; + \; \mathcal{L} G'(\chi)\Delta H \; + \; \mathcal{L}'\Delta H) H^*$$

The first term on the right hand side is a commutator, the second, a redundant perturbation.

B) Wegner's first and second invariance theorems [3] .

Theorem 1. If we perturb a linearized renormalization transformation by an operator which always produces a redundant perturbation only eigenvalues of redundant perturbations can change.

Theorem 2. (Also Jona-Lasinio) [4] . The similarity transformation generated by the infinitesimal change in coordinates

$$\sigma_k' = \sigma_k + \alpha \chi_k$$

changes no eigenvalue y_i . This similarity transformation manifests itself in L' as a commutator.

C) Invariance of scaling eigenvalues in case a) and b).

The form L' given above is the sum of a commutator and an operator which produces only redundant perturbations. The first part changes no eigenvalues by Theorem 2 ; the second changes only the eigenvalues of the redundant operators by Theorem 1. Eigenvalues of scaling ope-

rators are invariant to both.
In case c) L' cannot be put into above form. No invariance theorem can be proven. We note only that in this case exponents of scaling operators may depend on initial Hamiltonian.

VIII. Conclusions

If a renormalization group of Wegner's type is slightly perturbed the new group will have a fixed point close to that of the original group if the former has no marginal operator. If the original group has a redundant marginal operator the new group will have no fixed point near the original fixed point unless the perturbation to the group satisfies certain conditions. In either of these two cases the new fixed point, if it exists, will be related to the old by a redundant perturbation and the eigenvalues corresponding to scaling operators will be unchanged. If the original group has a non-redundant marginal operator, which is the case in which critical exponents may depend on initial Hamiltonian, no invariance theorem can be proved.

+ Work supported in part by U.S. National Science Foundation.

REFERENCES

1) P. SHUKLA and M.S. GREEN, Phys.Rev. Lett. 33, 1263 (1974), 34, 436 (1975), J. RUDNICK, 34, 438 (1975).

2) M.S. GREEN and P. SHUKLA, Julius Jackson Memorial Volume, American Institute of Physics (1976).

3) F. WEGNER, J. Phys. C 7 , 2098 (1974).
Phase transitions and critical phenomena, Vol. 6.
C. DOMB and M.S. GREEN, eds., Academic Press London (1976).

4) G. JONA-LASINIO, Nobel Symposium 24, B. LUNDQUIST and S. LUNDQUIST, eds., Academic Press N.Y. (1973).

SCALE TRANSFORMATIONS IN DYNAMIC MODELS

SHANG-KENG MA

Department of Physics
University of California
San Diego, La Jolla, California, USA

Service de Physique Théorique
CEN Saclay
91190 Gif-sur-Yvette, France

Scale Transformations in Dynamic Models

SHANG-KENG MA

Department of Physics
University of California
San Diego, La Jolla, California, USA

Service de Physique Théorique
CEN Saclay
91190 Gif-sur-Yvette, France

I. Review of Basic Ideas and Terminology

In these lectures I shall introduce some basic techniques used often in the recent literature on critical dynamics, and then discuss some work which I am currently involved in. Before we begin, a quick review of elementary concepts and terminology is appropriate.[1] I shall assume that you are familiar with basic aspects of static critical phenomena.

1.1. Scale Transformation and the Renormalization Group

The use of symmetry transformations is familiar in all branches of physics. Rotations in atomic physics, translations in solid state physics, and Isotopic spin rotations in nuclear physics are outstanding examples. Observations of systems near their critical points show the existence of approximate symmetry under scale transformation, i.e. a change of length scale, or the reference standard of length. These observations are qualitatively summarized in terms of the scaling hypothesis, which states that a system at its critical point is invariant under scale transformations, (or simply scale invariant) and near its critical point this invariance is broken and a preferred length scale, the correlation lenth ξ, appears.

A system which is rotationally invariant looks the same when the reference axes are rotated (or the system is rotated the opposite way). A system which is scale invariant looks the same when the unit of length is changed (or the system is shrunk or enlarged). A simple scale invariant system is shown in Fig. 1.

Fig. 1

When you back up along the tracks the picture you see is shrunk. If you back up for a distance equal to that between two poles, the picture would look identical.

For a system, a ferromagnetic material to be specific, at its critical point, the probability distribution of spin fluctuations remains unchanged when one changes the unit of length, analogous to the case in the picture.

A very important observation on Fig. 1 is that in order to have scale invariance, one must have a limited eye sight, i.e., one must not see objects that are infinitely far. In other words, the resolving power of the eyes are assumed to be finite and constant. Otherwise, one would see the station far away, and,when one backs up, the station would look smaller and hence not the same.

The scaling hypothesis also requires that small scale details must be ignored in order that scale invariance can be possible.

Therefore, one must build in some procedure to "ignore" or "smear out" short distance details in addition to the change of length unit in order to have a scale transformation which is useful in the study of critical phenomena. Such an augmented scale transformation was formulated by Kadanoff and by Wilson, and is termed the "renormalization group", (RG for short).

In these lectures we want to see how dynamical equations transform under RG. From transformation properties of these equations we deduce properties such as critical exponents.

1.2. Phenomenological Formulation of Dynamics

Keeping in mind that we want to study critical behaviors and that they are nearly scale invariant, we are forced to abandon a purely microscopic formulation because it would be very inconvenient. Microscopic equations of motion have no dissipation term, for example. Once we shrink the system and smear out the short distance details, we immediately get dissipation terms. Therefore the microscopic equations cannot be scale invariant.

Of course, one could in principle derive equations with dissipation from purely microscopic equations, but that is too hard. So let's just write phenomenological equations following general considerations like conservation laws plus plausible arguments. Kawasaki, Mori and other workers have put a lot of effort into setting up reasonable phenomenological equations. [2] Here we shall introduce the most important features of phenomenological equations by a simple example.

The simplest example is the harmonic oscillator. Let us

temporarily leave the subject of critical phenomena and study the familiar Hamiltonian

$$H = \frac{p^2}{2m} + \frac{1}{2} K x^2 \qquad (1.1)$$

Define the phase space coordinates q_1 and q_2 by

$$q_1 = p ,$$

$$q_2 = x \qquad (1.2)$$

which we shall call modes. The velocity (v_1, v_2) of the point (q_1, q_2) in the phase space is

$$v_1 = - K \ q_2$$

$$v_2 = q_1/m . \qquad (1.3)$$

The trajectory of (q_1, q_2) is an ellipse as shown in Fig. 2

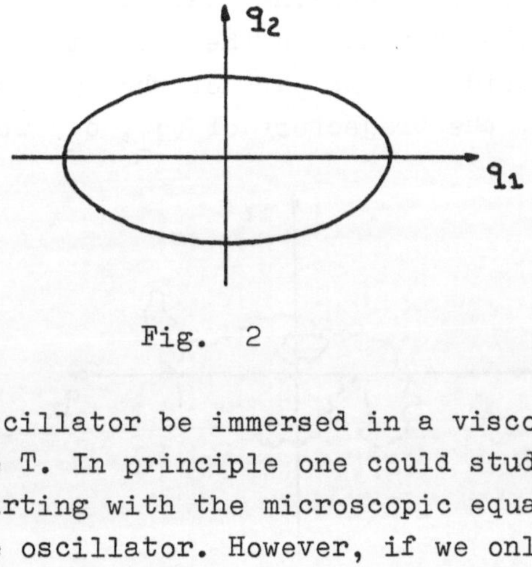

Fig. 2

Now let the oscillator be immersed in a viscous fluid at a certain temperature T. In principle one could study the oscillator-fluid system by starting with the microscopic equations for all the fluid atoms and the oscillator. However, if we only care about the motion of the oscillator, we can account for the effect of the fluid approximately by a damping on the oscillator and a random force. We write down the phenomenological kinetic equations

$$\frac{\partial q_1}{\partial t} = v_1 - \frac{\Gamma_1}{T} \frac{\partial H}{\partial q_1} + \zeta_1(t) , \qquad (1.4a)$$

$$\frac{\partial q_2}{\partial t} = v_2 \quad . \qquad (1.4b)$$

Here Γ_1/T is just the coefficient of viscosity apart from a constant factor. We write the damping term in the form of $-\partial H/\partial q_1$ to remind us that this term points in the direction of decreasing energy. The random force $\zeta_1(t)$, or "noise", is assumed to have zero average and a short correlation time, short compared to the time scale of the oscillator motion. We then write

$$\langle \zeta_1(t) \rangle = 0 \ ,$$

$$\langle \zeta_1(t) \zeta_1(t') \rangle = 2D_1 \ \delta(t-t') \ , \qquad (1.5)$$

where D_1 is another phenomenological constant. It is, however, related to Γ_1 of (1.4) via the Einstein relation which simply states that

$$\Gamma_1 = D_1 \ . \qquad (1.6)$$

The kinetic equations (1.4) are of course just the Langevin equations for the harmonic oscillator. There are two parts in the equations: first, the organized-motion part, i.e., (v_1, v_2), which shall be called "mode-mode coupling" terms; second, the disorganized-motion part, i.e., the damping and the noise terms, which reflect the effect of the fluid, or the part of the system not of interest.

As time passes, the trajectory of (q_1, q_2) would look like that shown in Fig. 3.

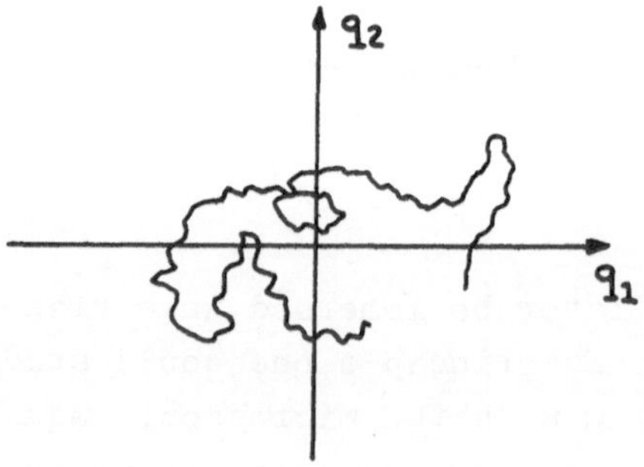

Fig. 3

The Langevin equations can only give the probability distribution of (q_1, q_2), not the exact trajectory. Eventually an equilibrium distribution is reached, namely, a Gaussian, as indicated by the cloud in Fig. 4.

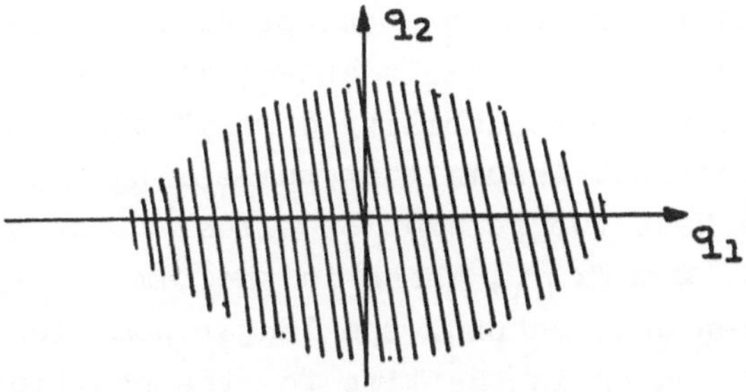

Fig. 4

The cloud is rotating by virtue of the mode-mode coupling term but the density of the cloud does not change.

Now we can generalize the above discussion. If we have a system described by the modes $q = (q_1, q_2, \dots q_n)$, then we can write the Langevin equations

$$\frac{\partial q_i}{\partial t} = v_i - \frac{\Gamma_i}{T} \frac{\partial H}{\partial q_i} + \zeta_i(t) ,$$

$$\langle \zeta_i(t) \zeta_j(t') \rangle = 2 \Gamma_i \delta_{ij} \delta(t-t') . \tag{1.7}$$

with further modifications if necessary. The probability distribution $P(q)$ generated by these equations follows a Fokker-Planck equation

$$\frac{\partial P}{\partial t} + \sum_i \frac{\partial J_i}{\partial q_i} = 0 , \tag{1.8a}$$

$$J_i = v_i P - \frac{\Gamma_i}{T} \frac{\partial H}{\partial q_i} P - \Gamma_i \frac{\partial P}{\partial q_i} \tag{1.8b}$$

Eq. (1.8a) is just the conservation of probability. The current in (1.8b) has three terms. The first one, $v_i P$, is the "streaming" term due to the mode-mode coupling. The second is the flux toward lower energy due to the dissipation and the last is the diffusion term which spreads out the probability as a result of the noise. If, in the absence of dissipation, H is conserved, (i.e., $\sum_i (\partial H/\partial q_i) v_i = 0$), and if the Liouville theorem is satisfied (i.e. $\sum_i \partial v_i / \partial q_i = 0$), then $P_o \propto \exp(-H/T)$ is a time independent solution of (1.8).

Finally we shall mention the concept of the relaxation time. For the harmonic oscillator example above, it is very easy to work out the time required for the probability distribution to approach the equilibrium distribution. Assume that the spring constant K is small for simplicity. We have two relaxation times, first the velocity relaxation time $\tau_1 \approx m\, T/\Gamma_1$, which is the time to approach a Maxwell distribution and, second, the longer position relaxation time $\tau_2 \approx \Gamma_1/T\, K$, which is the time for the position of the oscillator to approach a Gaussian of width 1/K. Note that $\tau_2 \propto 1/K$, the width.

One can obtain the equation for the slow mode q_2, the position of the oscillator, by eliminating the fast mode q_1. This is done by first solving (1.4a) for q_1 and then substituting the solution in (1.4b):

$$
\begin{aligned}
\frac{\partial}{\partial t} q_2 &= q_1/m \\
&= -\frac{K}{m}\int_{-\infty}^{t} dt' e^{-\gamma(t-t')} q_2(t') + \zeta'(t) \\
&\simeq -\frac{\Gamma'}{T}\frac{\partial H'}{\partial q_2} + \zeta'(t) , \qquad (1.9)
\end{aligned}
$$

where $\gamma = \Gamma_1/T$ and

$$
\zeta'(t) = \int_{-\infty}^{t} dt' \, e^{-\gamma(t-t')} \zeta_1(t')/m \qquad (1.10)
$$

is nothing but the random part of the velocity. In (1.9) the new damping constant Γ' is

$$
\Gamma'/T = T/\Gamma_1
$$

and $H' = \frac{1}{2} K q_2^2$. The new noise ζ' satisfies

$$
\begin{aligned}
\langle \zeta'(t)\, \zeta'(t') \rangle &= 2\, \Gamma'(\gamma/2)\, e^{-\gamma|t-t'|} \\
&\approx 2\, \Gamma' \delta(t-t') \qquad (1.11)
\end{aligned}
$$

provided we are not interested in a time scale much shorter than the position relaxation time $T/\Gamma' K$, which is assumed to be much larger than $1/\gamma$. Note that (1.9) and (1.11) have the same standard form as (1.7). The eliminated mode q_1 now appears through the noise.

The above example shows that smoothing out in the time scale should go along with the elimination of modes. Here the smoothing out in time is to ignore the time variation over a period $1/\gamma$. When we smear out short-distance details to define the RG transformation, we have to eliminate short-wavelength modes. This is the smoothing-out in space. They must be done together as subsequent examples will show in detail.

1.3. Van Hove Theory

We shall use the Van Hove theory as the simplest example in illustrating some of the above ideas. Let us start with the Ginzburg-Landau Hamiltonian for some spin system:

$$H/T = \frac{1}{2} \int d^d x \; ((\nabla\sigma)^2 + a\sigma^2 + b\sigma^4) \tag{1.12}$$

where $\sigma(x)$ is the spin density, and a, b are parameters. For simplicity, let us drop the last term. Then, in terms of the Fourier components σ_k of the spin density, H/T is

$$H/T = \frac{1}{2} \sum_{k<\Lambda} (k^2 + a) \; |\sigma_k|^2 \tag{1.13}$$

Modes with $k > \Lambda$ are excluded from the model. Now consider the dynamical model defined by the kinetic equations

$$\frac{\partial \sigma_k}{\partial t} = - \Gamma_k (k^2 + a)\sigma_k + \zeta_k , \tag{1.14}$$

$$\langle \zeta_k(t) \zeta_{k'}(t') \rangle = 2\Gamma\delta_{-k,k'} \; \delta(t-t')$$

which is of the form (1.7) without any mode-mode coupling. These are trivial to solve since each σ_k has an independent equation. The relaxation time for σ_k is clearly

$$\tau_k = \frac{1}{\Gamma_k (a+k^2)} . \tag{1.15}$$

Recall that in the mean field theory (or the Landau theory) we have $a \propto (T-T_c)$ and $\xi \propto |T-T_c|^{-1/2}$. Here ξ is the correlation length. Fig. 5 shows the H/T for a mode σ_k

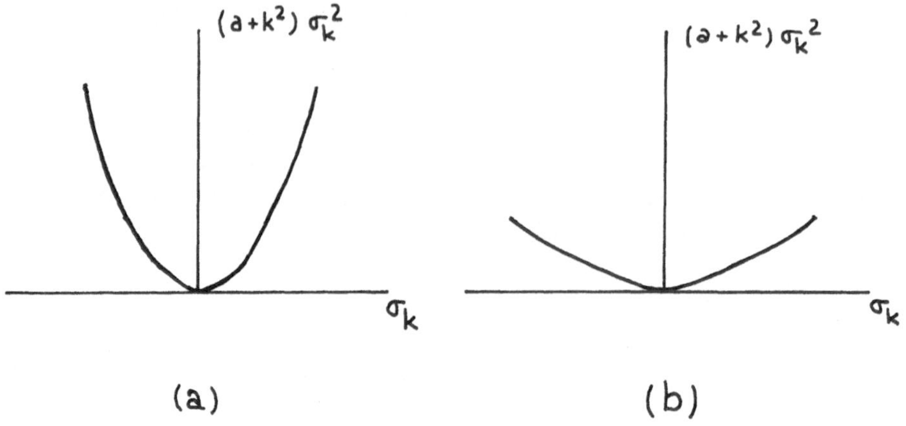

Fig. 5

For case (a) (large $(a+k^2)$, so the curve is steep,) by (1.15), the relaxation time is short. For case (b)(small $(a+k^2)$, i.e., for small $(T-T_c)$ and small k,) the relaxation time is long. The point being illustrated here is that the larger the fluctuation, (i.e., the flatter the curve), the longer the relaxation time. It takes longer to diffuse across a longer displacement. Near the critical point, the fluctuations have large amplitude and hence long relaxation time. This is often referred to as "the critical slowing down".

From (1.15), we see that

$$\tau_k \to (T-T_c)^{-1} \propto \xi^2 \qquad \text{for } k = 0$$
$$\to k^{-2} \qquad \text{for } T = T_c \tag{1.16}$$

if Γ is constant for $k \to 0$. The <u>dynamic exponent z</u> (or the characteristic time exponent) is defined by

$$\tau = \xi^z f(k\xi) \tag{1.17}$$

Therefore, we have $z = 2$ in this simple model. On the other hand, if $\Gamma_k \propto k^2$ for small k, we would have $z = 4$. This is the case where the total spin is conserved, i.e., the $k = 0$ component $\sigma_{k=0}$ is a constant of motion. The isotropic Heisenberg ferromagnet is an example.

Now let us try out our scale transformation or RG on the kinetic equations (1.14).

There are two steps. The first is to eliminate short wavelength modes, i.e., σ_k with $\Lambda/s < k < \Lambda$. Here $s \geq 1$ is the factor of scale change. We are shrinking the size of the system by a factor s. This step of elimination is trivial for this particular example. We simply ignore those equations in (1.14) with $\Lambda/s < k < \Lambda$. In general, this is the difficult step as our later examples will show.

The second step is to make the replacement

$$\sigma_k(t) \rightarrow s^{1-\eta/2} \quad \sigma_{sk}(t s^{-z}) \tag{1.18}$$

in the remaining equations. This is what one expects for the change of scale. Recall that under a change of scale, $x \rightarrow x/s$, $k \rightarrow sk$, and in general $A \rightarrow As^{\lambda}$, where λ is the <u>dimension</u> of A. Here we have defined η through the dimension of σ_k and $-z$ as the dimension of time. Note that the dimension of σ_k is in fact $-1+\eta/2$, not $1-\eta/2$. This is analogous to the rotation transformation (or any other transformation) where you change coordinates in one direction and the observables in the opposite direction. Now simply remember (1.18) as a rule, and η, z as adjustable constants in defining the RG.

After the replacement (1.18), Eq. (1.14) becomes

$$\frac{\partial}{\partial t} s^{1-\eta/2} \sigma_{sk}(t\, s^{-z}) = - \Gamma(k^2+a)\, s^{1-\eta/2} \sigma_{sk}(ts^{-z})$$

$$+ \zeta_k(t) \,, \quad 0 < k < \Lambda/s \quad . \tag{1.19}$$

Now we define

$$k' = sk \,,$$

$$t' = t\, s^{-z} \quad ,$$

$$\zeta'_{k'}(t') = s^{-1+\eta/2}\, \zeta_{k'/s}\, (t' s^{z}) \quad . \tag{1.20}$$

Thus we can write (1.19) in the form of (1.14)

$$\frac{\partial}{\partial t'} \sigma_{k'}(t') = - (\, \Gamma s^{z-2}) \ (k'^2+as^2)\ \sigma_{k'}(t')$$

$$+ \zeta'_{k'}(t') \quad , \tag{1.21a}$$

$$\langle \zeta'_{k'}(t') \, \zeta'_{k''}(t'') \rangle = 2(\Gamma s^{z-2+\eta}) \, \delta(t'-t'') \, \delta_{-k',k''} \,,$$

$$0 < k' < \Lambda . \qquad (1.21b)$$

The final expression for the transformation in the space of parameters $\mu = (\Gamma, a)$ is $\mu' = R_s \mu$,

$$\Gamma' = \Gamma s^{z-2} \quad ,$$

$$a' = a s^2 \quad . \qquad (1.22)$$

We have set $\eta = 0$, otherwise the correlation function of ζ'_k and the damping of $\sigma_{k'}$ would violate the Einstein relation.

Any point $\mu^* = (\Gamma^*, a^*)$ in the parameter space which is invariant under the transformation represents a scale invariant set of equations. Such a point is a <u>fixed point</u>. From (1.22), it is obvious that in order to have a fixed point we must choose

$$z = 2 \qquad (1.23)$$

in order that $\Gamma^* \neq 0$. It is also clear that $a^* = 0$, or $a^* = \infty$. Of course there are fixed points with $\Gamma^* = 0$ or ∞ , but they are not very interesting.

For $z = 2$, and $a = 0$, we have a fixed point for every value of Γ.

From the transformation properties of the parameters, we can learn something about the solution of the equations without solving them explicitly. This is of course a major motivation for studying the transformations. The quantities of interest are often the correlation function C defined by

$$C(k, \omega) = \int dt \, e^{i\omega t} \langle \sigma_k(t) \, \sigma_{-k}(0) \rangle \qquad (1.24)$$

and the response function G defined by adding a small external field h:

$$\langle \sigma_k(\omega) \rangle = G(k,\omega) \, h_k(\omega) + O(h^2). \qquad (1.25)$$

C and G are related by

$$C(k,\omega) = \frac{2}{\omega}\,\mathrm{Im}G(k,\omega) \quad ,$$

$$G(k,\omega) = -\int \frac{d\omega'}{2\pi} \; \frac{C(k,\omega')\omega'}{\omega-\omega'+i0^+} \tag{1.26}$$

For those modes $\sigma_k(t)$ which are not eliminated in the RG transformation, the content of their equations of motion is not changed except for a change of name defined by (1.18). Therefore

$$\langle \sigma_k(t)\ \sigma_{-k}(0)\rangle_\mu = s^{2-\eta} \langle \sigma_{sk}(ts^{-z})\,\sigma_{-sk}(0)\rangle_{\mu'} \tag{1.27}$$

which implies

$$C(k,\ \omega, \mu) = s^{2-\eta+z}\ C(sk,\ \omega s^z, \mu'), \tag{1.28a}$$

$$G(k,\ \omega, \mu) = s^{2-\eta}\, G(sk,\ \omega s^z,\ \mu'). \tag{1.28b}$$

For our simple example, we have $z = 2$, $\eta = 0$ and

$$C(k,\omega,a) = s^4\ C(sk,\ \omega s^2,\ as^2)$$

$$G(k,\omega,a) = s^2\ G(sk,\ \omega s^2,\ as^2) \tag{1.29}$$

Let us write $\xi = a^{-1/2}$, and set $s = \xi$. Then

$$C(k,\omega,a) = \xi^4\ C(k\xi,\ \omega\xi^2,\ 1)$$

$$G(k,\omega,a) = \xi^2\ G(k\xi,\ \omega\xi^2,\ 1) \tag{1.30}$$

These are of course the statement of the dynamic scaling hypothesis for T very close to T_c. If $\xi \to \infty$, we set $s = k^{-1}$ to obtain

$$C(k,\omega,0) = k^{-4}\ C(1,\ \omega/k^2,\ 0)$$

$$G(k,\omega,0) = k^{-2}\ G(1,\ \omega/k^2,\ 0) \tag{1.31}$$

This is everything that the RG transformation tells us. Note that we have not solved any equation. Nor have we spoken of above or below T_c. We simply transformed one set of equations into another.

II. Simple Examples

We shall apply the RG transformation to a couple of less trivial examples as further illustrations of the ideas introduced earlier. In intermediate stages of our discussion, we shall skip the detailed algebra.

2.1. Weak Impurity

There are always impurities in real materials. The study of critical behavior must pay attention to their effects. There are many different forms of impurities, random distortions of the crystal lattice, non-magnetic atoms replacing a fraction of spins, etc.. The effects of impurities(even though a minute amount) on critical behavior are expected to be very important for the following reason. A small amount of impurity acts as a small perturbation. The response of the system to the perturbation is given by various susceptibilities. Near T_c, some of them are singular functions of $T-T_c$ and may even diverge at T_c. This means that a small amount of impurity can produce a large effect near the critical point. The critical exponents may be modified. The critical point itself may disappear.

Now we shall consider a model with weak non-magnetic impurity. This is the kind which does not change under the inversion of spins (or rotations of spins). We want to see how the relaxation time is affected.

Let us write the equation of motion in coordinate space

$$\frac{\partial \sigma_i}{\partial t} = -\Gamma \frac{\delta \mathcal{H}}{\delta \sigma_i} + \zeta_i \quad , \qquad i = 1,2,\ldots n \ ,$$

$$\mathcal{H} = \frac{1}{2} \int d^d x \left[(\nabla \sigma)^2 + r_0 \sigma^2 + \frac{1}{4} u \sigma^4 + \varphi \sigma^2 \right] \ ,$$

$$\langle \varphi(x) \rangle = 0 \ ,$$

$$\langle \varphi(x) \varphi(x') \rangle = \Delta \delta(x-x') \ ,$$

$$\langle \zeta_i(x,t) \zeta_j(x',t') \rangle = 2 \Gamma \delta_{ij} \ \delta(x-x') \ \delta(t-t') \qquad (2.1)$$

This is just the Time-Dependent Ginzburg-Landau model of an n-component spin density σ_i . In addition to the noise ζ_i, there is the fixed random field coupled to σ^2. Note that φ is fixed in time, or "quenched". It has a short range correlation. We have not included any mode-mode coupling.

This model is defined by the set of parameters

$$\mu = (r_o, u, \Delta, \Gamma) \quad . \tag{2.2}$$

Now we want to carry out the transformation $R_s \mu = \mu' = (r_o', u', \Delta', \Gamma')$. Before doing that, we ask first whether Δ' would grow or diminish as s increases. If Δ' diminishes, then, if Δ is unimportant, the critical behavior would be that of a pure system. (Here we are assuming there is a fixed point $\mu_o^* = (r_o^*, u^*, 0, \Gamma^*)$.) If Δ' grows, then we expect the critical behavior of the pure system to be modified. We can answer this question by a simple dimensional analysis.

From statics, we know that r_o transforms into $r_o s^{2-\eta}$. Eq. (2.1) shows that φ has the same dimension as r_o. Thus, Δ will transform to $\Delta' = \Delta s^{2(2-\eta)-d}$ where the -d comes from the $\delta(x-x')$ defining Δ. Of course, this argument is correct only if r_o and Δ are small and near the fixed point with $r_o^* = 0$. Otherwise u and non-linear terms in Δ will enter. Anyway, we already see that for $d > 4$, Δ' will diminish. For $d < 4$, it is not clear. Therefore, one considers $d = 4 - \epsilon$.

The static RG for small ϵ was worked out to $O(\epsilon^2)$ by Lubensky [1] who determined various fixed points. His results conclude that there is a fixed point with $\Delta^* \neq 0$ if the specific heat exponent α for the pure system is positive. If $\alpha < 0$, then $\Delta^* = 0$ and the pure fixed point is stable. This conclusion is in fact general and easy to understand. The random field φ in (2.1) acts like a local change of r_o, or a "temperature fluctuation". The specific heat is just the temperature susceptibility giving the response to a small change in temperature. If $\alpha > 0$, the response is thus infinite at the critical point and we expect the fixed point describing critical behavior of the pure system to be modified substantially [2] .

Lubensky's results also show that $\Delta^* = O(\epsilon)$, i.e., Δ^* is small when ϵ is small. In the following, we shall take the static results for granted and concentrate on the dynamical aspect.

For simplicity, let us forget about the u term for the moment and write (2.1) in Fourier component form:

$$\frac{\partial \sigma_k}{\partial t} = - \Gamma (r_o + k^2)\ \sigma_k + \zeta_k$$

$$- \Gamma \sum_{q'} \varphi_{k-q'}\ \sigma_{q'} \quad , \qquad k, q' < \Lambda. \tag{2.3}$$

The spin index i is suppressed.

We want to eliminate the shortwave modes σ_q, $\Lambda/s < q < \Lambda$. So let us solve (2.3) for σ_q to obtain

$$\sigma_q(t) = \Gamma \int_{-\infty}^{t} dt'\ e^{\Gamma (r_o + q^2)\ (t'-t\ )}$$

$$\times \left[\zeta_q(t') - \sum_{k'} \varphi_{q-k'}\ \sigma_{k'}\ (t') \right] \tag{2.4}$$

Now substitute this in (2.3) for $k < \Lambda/s$. The sum in (2.3) has a part with $\Lambda/s < q' < \Lambda$. Let us concentrate on the term involving φ :

$$\ldots \sum_{q'} \varphi_{k-q'} \sum_{k'} \varphi_{q'-k'}\ \sigma_{k'}(t') \tag{2.5}$$

The product of two φ's has a net average over the probability distribution of φ.

$$\langle \varphi_{k-q'}\ \varphi_{q'-k'} \rangle = \Delta \delta_{kk'}\ . \tag{2.6}$$

This makes (2.3) look like

$$\frac{\partial \sigma_k}{\partial t} = - \Gamma (r_o + k^2)\ \sigma_k$$

$$+ \Gamma^2 \Delta \sum_q \int_{-\infty}^{t} dt'\ e^{- \Gamma (r_o + q^2)(t-t')}\ \sigma_k(t')$$

$$+ \ldots \tag{2.7}$$

where q is summed over the shell $\Lambda/s < q < \Lambda$. This is already an integral equation for σ_k. To reduce it to the form of (2.1), we smear it out in time. Write

$$\sigma_k(t') = \sigma_k(t) + (t'-t)\ \frac{\partial \sigma_k(t)}{\partial t} + \ldots \tag{2.8}$$

and substitute it in (2.7). The integral in t′ is then easily done:

$$\frac{1}{\Gamma}\frac{\partial\sigma_k}{\partial t} = -(r_0 + k^2)\,\sigma_k$$
$$+\,\Delta\sum_q\left(\frac{1}{r_0+q^2}\,\sigma_k - \frac{1}{\Gamma(r_0+q^2)^2}\,\frac{\partial\sigma_k}{\partial t}\right)$$
$$+ \dots \tag{2.9}$$

Move the $\frac{\partial\sigma_k}{\partial t}$ on the right to the left to obtain

$$\frac{1}{\Gamma}(1+\Delta K_4 \ln s)\,\frac{\partial\sigma_k}{\partial t} = -(r''_0 + k^2)\,\sigma_k + \dots \tag{2.10}$$

since

$$\sum_q \frac{1}{(r_0+q^2)^2} = (2\pi)^{-4}\int d^4q\,\frac{1}{(r_0+q^2)^2} = K_4 \ln s \tag{2.11}$$

and where $K_4 \equiv 1/(8\pi^2)$. We take d=4 in doing the integral as a first approximation. After going through the replacement (1.18) and the algebra (1.19) - (1.21), we simply find

$$\Gamma' = \Gamma(1-\Delta K_4 \ln s)\,s^{z-2} \tag{2.12}$$

which differs from (1.22) by the Δ term.

So far we have kept track of only a part of the transformed equation of motion. The rest of the transformation will give the static results, which we simply state here:

$$r'_0 = s^2\left\{r_0 + \left[\left(\frac{n}{2}+1\right)u - \Delta\right]\right.$$
$$\left.\times\,K_4\left(\frac{\Delta^2}{2}(1-s^{-2}) - r_0 \ln s\right)\right\},$$
$$u' = s^{\varepsilon}\,u\left\{1-\left[\left(\frac{n}{2}+4\right)u - 6\Delta\right]K_4 \ln s\right\},$$
$$\Delta' = s^{\varepsilon}\Delta\left\{1-\left[(n+2)u - 4\Delta\right]K_4 \ln s\right\}, \tag{2.13}$$

good to $O(\varepsilon)$. The fixed point value of Δ is

$$\Delta^* = \frac{\varepsilon(4-n)}{8(n-1)K_4}. \tag{2.14}$$

Using this value in (2.12), we obtain

$$z = 2 + K_4 \Delta^*$$

$$= 2 + \frac{\varepsilon(4-n)}{8(n-1)} + O(\varepsilon^2) \ , \ 1 < n < 4 \quad . \qquad (2.15)$$

The $O(\varepsilon^2)$ is available in a paper by Grinstein, Mazenko and the lecturer. [3]

The factor $(4-n)\varepsilon$ in (2.14) shows that for $n > 4$ Δ^* must be zero since Δ is always positive. To $O(\varepsilon)$, the specific heat exponent α is proportional to $(4-n)\varepsilon$. Thus the 4-n in (2.14) should be interpreted as reflecting the fact that Δ^* is non-zero only if $\alpha > 0$. For n=1, (2.14) evidently breaks down. The formula (2.13) is in fact inadequate. It was shown by Khmelnitsky [4] that for n = 1,

$$\Delta^* = \left(\frac{6}{53} \varepsilon \right)^{1/2} / K_4 + O(\varepsilon) \ . \qquad (2.16)$$

Thus the exponent z is

$$z = 2 + \left(\frac{6}{53} \varepsilon \right)^{1/2} + O(\varepsilon) \ , \ n = 1 \ . \qquad (2.17)$$

For the purpose of extrapolating down to d = 3, the results involving α are usually not very helpful. While $\alpha > 0$ for $n > 4$ to $O(\varepsilon)$, the value of α for d = 3 appears to be negative for $n > 1$. For n = 1, d = 3, our knowledge about α seems quite incomplete.

What we have learned from the RG study is, among other things, how the effect of the impurities are modified by the eliminated modes. Another way of saying it is that we considered the way impurities are "dressed" upon coarse graining. The important thing is to find out whether the impurity effect is amplified or diminished upon dressing.

2.2. Ferromagnet

Let us consider a dynamical model of a 3-component isotropic ferromagnet as another example of the application of the RG transformation. [6]

There are two important features of the isotropic ferromagnet, the conservation of the total spin, and the precession of spins. The precession is described by the equation

$$\frac{\partial \vec{\sigma}}{\partial t} = \lambda\, \vec{\sigma} \times \vec{B} \tag{2.18}$$

where λ is a constant, and $\vec{B}$ is the local magnetic field seen by $\vec{\sigma}$:

$$\vec{B} = \nabla^2 \vec{\sigma} + \vec{h} \quad . \tag{2.19}$$

Note that $\nabla^2 \vec{\sigma}(x)$ is the mean of the neighboring spins minus $\vec{\sigma}(x)$ itself. Putting in the damping and the noise, we have

$$\frac{\partial \sigma_{ik}}{\partial t} = v_{ik} - \gamma k^2 \frac{\partial \mathcal{H}}{\partial \sigma_{i-k}} + \zeta_{ik} \quad . \tag{2.20}$$

$$\langle \zeta_{ik}(t)\, \zeta_{jk'}(t') \rangle = 2\gamma k^2\, \delta_{-k,k'}\, \delta(t-t')\, \delta_{ij} \quad ,$$

$$k < \Lambda, \; i = 1, 2, 3,$$

$$\vec{v}_k = \lambda L^{-d} \sum_{k'} \vec{\sigma}_{k+k'} \times \vec{B}_{-k'} \tag{2.21}$$

The mode-mode coupling term v_{ik} describes the precession. The γk^2 factor reflects the conservation of total spin. The set of parameters defining the model is

$$\mu = (\lambda, \gamma, r_0, u) \tag{2.22}$$

There is a fixed point μ_0^* with $\lambda^* = 0$, i.e., the TDGL model with conserved spin [5] giving $z = 4 - \eta$. The first thing one should decide is whether μ_0^* is stable when a small λ is introduced, i.e., whether λ' in $\mu' = R_s \mu$ would diminish as s increases. This is easily checked through the scale change operation (1.18) - (1.20) on the part involving λ , i.e.,

$$\frac{\partial \sigma}{\partial t} = - \lambda\, \sigma \times \nabla^2 \sigma \tag{2.23}$$

which goes to

$$s^{-z} \frac{\partial \sigma}{\partial t'}(x',t') = \lambda \sigma \times s^{-2}\, \nabla'^2\, s^{-\frac{d}{2}+1-\eta/2}\, \sigma(x',t')$$

or

$$\lambda' = \lambda\, s^{-1-\frac{d}{2}+4} \tag{2.24}$$

where we have set $z = 4$ and $\eta = 0$. Clearly, if $d > 6$, λ' will diminish and for $d < 6$, it will grow . For $d = 6 - \varepsilon$ with small ε , we can use perturbation methods to perform the RG transformation. We shall quote the result here. The details will appear later. One finds

$$\lambda' = \lambda\, s^{z-4+\epsilon/2}$$

$$\gamma' = s^{z-4}\, \gamma(1 + \tfrac{1}{3}(\lambda/\gamma)^2 K_6 \ln s) \,, \tag{2.25}$$

with $K_6 = (64\pi^3\varepsilon)^{-1}$. (2.26)

This formula gives two fixed points with

$$\tilde{\lambda}^* = \pm(96\pi^3\varepsilon)^{1/2} \,,$$

$$z = 4 - \varepsilon/2 \,, \tag{2.27}$$

where $\tilde{\lambda} \equiv \lambda/\gamma$. The $\pm$ sign simply gives the direction of precession (right-hand or left-hand).

III. Monte Carlo Study of the Renormalization Group

Numerical study of the renormalization group has given a great deal of insight into the foundation of the RG concept. [1] The fact that the numerical program is still at its developing stage shows that our understanding of the RG is still immature, and the introduction of new techniques should be helpful. The Monte Carlo method has been a very successful numerical tool for many branches of science [2] . It seems entirely evident that one ought to try it on the study of the RG. Here I shall report the basic methods and preliminary results of such a study.

The reader should be reminded that the RG is meant to be a transformation on a scale much shorter than the correlation length, (we always assume short range interactions), and the quantities being transformed are the parameters describing the interactions. In other words, the RG deals with smooth "local" quantities, (in contrast to the global quantities like the specific heat and susceptibility which are singular at the critical point). This is of course the basic motivation of using the RG to study critical phenomena.

An advantage of the Monte Carlo Method is its flexibility. One can study the RG with different block sizes and shapes, and lattices in different dimensions with basically the same numerical program. Another advantage is the "experimental" nature of the method. These advantages in particular allow the study of the dynamic RG which so far has not been done except using perturbation theory in specialized dimensions, [3] as we have discussed earlier. The major disadvantage so far is the poor numerical precision (5% for statics and 25% for dynamics) which can be improved easily by more skillful programming and longer computing time. The results reported here are obtained in 1 minute (static results) to 3 minutes (dynamic results) on a CDC 7600 computer at Saclay.

3.1. The Kinetic Ising Model

Before outlining the program, we define the kinetic Ising model which we use to demonstrate the method. Consider Ising spins $\sigma_r = \pm 1$, (r is the coordinate vector), on a 2-dimensional square lattice. The dynamics (and hence the statics) is defined by the flip probability

$$w(\sigma_r, B_r)\,dt = \Gamma e^{-\sigma_r B_r}\,dt , \tag{3.1}$$

for each spin, which is the probability that the spin at r flips from σ_r to $-\sigma_r$ in the time interval dt. B_r is the local magnetic field seen by σ_r, and is the sum of the applied field h and the field produced by neighbors. For example, if there is only nearest neighbor interaction, we have

$$B_r = h + J \sum_{\delta} \sigma_{r+\delta} \tag{3.2}$$

where J is interaction strength and δ is summed over nearest neighbors. In general, we can write formally

$$B_r = \partial \mathcal{H}[\sigma] / \partial \sigma_r , \tag{3.3}$$

where $\mathcal{H}[\sigma]$ is the -(Hamiltonian/temperature) for the spin configuration σ . The quantity Γ in (3.1) is independent of σ_r but may depend on the neighboring spins. The transition rate $w(\sigma_r, B_r)$ of the process $\sigma_r \rightarrow -\sigma_r$ and that of the inverse process $-\sigma_r \rightarrow \sigma_r$ clearly satisfy the condition of detailed balance, namely

$$\frac{w(-\sigma_r, B_r)}{w(\sigma_r, B_r)} = e^{2B_r\sigma_r} = e^{\mathcal{H}[\sigma] - \mathcal{H}[\sigma']} \tag{3.4}$$

where the configuration σ' is obtained from σ by reversing σ_r . This condition implies that the probability distribution for the spin configurations will approach exp ($\mathcal{H}[\sigma]$) after a long time.

Let μ denote the set of parameters specifying the model

$$\mu = (J, K, L, \Gamma) \tag{3.5}$$

where J, K, L refer to respectively the nearest-neighbor, next-nearest-neighbor and four-spin interaction parameters. Of course, one can and should try to study the difference in results when more parameters are included. What we report here is all based on (3.5).

The basic steps of the program are shown in the following chart.

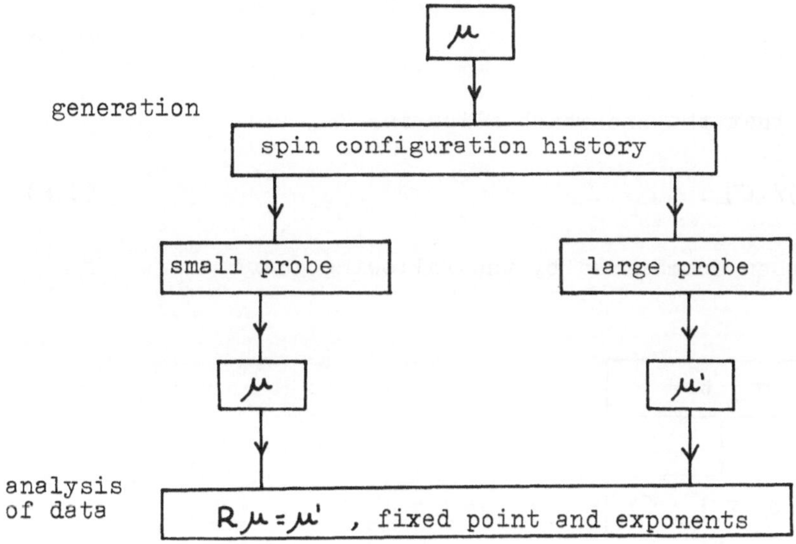

Given μ, we generate a sequence of spin configurations or a "history" according to the probability (3.1). Then we can "measure" the interaction parameters which generate these configurations. The small probe is to measure μ . The results should be just the μ we start with. The purpose is to verify this and to see how accurately the generated history represents the intended distribution. The large probe is designed to measure the effective interaction parameters $\mu' = (J', K', L', \Gamma')$ among block spins. The transformation from μ to μ' is the RG transformation R. It is then a matter of data analysis to obtain the fixed point and critical exponents. We now proceed to explain these steps in more detail.

3.2. Generation of Spin Configurations

The spins are flipped one at a time as follows. Given a configuration σ at time t, the probability that no spin shall flip during the subsequent interval t' is $\exp(-\Omega t')$ where

$$\Omega = \sum_r w(\sigma_r, B_r) \tag{3.6}$$

is the probability per second that one of the spins flips. Thus, the probability that nothing happens for t' and then one spin flips during the subsequent interval dt' is

$$e^{-\Omega t'} \Omega dt' \quad . \qquad (3.7)$$

The probability that the one which flips is σ_r is

$$w(\sigma_r, B_r)/\Omega . \qquad (3.8)$$

The program is then summarized by the following chart.

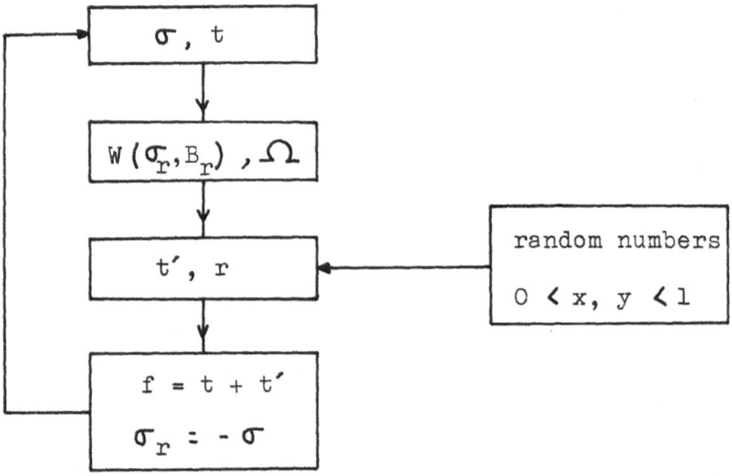

The random number generator determines the time t' and which spin to flip according to (3.7) and (3.8) respectively. In practice we take

$$t' = -(\ln x)/\Omega . \qquad (3.9)$$

Divide the interval (0,1) into N portions of length given by (3.8), where N is the total number of spins. The random number y falls in one of the portions. The spin belonging to that portion gets flipped. The procedure is then repeated.

3.3. Determination of Parameters

The parameters are determined by observing how a spin behaves in a given environment provided by its neighbors. When the range of interactions is no longer than the next nearest neighbor, the environment of a spin is defined by the configuration of its

8 neighbors. What we measure are τ_+ , the length of time that a spin spends in +1 state before flipping and τ_- that it spends in -1 state for a given environment. The ratio

$$\tau_+/\tau_- = e^{2B_r} \qquad (3.10)$$

gives an equation for J, K and L since B_r is a linear combination of these parameters. For example, if all 8 neighboring spins are + 1, then $B_r = 4 (J + K + L)$. By measuring the ratio τ_+/τ_- for three or more different environments, one can solve for J, K, and L. The geometrical mean $(\tau_+\tau_-)^{1/2}$ gives Γ . For determining the static parameters J, K and L, one can simply accumulate the total time spent in the + state and that in the - state, and then take the ratio. For determining the dynamic parameter Γ, one has to measure the mean time that a spin stays without flipping in the + 1 or in the - 1 state. One must also make the correction due to the fact that the spins in the neighborhood are also allowed to flip, (division by the branching ratio). These measurements are what we called "small probe" in the first chart. The small probe looks at 3 x 3 spins. The probe scans through the system of 15 x 15 spins to observe the local configurations. We fix the 56 spins on the boundary to + 1 and the probe is kept away from the boundary. Note that our measurement is local, i.e., we always look at one spin at a given environment, never any collective property of the whole system. The role of the boundary spins in our calculation is entirely different from that in the calculations cited in Ref.[1]. The purpose of fixing the boundary spins here is technical.

The large probe is the same as the small probe except that it looks at 3 x 3 block spins at a time. Each block spin is a set of 2 x 2 spins. The value of the block spin σ' is chosen as + 1 if the majority of the constituents (excluding one spin arbitrarily to assure majority rule)of the block is positive, and - 1 if negative. The results of measurements of the large probe gives $\mu' = (J', K', L', \Gamma')$. A few remarks are in order.

Of course there are truncations involved here just as there were in earlier numerical works [1], even though they are of somewhat different nature. First of all, earlier work showed that there are, although small, interactions between block spins with a range longer than next-nearest-neighbor. There are also other kinds of interactions within the 3 x 3 block spins. In short, we should measure more than just J',K',L'and should use a bigger probe.

Fitting the data with just J', K', and L' is a truncation. The accuracy of the computation so far does not allow a study of the effect of truncation at the level of the four-spin interaction, i.e., if we make a two-parameter fit (J', K'), the results of exponents are within our accuracy the same as a three-parameter fit with L'.

The measurement of Γ' is more complicated. Intuitively one expects that the block spins would not flip so easily as the original spins and hence have a longer relaxation time. This picture is incomplete. When a block spin flips over, there is an intermediate period during which the block spin flips rapidly, (as rapidly as the original spin). This is the period when the block spin is determined by a single spin because the other two have opposite signs. Thus one must smooth out this rapid fluctuation in order to obtain the Γ' which fits the longer time behavior of a block. This can be done by constructing from the data the histogram which gives the probability distribution of observed times. One plots this distribution on a semi-log paper and indeed sees two time scales. Fitting a straight line to the long-time part and determining the slope, one then deduces Γ'. As we mentioned before, Γ and also Γ' may depend on the neighboring spins. We need to introduce more parameters to account for this dependence. We have not included more parameters. The dynamic part is so far only a one-parameter fit.

There is another way of determing μ'. Instead of letting spins flip according to (3.1), we impose on the spins another field $B_r[\sigma']$. This new field is prescribed by a particular block configuration σ' and is so strong that only those spin configurations which are consistent with the block configuration σ' can occur. This approach is closer to the spirit of earlier work cited in Ref. [1] . One can also fix the 8 neighbor block spins of an unfixed block spin and observe how this unfixed one behaves. Such calculations are also done, with 12 x 12 spins and with periodic boundary condition. It is not entirely clear to what extent this approach is equivalent to the one described above. No significant difference between the exponents calculated with the two approaches are observed. We also tried larger size blocks (3 x 3 per block instead of 2 x 2) and obtained the same results within the accuracy of our calculation.

3.4. Extraction of Static and Dynamic Exponents

The static exponents are obtained through the eigenvalues of the linearized RG transformation evaluated at the fixed point μ^*. The "temperature eigenvalue" $\lambda_T = 2^{1/\nu}$ is the largest eigenvalue of the 3 x 3 matrix $\partial J'_\alpha / \partial J_\beta$, ($J_1$ = J, J_2 = K, J_3 = L). The "magnetic eigenvalue" $\lambda_H = 2^{2-\eta/2}$ taken as the larger of the eigenvalues of the 2 x 2 matrix $\partial h'_\alpha / \partial h_\beta$, ($h_1$=h, h_2 = 3-spin interaction parameter). These matrices are calculated directly through various average values at given block configurations when the large probe scans through the system. For example

$$\frac{\partial}{\partial J_\alpha} (J' + K' + L') = \frac{\partial}{\partial J_\alpha} \frac{1}{8} \ln(\tau'_+ / \tau'_-)$$

$$= \left\langle \frac{\partial \mathcal{H}}{\partial J_\alpha} \right\rangle_+ - \left\langle \frac{\partial \mathcal{H}}{\partial J_\alpha} \right\rangle_- \tag{3.11}$$

where $\tau'_\pm$ is the time that a block spin spends in ± state given all 8 neighboring block spins in + 1 state and

$$\langle A \rangle_\pm \equiv \sum_\sigma^\pm e^{\mathcal{H}} A \Big/ \sum_\sigma^\pm e^{\mathcal{H}} . \tag{3.12}$$

This means taking the average over the original spins at fixed block spins, the center one fixed at ± 1 and the 8 neighbors at +1. Eq. (3.11)follows from the fact that

$$\tau_\pm \propto \ln\left(\sum_\sigma^\pm e^{\mathcal{H}} \right) . \tag{3.13}$$

Average values like (3.11) converage much faster than the interaction parameters themselves in our calculation.

In our calculations so far the eigenvalues λ_T and λ_H converged to the ideal values 2 and 3.668 respectively, (For ν = 1, η = 1/4), within 10% after about 50 flips per spin and stablized to 1.9 and 3.4 respectively after 200 flips/spin with a fluctuation of 5%. For the programs which used fixed block spins, more flips are needed (1000 flips/spin) . However the scanning process takes time so that both approaches took about the same total computing time. To obtain the dynamic results, a longer time is needed because we need to obtain the histogram for the time between flips. Using three times as long as the static program, we obtain the ratio $\Gamma / \Gamma' = 2^z$ ranging from 2.6 to 3.0. This puts z between 1.35 and 1.6. This result may seem crude, but so far we have known no other non-perturbative dynamic RG calculation.

To obtain more precise results, the program needs to be improved. So far little attention has been paid in minimizing computer time and round-off errors. We have merely demonstrated that the Monte Carlo method is indeed very useful in studying the RG, both in statics and in dynamics. It is ready for applications to different models.

IV. Perturbation Methods

We shall briefly review the graphic expansion and then apply it to the study of the spin glass.

4.1. Graph Expansion

Let us start with the model defined by (2.1). namely, a TDGL model with a random field φ added. Let us write the equation of motion in the frequency-wavevector representation

$$- i\ \omega \sigma_k(\omega) = - \Gamma(k^2 + r_o)\ \sigma_k(\omega) + \Gamma h_k(\omega)$$

$$- \quad \Gamma \sum_{k'} \varphi_{k-k'}\ \sigma_{k'}(\omega) + \zeta_k\ (\omega) \tag{4.1}$$

An external field h is added. We omit the $u\sigma^2\sigma$ term for the moment. We now rewrite (4.1) as

$$\sigma_k(\omega) = \sigma_k^o(\omega) - G_o(k,\omega) \sum_{k'} \varphi_{k-k'}\ \sigma_{k'}(\omega)$$

$$+ G_o\ (k,\omega)\ h_k(\omega)\ . \tag{4.2}$$

where

$$G_o^{-1}(k,\omega) = - \frac{i\,\omega}{\Gamma} + k^2 + r_o\ , \tag{4.3}$$

$$\sigma_k^o(\omega) = G_o(k,\omega)\ \frac{1}{\Gamma}\ \zeta_k(\omega)\quad . \tag{4.4}$$

Eq. (4.2) can be solved by iteration in powers of φ . Let us draw a line for G_o, a wavy line for $-\varphi$, a cross for h and a short line ending for $\sigma_k^o(\omega)$. Then the solution of (4.2) is represented by a series of graphs shown in Fig. 6

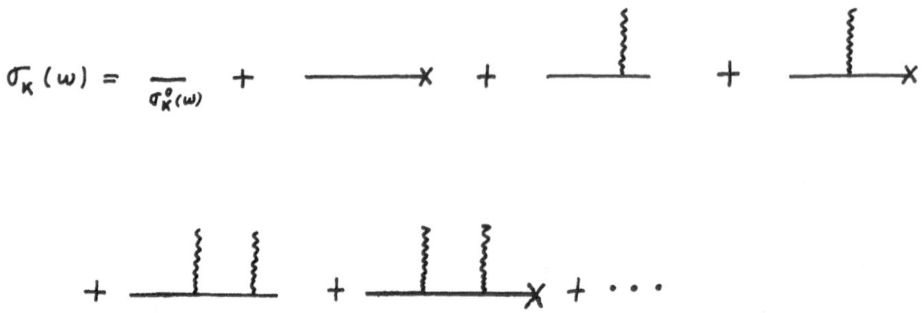

Fig. 6

Assuming that φ has a Gaussian distribution of zero average, the graphs with odd number of wavy lines would not contribute when we average over φ. For graphs with even numbers of wavy lines, we join the wavy lines by pairs when averaging. By definition, the response function G is given by

$$\langle \sigma_k(\omega) \rangle = G(k, \omega)\, h_k(\omega) + O(h^2) . \qquad (4.5)$$

Since $\langle \sigma_k^0 \rangle = 0$, (assume that the temperature is above T_c,) we have the graphs for $G(k, \omega)\, h_k(\omega)$ in Fig. 7

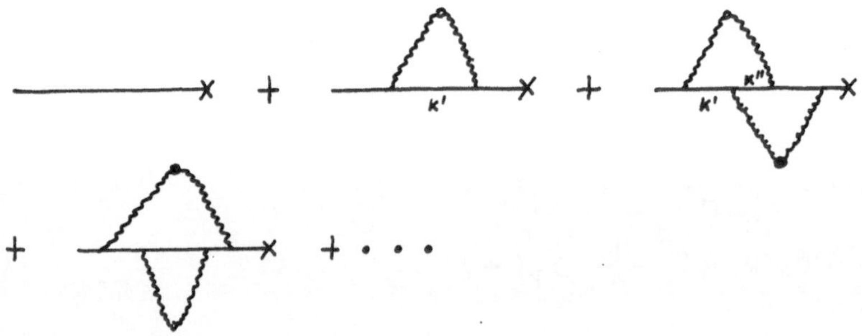

Fig. 7

The small circles are to remind you that an average over the probability distribution of φ is taken.

The analytic expression for these graphs is

$$G(k,\omega) = G_0(k,\omega) + G_0^2 \int \frac{d^dk'}{(2\pi)^d} G_0(k',\omega)\,\Delta$$

$$+ G_o^2 \int \frac{d^dk'}{(2\pi)^d} \frac{d^dk''}{(2\pi)^d} G_0(k',\omega)\, G_0(k'',\omega)$$

$$\times\, G_0(k-k'-k'',\omega)\,\Delta^2$$

$$+ \dots \tag{4.6}$$

We have used the formula

$$\langle \varphi_k \; \varphi_{-k'} \rangle = \Delta\, \delta_{kk'} \; . \tag{4.7}$$

Note that the random field φ is static, that is why the frequency variable ω of G_0 is never changed by φ. The component label of the line is also unchanged and has been suppressed.

We can now easily generalize the above graph expansion to include the $u\,\sigma^2\sigma$ term in the kinetic equation. We add a term

$$-\frac{1}{2}u\, G_0(k,\omega) \int \frac{d\omega'}{2\pi} \frac{d\,\omega''}{2\,\pi} \sum_j \sigma_{jk'}(\omega')\; \sigma_{jk''}(\omega'')\; \sigma_{ik-k'-k''}(\omega-\omega'-\omega'') \tag{4.8}$$

to (4.1). This is of course just the Fourier transform of $-\frac{1}{2}u\,\sigma^2(x,t)\,\sigma_i(x,t)$, $(\sigma^2 \equiv \Sigma_j\; \sigma_j^{\,2}$, $j = 1,\dots n)$.
We have written the sum over components explicitly.

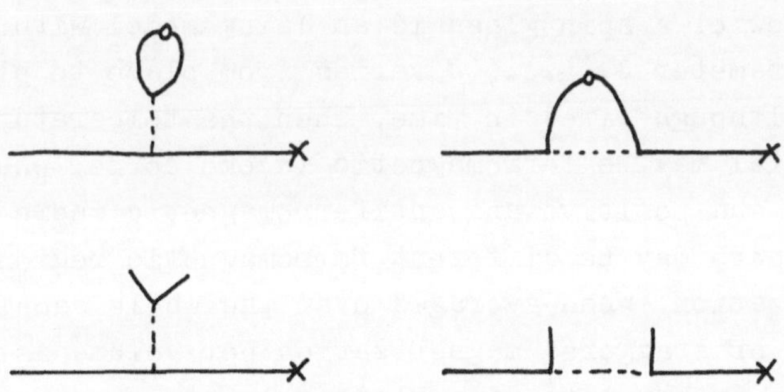

Fig. 8

Fig. 8 shows some terms in $G(k, \omega)h_k$ generated by the u term. We use $G(k, \omega)h_k$ generated by the u term. We use a dashed line to represent u and to separate the $\sum_j \sigma_j^2$ part and σ_i in (4.8). The small circle again reminds us of averaging. This time it gives the averaging over the noise ζ. Fig. 8b shows the iteration terms whose averages give Fig. 8a. The short ends represent $\sigma_k^o(\omega)$. By (4.4) we have

$$\langle \sigma_{ik}^o(\omega)\, \sigma_{jk'}^o(\omega') \rangle = \delta_{ij}\, \delta_{-kk'} 2\pi\, \delta(\omega + \omega') C_o(k, \omega)$$

$$C_o(k, \omega) = \frac{2}{\Gamma} G_o(k, \omega) G_o(-k, -\omega)$$

$$= \frac{2}{\omega} \text{ Im } G_o(k, \omega) \quad . \tag{4.9}$$

The analytic expression for Fig. 4.3(a) is

$$-u\, G_o(k, \omega)^2 \int \frac{d^d k'}{(2\pi)^d} \frac{d\omega'}{2\pi} C_o(k', \omega') \left(\frac{n}{2} + 1\right)$$

$$= -u\, G_o(k, \omega)^2 \int \frac{d^d k'}{(2\pi)^d} \frac{1}{r_o + k'^2} \quad . \tag{4.10}$$

The $\frac{n}{2}$ goes with the closed loop of the solid line.

It is a matter of experience to deal with more complicated graphs and to calculate the integrals. You might find some useful hints in my book [1].

4.2. The Spin Glass

The simple graphic expansion outlined above is sufficient for a crude study of the spin glass, which received much attention recently. [2 , 3 , 4 , 5]

The simplest view of a spin glass is an Ising model with a random interaction parameter J, i.e., J varies from place to place in a disordered manner although fixed in time. When the temperature is low enough, the material may be ferromagnetic in one corner where J happens to be large and positive and antiferromagnetic where J is large and negative. There may be different ferromagnetic regions but with opposite magnetization. When averaged over the whole sample, the net magnetization or staggered magnetization per volume is zero, although the system is frozen into some kind of low temperature ordered phase, which is called the "spin glass phase".

To get a simple picture we use the model (4.1), i.e., a ferromagnet with non-magnetic impurities represented by φ. Of course, when φ is too weak, we would not get anything more than we discussed earlier. But if φ is not very weak, there is no perturbation expansion which is useful. What we can do here is to sum a subset of graphs to obtain a mean-field type of description.

First, we have to identify the order parameter if there is a low temperature ordered phase. There are two averagings involved, the time averaging, (or canonical averaging at a given φ), and space averaging, (or averaging over the probability distribution of φ). It is the second averaging that makes the magnetization average to zero. At a given point, the magnetization averaged over time is not zero in the spin glass phase. So let us define the order parameter q as

$$q = \langle\langle\sigma\rangle_t^2\rangle \tag{4.11}$$

where the subscript t refers to time average. For high temperatures, we expect q = 0. So the first thing to look for is some instability which would occur when temperature is lowered. To see the instability, we switch on a very weak random static magnetic field h. The time average of σ_k is just $G_\varphi(k,0)\ h_k$, i.e., the zero-frequency Fourier component. Here G_φ is the response function with a fixed φ. Fig. 9a shows a term in the expansion of G_φ hk.

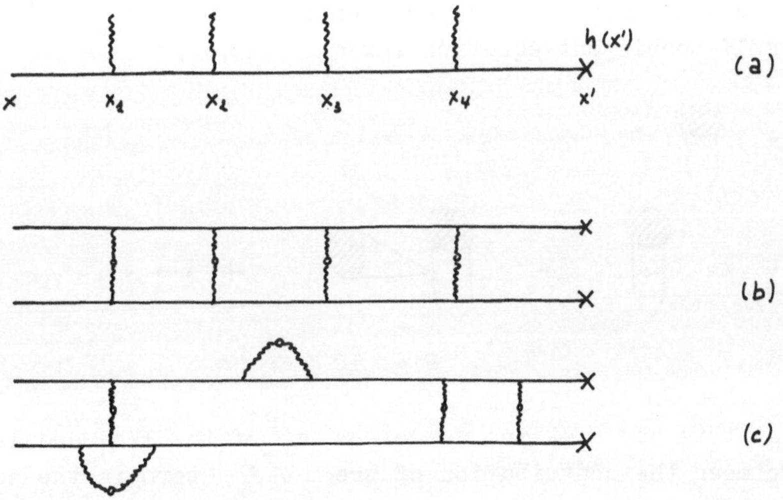

Fig. 9

It is more appropriate to think in terms of position space than in terms of wavevector space. Fig. 9a represents the effect of the random field φ at several different locations on the average $\langle\sigma(x)\rangle_t$. To calculate $\langle\langle\sigma\rangle_t^2\rangle$ we multiply the two identical series for $\langle\sigma(x)\rangle_t$ and average over φ. Since $\langle\varphi(x)\ \varphi(y)\rangle = \Delta\ \delta(x-y)$, the φ from one series can pair up only with the φ at the same position of the other series. A term of the product is shown in Fig. 9b. Clearly there are other terms besides the ladder type of graphs. The wavy lines can pair up in many other ways. Fig. 9c shows an example. In general, $\langle\langle\sigma\rangle_t^2\rangle$ is the sum of graphs which fall into two pieces upon removing all small circles.

Let us just sum the ladder graphs of the type shown in Fig. 9b. Then we get a geometrical series which is easily summed:

$$\langle\langle\sigma\rangle_t^2\rangle \propto \frac{h^2}{1 - \Pi\Delta} \quad , \tag{4.12}$$

where

$$\Pi = \int \frac{d^d p}{(2\pi)^d} \frac{1}{(r_o+p^2)^2} \quad . \tag{4.13}$$

Since Π increases as r_o decreases, there will be a value r_{og} at which the denominator of (4.12) vanishes provided that Δ is sufficiently large. Thus, in the limit $\langle h^2\rangle \to 0$, the order parameter q may not be zero when $r_o < r_{og}$. There is some resemblence between these arguments and those in the theory of superconductivity.

Now we want to know how q depends on r_o when $r_o < r_{og}$. Fig.10 shows the self-consistent equation for q

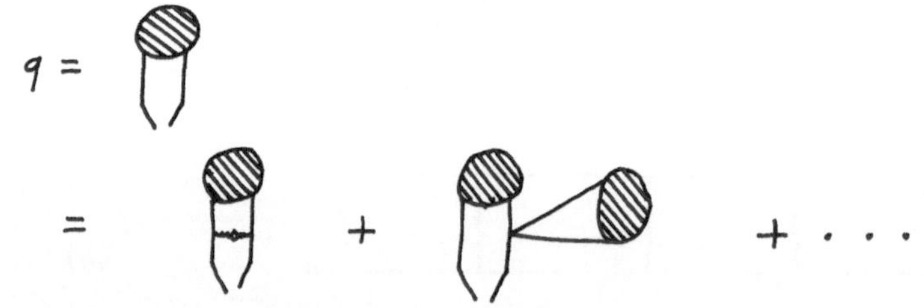

Fig. 10

The last term has a 4-line vertex on one of the legs and is intended to mean the contribution of the u $\sigma^2\sigma$ term in the equation of motion, i.e., the sum of the various permutations of the legs in the interaction represented by the dashed lines earlier. The equation represented by Fig. 10 is

$$q = \Pi\Delta q - \frac{3}{2} u \left(-\frac{1}{\Pi}\frac{\partial}{\partial r_o}\Pi\right) q^2 \quad . \tag{4.14}$$

Without u, we cannot have stability. This is the case for any mean field theory. Note that adding a vertex on a line is the same as applying $-\partial/\partial r_0$ to the G_0 for that line. Since

$$\Pi(r_0) \approx \Pi(r_{og}) + (\frac{\partial}{\partial r_0}\Pi)(r_0 - r_{og}) ,$$

$$1 = \Delta\,\Pi(r_{og}) , \tag{4.15}$$

we have q = 0, or

$$0 = \frac{\partial \Pi}{\partial r_0}\ \Delta\,((r_0 - r_{og}) + \frac{3}{2}u\,q) \tag{4.16}$$

which implies that $q \propto (r_{og} - r_0)$ for small enough $(r_0 - r_{og})$. In the language of exponents, we have $\beta = 1$.

The natural question to ask is what happens if more graphs are included. The answer is that the perturbation series will diverge unless the dimension is greater than 6. We shall stop here and refer further detailed discussion to Ref. [5].

The dynamical behavior of the spin glass is still more complicated and more interesting. At this stage, one does not understand very well the precise nature of the spin-glass phase. The meta-stable configurations play a very important role. It appears that one would not be able to organize and understand the experimental results unless one understands better both the dynamics and the statics.

ACKNOWLEDGEMENT

I thank Professor L. Garrido for his warm hospitality at Sitges and Dr. R.B. Jones and Mr. V. Canivell for helpful comments on the manuscript. My knowledge on the dynamics of random systems, the ferromagnet and the spin glass is due to collaboration with G. Grinstein, G.F. Mazenko and J. Rudnick.

REFERENCES

SECTION I

1) For a more detailed introduction, see S. MA: Modern Theory of Critical Phenomena, W.A. Benjamin, Inc., Reading, Mass. 1976 Chapters XI - XIV.

2) K. KAWASAKI, Ann. Phys. 61, 1 (1970). H. MORI, H. FUJISAKA, and H. SHIGEMATSU, Prog. Theor. Phys. 51, 109 (1974). M.S. GREEN J. Chem. Phys. 20, 1261 (1952), 22, 398 (1954).

SECTION II

1) T.C. LUBENSKY, Phys. Rev. B11, 3573 (1975).

2) For a detailed discussion of this point and generalization see pp. 371-383 of Ref. I.1.

3) G. GRINSTEIN, G.F. MAZENKO and S. MA, to be published.

4) D.E. KHMELNITSKY, preprint 1975.

5) See pp. 464-472 of Ref. I.1 for more discussion of TDGL models. The isotropic ferromagnet was studied using RG by S. MA and G.F. MAZENKO. Phys. Rev. B11, 4077 (1975).

SECTION III

1) Th. NIEMEIJER and J.M.J. van LEEUWEN, Physica, 71, 17 (1974), and their review article in Phase Transition and Critical Phenomena, Vol. 6, C.DOMB and M.S. GREEN, Editors (Academic Press, 1976, to be published). L.P. KADANOFF's lectures in this volume; Chapter VIII of Ref. I.1.

2) J.M. HAMMERSLEY and D.C. HANDSCOMB, Monte Carlo Methods, Methuen, London, 1964; K. BINDER in Phase Transition and Critical Phenomena, C. DOMB and M.S. GREEN, Editors (Academic Press, 1976, to be published), Vol. 5.

3) See chapters XI - XIV of Ref. I.1.

SECTION IV

1) See chapter IV of Ref. I.1.

2) S.F. EDWARDS and P.W. ANDERSON, J. Phys. F 5, 965 (1975).

3) S. KIRKPATRIK and D. SHARRINGTON, Phys. Rev. Lett. 36, 69 (1976).

4) T.C. LUBENSKY, Phys. Rev. Lett. 36,415 (1976).

5) S. MA and J. RUDNICK, to be published.

CRITICAL DYNAMICS IN FOKKER-PLANCK FORMALISM

C.P. ENZ

Département de Physique Théorique
Université de Genève
CH-1211 GENEVE 4

Critical Dynamics in Fokker-Planck Formalism

C.P. ENZ

Département de Physique Théorique
Université de Geneve
CH-1211 GENEVE 4

I. Classification of Models

1.1. Introduction

Dynamics in the vicinity of a critical point is characterized by the softening of one of the long-wave modes of excitation, which means that the complex soft mode frequency ω_q goes to zero at $T = T_c$ and $q = 0$ [1]. The $q = 0$ component of the dynamical variable describing this mode is called the order parameter.

A first general classification results from the distinction between real and imaginary ω_q, corresponding to a propagating and a diffusive (or relaxational) soft mode, respectively. In the first case the order parameter response function $\chi(q,\omega)$ has two poles $\omega = \pm\,\omega_q - i\,\Gamma$ where Γ is the width of the soft mode peak (we suppress the q-dependence of Γ); $\Gamma \gg \omega_q$ is an overdamped mode. In the second case there is a single pole on the imaginary ω-axis, and the softening is called critical slowing down.

This classification is the result of the most fundamental symmetry operation of dynamics, namely time reversal $\mathcal{T}$. In the undamped limit $\Gamma = 0$ of the propagating case the order parameter dynamics is canonical and hence $\mathcal{T}$- invariant. In the diffusive case the order parameter dynamics violates $\mathcal{T}$ and hence is dissipative. This is the case of Ginzburg-Landau (GL) dynamics which historically was the first model of critical dynamics to be analyzed with renormalization group (RG) techniques [2].

1.2. Ginzburg-Landau versus cluster dynamics

GL-dynamics is determined by the GL free energy functional

$$F_{GL} = \int d^d x \left\{ \frac{1}{2} r |\Psi|^2 + \frac{1}{2} |\nabla \Psi|^2 + u |\Psi|^4 \right\} \tag{1.1}$$

where $\Psi = (\Psi_1, \dots \Psi_n)$ is the order parameter, $|\Psi|^2 \equiv \sum_{\alpha=1}^{n} \Psi_\alpha^2$,

$|\nabla\Psi|^2 \equiv \sum_{\alpha=1}^{n} |\nabla \Psi_\alpha|^2$ and $r = (T - T_c)\, a$.

In critical phenomena spatial dimension d and order parameter dimension n are classification parameters and the coefficient of $|\nabla\Psi|^2$ in (1.1) is normalized to one thus fixing a length scale relative to r. In the original version of superconductivity [3] $d = 3$ and $n = 2$, corresponding to a complex Ψ , Ginzburg and Landau only considered equilibrium, as determined by

$$0 = \frac{\delta F_{GL}}{\delta \Psi^*} = r \Psi + 4 u |\Psi|^2 \Psi - \nabla^2 \Psi \tag{1.2}$$

This equation has solutions $\Psi = 0$ and for $T < T_c$,

$$|\Psi|^2 = - \frac{r}{u} = \frac{a}{u} (T_c - T) = n_c \tag{1.3}$$

where n_c is the condensate density. The particular inhomogeneous solutions which reach $\pm \sqrt{n_c}$ asymptotically are

$$\Psi(x) = \pm \sqrt{n_c} \tanh \left(\frac{x - x_0}{\sqrt{2}\, \xi_{GL}} \right) \tag{1.4}$$

where

$$\xi_{GL}(T) = r^{-1/2} \propto (T_c - T)^{-1/2} \tag{1.5}$$

is the GL correlation length.

The generalization of eq. (1.2) to dynamics is not unique. One possibility is (for arbitrary n) the GL dynamics of the Introduction or what is called the time-dependent Ginzburg-Landau (TDGL) equations

$$\dot{\Psi}_\alpha = - \Gamma' \frac{\delta F_{GL}}{\delta \Psi_\alpha} \quad ; \quad \alpha = 1, \dots n \tag{1.6}$$

These equations are manifestly $\mathcal{T}$-violating and diffusive.

Quite a different possibility is given by the Newtonian equations of motion

$$\ddot{\Psi}_\alpha = - c^2 \frac{\delta F_{GL}}{\delta \Psi_\alpha} \quad ; \quad \alpha = 1, \dots n \tag{1.7}$$

which are manifestly $\mathcal{T}$-invariant and can be obtained from canonical dynamics. Since an example of eq.(1.7) is supplied by anharmonic elasticity [4, 5] or cluster dynamics [6] it is not surprising that these equations have soliton solutions

$$\Psi(x,t;v) = \pm \sqrt{n_c} \tanh \left(\frac{x \pm v t - x_o}{\sqrt{2}\, \xi_{GL} \sqrt{1 - v^2/c^2}} \right) \tag{1.8}$$

This shows that c in eq. (1.7) has the meaning of a limiting velocity whereas v is the velocity of propagation of cluster walls [6]. For $v = 0$ eq. (1.4) is recovered which shows that in the sense of these particular solutions eq. (1.7) is a more natural generalization of eq. (1.2) than eq. (1.6).

1.3. Schrödinger-Gross-Pitaevskii dynamics

Apart from a factor i eq. (1.6) has, for $n = 2$, the form of a Schrödinger equation

$$\dot{\Psi} = - \frac{i}{\hbar} \frac{\delta H_{SGP}}{\delta \Psi^*} \tag{1.9}$$

which describes a probability flow and is $\mathcal{T}$-invariant and hence non-dissipative. In this equation

$$H_{SGP} = H_S + H_{GP} \tag{1.10}$$

where

$$H_S = \int d^d x \; \Psi^* \left(- \frac{\hbar^2}{2m} \nabla^2 + V_1(x)\right) \Psi \tag{1.11}$$

is the ordinary Schrödinger Hamiltonian while

$$H_{GP} = \frac{1}{2} \int d^d x \int d^d x' \, |\Psi(x)|^2 \, V_2(x - x') \, |\Psi(x')|^2 \tag{1.12}$$

is the Gross-Pitaevskii interaction introduced to describe Bose fluids [7,8]. If we put in these equations

$$\frac{\hbar}{m} = \Gamma'' \, , \; V_1(x) = \frac{r}{2} \hbar \Gamma'' \, , \quad V_2(x) = 2 u \hbar \Gamma'' \delta(x) \tag{1.13}$$

then $H_{SGP} = \hbar \Gamma'' F_{GL}$ and eq. (1.9) can be combined with eq. (1.6) to give mixed GL + SGP equations

$$\dot{\Psi}_\alpha = -\Gamma \frac{\delta F_{GL}}{\delta \Psi_\alpha} \quad , \quad \alpha = 1,\ldots n \tag{1.14}$$

where

$$\Gamma = \Gamma' + i\,\Gamma'' \tag{1.15}$$

gives rise both to flow and to diffusion.

1.4. Generalized Ginzburg-Landau dynamics

It is also of interest to couple the order parameter to a conserved field ε (e.g. the energy density or the magnetization density in the case of an anti-ferromagnet) via the Hamiltonian

$$H' = \int d^d x \left\{ \frac{1}{2} \; C^{-1} \varepsilon^2 + \gamma |\Psi|^2 \varepsilon \right\} \tag{1.16}$$

Taking into account the effect of the large number of degrees of freedom not explicitly described by the equations of motion with the help of random variables (see Section 2) η_α and ζ we arrive at the set of equations [2,9,10]

$$\left.\begin{aligned} \dot{\Psi}_\alpha &= -\Gamma \frac{\delta H}{\delta \Psi_\alpha} - i\,g\,\Psi_\alpha \frac{\delta H}{\delta \varepsilon} + \eta_\alpha \\ \dot{\varepsilon} &= \lambda \nabla^2 \frac{\partial H}{\partial \varepsilon} + g \sum_{\alpha,\beta} \nabla \cdot (\Psi_\alpha D^o_{\alpha\beta} \nabla \Psi_\beta) + \zeta \end{aligned}\right\} \tag{1.17}$$

where

$$H = F_{GL} + H' = H_0 + H_1 \tag{1.18}$$

and

$$\left.\begin{aligned} \langle \eta_\alpha(x,t)\, \eta_{\alpha'}(x',t') \rangle &= 2\,\Gamma' \delta_{\alpha\alpha'} \cdot \delta(x-x')\, \delta(t-t') \\ \langle \zeta(x,t)\, \zeta(x',t') \rangle &= -2\lambda \nabla^2 \delta(x-x')\, \delta(t-t') \end{aligned}\right\} \tag{1.19}$$

The terms proportional to g are needed in case there exist kinematical constraints ("Poisson bracket relations") due to an internal symmetry of the order parameter for which $\int d^d x\, \varepsilon$ is a generator [9,10,11]). These terms establish the covariance of eq. (1.17) under the internal symmetry, and D^0 is a (skew symmetric) representation matrix.

Writing

$$\Gamma = \Gamma_0 + \Gamma_1 \nabla^2 \tag{1.20}$$

$\Gamma_o = 0$ describes a conserved order parameter while for $\Gamma_o \neq 0$ the order parameter is not conserved [2]. The Γ''-part together with the interaction H_1 of the Hamiltonian (1.18) describes mode-mode coupling terms [12]. We have thus arrived at a wide class of models for critical dynamics. However, Newtonian dynamics is still not contained in it.

1.5. Mixed canonical and dissipative dynamics

A very general class of dynamical systems is defined by the generalized Langevin equations [13]

$$\dot{\Psi}_\mu = f_\mu + \xi_\mu \tag{1.21}$$

in which the deterministic forces f_μ are given by [14]

$$f_\mu = \frac{\partial H}{\partial \Psi_\nu} D_{\nu\mu} = f^o_\mu + f'_\mu \tag{1.22}$$

and the matrix D consists of a skew part D^o and a symmetric part D'

$$D = D^o - D' \tag{1.23}$$

The matrix D^o determines the non-dissipative part f^o_μ [14] which gives rise to canonical dynamics and describes a flow. The matrix D' determines the dissipative part f'_μ [14] which gives rise to GL-dynamics and describes diffusion. D' also determines the fluctuating random variables ξ_μ (see Section 2)

$$\langle \xi_\mu(t)\, \xi_\nu(t') \rangle = 2D'_{\mu\nu}\, \delta(t - t') \tag{1.24}$$

f^o_μ together with the interaction H_1 of the Hamiltonian

$$H = H_o + H_1 \tag{1.25}$$

gives rise to mode-mode coupling terms [12]. Note that the Ψ's in eq. (1.21) describe all relevant dynamical variables, not only the order parameter.

A model which belongs to the class of systems defined by eqs. (1.21), (1.22) is the thermo-elastic coupling introduced earlier [15] to describe critical dynamics near structural phase transitions (see Section 4). In this model the identification

$$\Psi = (\vec{u}_q\, , \vec{p}_q\, , \varepsilon_q) \tag{1.26}$$

is made where $\vec{u}_q$, $\vec{p}_q$ and ε_q are the Fourier components of the displacement field, the canonically conjugate momentum and the energy density, respectively, and the order parameter is $\vec{u}_0$ at $q = 0$. The Hamiltonian is given by [14,15]

$$H_0 = \frac{1}{2} \sum_q \left\{ \frac{1}{\rho} \vec{p}_q \cdot \vec{p}_{-q} + \rho \vec{u}_q \cdot \omega_q^2 \vec{u}_{-q} + \frac{1}{Q} \varepsilon_q \varepsilon_{-q} \right\} \tag{1.27}$$

where ρ is the mass density, ω_q the soft-mode frequency and $Q = T\rho c_v$, c_v being the specific heat, and

$$\begin{aligned} H_1 &= \gamma V^{-\frac{1}{2}} \sum_{qq'} \vec{u}_q \cdot \vec{u}_{-q'} \varepsilon_{q'-q} \\ &+ v V^{-1} \sum_{qq'q''} \vec{u}_q \cdot \vec{u}_{-q'} \vec{u}_{q''} \cdot \vec{u}_{q'-q-q''} \end{aligned} \tag{1.28}$$

The associated D-matrix is found to be [14]

$$D = \begin{pmatrix} \mathbb{0} & -\mathbb{1} & \vec{0} \\ +\mathbb{1} & -\rho\Gamma\mathbb{1} & \vec{0} \\ \vec{0} & \vec{0} & -T\lambda q^2 \end{pmatrix} \delta_{q',-q} \tag{1.29}$$

where Γ (which is the reciprocal of the Γ in eq. (1.15)) is the soft-mode width and λ the heat conductivity.

II. Fokker-Planck Formalism

2.1. Introduction

The objects we are finally interested in are response functions and correlation functions of the dynamical variables Ψ_μ. Therefore we need some kind of averaging procedure. Since canonical dynamics was, so to speak, our zeroth order approximation (see ref. [15], Appendix A) we could think of using density matrices. The dissipation built into dynamics would not forbid such a choice, although Liouville's theorem of course breaks down : Instead of an incompressible distribution in Ψ-space (the phase space) we would have a compressible one [15].

However, since the Ψ's describe the dynamics of critical systems only incompletely we had to introduce fluctuating random forces. This immediately rules out a density matrix description because of its deterministic character which implies that averaging can be done at any moment of time (the initial time). Dynamics with random fluctuations, on the other hand, is Markovian, i.e. it suffers from loss of memory. This means that the averaging has to be "fresh", i.e. has to be done at the time one is interested in. This implies a description in terms of a probability distribution $P(\Psi,t)$ which indicates the likelihood of the dynamical variables $\Psi_\mu(t)$ to take values Ψ_μ at time t.

The problem then is to determine $P(\Psi,t)$ for a given set of generalized Langevin equations (1.21). This problem is answered by the Fokker-Planck (FP) equation for $P(\Psi,t)$ associated to the Langevin equations. This subject is well developed [16], in particular in its application to lasers and non-linear optics [13,17]. This application shows the wide scope of Fokker-Planck theory since lasers are open systems while ordinary phase transitions occur in closed systems.

2.2. Analytical Derivation of the FP Equation [16]

In the application of FP theory to critical dynamics [18,19,20] we are interested in averages of multi-time observables $A(\Psi(t),\Psi(t'),..$

$$\langle A(\Psi(t)\ ,\ \Psi(t')\ ,\ \dots\ )\rangle =$$

$$\int d\Psi\ d\Psi' \dots P(\Psi,\ t;\ \Psi';\ t';\ \dots)\ A\,(\Psi,\Psi';\ \dots) \qquad (2.1)$$

where $P(\Psi,\ t;\ \Psi';\ t';\dots\ )$ is a multi-time probability distribution and $\int d\Psi$ means integration over all Ψ_μ . If we choose in particular $A(\Psi) = \delta(\Psi-\Psi')$ we have formally

$$P(\Psi';\ t) = \langle \delta\,(\Psi(t) - \Psi')\rangle \qquad (2.2)$$

where the δ-function is a product of all $\delta(\Psi_\mu(t) - \Psi_\mu)$ and the meaning of the average will become clear later.

Choosing initial conditions such that the Ψ's take precise values,

$$\Psi_\mu(0) = \Psi_\mu \qquad (2.3)$$

any function $A(\Psi(0))$ is non-fluctuating, in particular eq. (2.2) becomes

$$P(\Psi';\ 0) = \delta(\Psi - \Psi') \qquad (2.4)$$

The time-evolution may then be written as

$$P(\Psi,t) = V(\Psi,t)\ P(\Psi,0) \qquad (2.5)$$

with $V(\Psi,0) = 1$. Eliminating the initial distribution from eq. (2.5) and

$$\dot{P}(\Psi,t) = \dot{V}(\Psi,t)\ P(\Psi,0) \qquad (2.6)$$

we obtain the FP equation

$$\dot{P}(\Psi,t) = \mathcal{F}(\Psi,t)\ P(\Psi,t) \qquad (2.7)$$

where the FP operator $\mathcal{F}$ is defined through

$$\dot{V}(\Psi,t) = \mathcal{F}\ (\Psi,t)\ V(\Psi,t) \qquad (2.8)$$

We will now determine $\mathcal{F}$ by expressing the evolution operator V as a formal power series in the derivatives ∂_μ. Since according to eqs. (1.22) and (1.27), (1.28) the non-linearities in the Langevin equations are in general of low order (higher orders are generated by the renormalization group recursions, see Section IV) increasing powers of derivatives are expected to be of decreasing importance. Although the validity of this argument in the critical region may be questioned in general [18] Markovicity is likely to save it.

We start by writing eq. (2.2) in the form

$$P(\Psi,t) = \int d\Psi' < \delta(\Psi(t) - \Psi)\, \delta(\Psi(t') - \Psi') > \qquad (2.9)$$

valid for arbitrary t'. Making use of the identity

$$\delta(\Psi_\mu(t) - \Psi_\mu) = \int \frac{ds_\mu}{2\pi}\, e^{is_\mu(\Psi_\mu(t) - \Psi'_\mu)}\, e^{is_\mu(\Psi'_\mu - \Psi_\mu)}$$

$$= \sum_{n_\mu=0}^{\infty} \frac{1}{n_\mu!} (-\partial_\mu)^{n_\mu} (\Psi_\mu(t) - \Psi'_\mu)^{n_\mu}\, \delta(\Psi_\mu - \Psi'_\mu)$$

eq. (2.9) becomes

$$P(\Psi,t) = \sum_{\{n_\lambda\}=0}^{\infty} \prod_\mu \left\{\frac{1}{n_\mu!} (-\partial_\mu)^{n_\mu}\right\} \times$$

$$\times < \prod_\nu \{(\Psi_\nu(t) - \Psi_\nu(t'))^{n_\nu}\, \delta(\Psi_\nu(t') - \Psi_\nu) \qquad (2.10)$$

Taking here $t' = 0$ and making use of eqs. (2.3), (2.4) we find by comparison with eq. (2.5)

$$V(\Psi,t) = \sum_{\{n_\lambda\}=0}^{\infty} \prod_\mu \left\{\frac{1}{n_\mu!} (-\partial_\mu)^{n_\mu}\right\} < \prod_\nu (\Psi_\nu(t) - \Psi_\nu)^{n_\nu} >$$

$$= 1 - \partial_\mu < \Psi_\mu(t) - \Psi_\mu > \qquad (2.11)$$

$$+ \frac{1}{2} \partial_\mu \partial_\nu < (\Psi_\mu(t) - \Psi_\mu)(\Psi_\nu(t) - \Psi_\nu) > + \mathcal{O}(\partial^3)$$

Here we eliminate the variables $\Psi_\mu(t)$ successively in favour of the initial values. Integration of the Langevin equations (1.21) with the initial conditions (2.3) leads to

$$\Psi_\mu(t) - \Psi_\mu = \int_0^t dt' \left\{ f_\mu(\Psi(t'), t') + \xi_\mu(t') \right\} \qquad (2.12)$$

Making use of the Taylor expansion

$$f_\mu(\Psi(t), t) = f_\mu(\Psi,t) + (\Psi_\nu(t) - \Psi_\nu)\, \partial_\nu f_\mu(\Psi,t)$$

$$+ \frac{1}{2} (\Psi_\nu(t) - \Psi_\nu)(\Psi_\lambda(t) - \Psi_\lambda)\, \partial_\nu \partial_\lambda f_\mu(\Psi,t) + \mathcal{O}(\partial^3)$$

iteration of eq. (2.12) yields

$$\Psi_\mu(t) - \Psi_\mu = g_\mu(\Psi,t) + \zeta_\mu(t) + \int_0^t dt' \left\{ g_\nu(\Psi,t') + \zeta_\nu(t') \right\} \partial_\nu f_\mu(\Psi,t') + \mathcal{O}(\partial^2) \tag{2.13}$$

where

$$g_\mu(\Psi,t) \equiv \int_0^t dt' \, f_\mu(\Psi,t')$$

$$\zeta_\mu(t) \equiv \int_0^t dt' \, \xi_\mu(t')$$

We now insert eq. (2.13) and its derivative into the equation

$$\mathcal{F} = \dot{V} V^{-1} = - \partial_\mu \langle \dot{\Psi}_\mu(t) \rangle + \partial_\mu \partial_\nu \langle \dot{\Psi}_\mu(t) \, (\Psi_\nu(t) - \Psi_\nu) \rangle - \partial_\mu \langle \dot{\Psi}_\mu(t) \rangle \partial_\nu \langle \Psi_\nu(t) - \Psi_\nu \rangle + \mathcal{O}(\partial^3)$$

which follows from eqs. (2.8), (2.11), noting that <u>derivatives always act on everything on the right, or everything within a bracket.</u> Since $\langle f_\mu(\Psi,t) \rangle = f_\mu(\Psi,t)$, $\langle \xi_\mu(t) \rangle = 0$ and

$$\langle \xi_\mu(t) \, \xi_\nu(t') \rangle = 2D'_{\mu\nu} \, \delta(t - t') \tag{2.14}$$

(actually, we need $\int_0^t dt' \langle \xi_\mu(t) \, \xi_\nu(t') \rangle = D'_{\mu\nu}$ which is an integral over the half δ-function only) we obtain

$$\mathcal{F}(\Psi,t) = - \partial_\mu f_\mu(\Psi,t) + \partial_\mu \partial_\nu D'_{\mu\nu}(\Psi,t) + \mathcal{O}(\partial^3) \tag{2.15}$$

2.3. <u>Functional Form of FP Theory [13,20]</u>

This form which is due to Onsager and Machlup [21] starts from an entropy functional expressed as a quadratic form in the fluctuating random variables

$$S_\tau \left\{ \xi(t) \right\} = S_0 - \frac{1}{2} \int_{-\infty}^{\tau} dt \, \xi(t) \, (2D')^{-1} \, \xi(t) \tag{2.16}$$

This gives rise to a probability functional in the Boltzmann sense

$$W_\tau \left\{ \xi(t) \right\} = \exp \left[S_\tau \left\{ \xi(t) \right\} \right] \tag{2.17}$$

If we define a trace by the functional integral

$$\mathrm{Tr}_\tau (WA) = \int d \left\{ \xi(t) \right\} \, W \left\{ \xi(t) \right\} \, A \left\{ \xi(t) \right\} \tag{2.18}$$

so that

$$Z_\tau \quad = \quad Tr_\tau \; W \tag{2.19}$$

is the partition function we may define averages by

$$\langle A \rangle_\tau \quad = \quad Z_\tau^{-1} \; Tr_\tau \, (WA) \tag{2.20}$$

We first have to show that the matrix $(2D')^{-1}$ of the quadratic form in eq. (2.16) is correctly chosen to give the result (2.14). For this we discretize the time by dividing it into sufficiently small intervals $t_{i+1} - t_i = \varepsilon$ such that $\xi_{\mu i} = \xi_\mu(t_i)$ are representative values. Then

$$\begin{aligned} \langle \xi_{\mu i} \xi_{\nu j} \rangle_{t_0} &= Z_{t_0}^{-1} \int d\{\xi\} \, \xi_{\mu i} \, \xi_{\nu j} \\ &\quad e^{-\frac{1}{2}\sum_{\ell<0} \xi_\ell (2D'/\varepsilon)^{-1} \xi_\ell} \\ &= \delta_{ij} \int d\xi \, \xi_\mu \xi_\nu e^{-\frac{1}{2}\xi(2D'/\varepsilon)^{-1}\xi} \\ &= \frac{2}{\varepsilon} D'_{\mu\nu} \delta_{ij} \xrightarrow{\varepsilon \to 0} 2D'_{\mu\nu} \; \delta(t_i - t_j) \end{aligned} \tag{2.21}$$

which is eq. (2.14).

Since we are interested in the dynamical variables Ψ_μ rather than in the ξ_μ we seek a transformation $\{\xi\} \rightarrow \{\Psi\}$ such that probability is conserved,

$$W\{\xi\} d\{\xi\} \quad = \quad W\{\Psi\} \, d\{\Psi\} \tag{2.22}$$

This transformation is determined by the discretized Langevin equations which in the notation $\Psi_{\mu i} \equiv \Psi_\mu(t_i)$, $f_{\mu i} \equiv f_\mu(\Psi_i, t_i)$ are

$$\Psi_{\mu i} - \Psi_{\mu i-1} \quad = \quad \varepsilon\lambda \, f_{\mu i} + \varepsilon(1-\lambda) \, f_{\mu\, i-1} + \varepsilon \, \xi_{\mu i} \tag{2.23}$$

Here the ambiguity of choice between $f_{\mu i}$ and $f_{\mu\, i-1}$ is expressed by a free parameter λ . From eq. (2.33) we obtain

$$\varepsilon \frac{\partial \xi_{\mu i}}{\partial \Psi_{\nu j}} \quad = \quad A^{(i)}_{\mu\nu} \; \delta_{ij} \; - \; B^{(i-1)}_{\mu\nu} \; \delta_{i-1,j} \tag{2.24}$$

with

$$A^{(i)}_{\mu\nu} = \delta_{\mu\nu} - \varepsilon\lambda \left(\frac{\partial f_\mu}{\partial \Psi_\nu}\right)_i$$

$$B^{(i)}_{\mu\nu} = \delta_{\mu\nu} + \varepsilon(1-\lambda) \left(\frac{\partial f_\mu}{\partial \Psi_\nu}\right)_i$$

Since the matrix (2.24) is triangular in the indices i,j the functional determinant becomes

$$\mathrm{Det} \left\| \varepsilon \frac{\partial \xi_{\mu i}}{\partial \Psi_{\nu j}} \right\| = \prod_i \mathrm{Det} \left\| A^{(i)} \right\|$$

$$= \exp\left\{ -\varepsilon\lambda \sum_i \left(\frac{\partial f_\mu}{\partial \Psi_\mu}\right)_i + \mathcal{O}(\varepsilon^2) \right\} \tag{2.25}$$

Insertion of eq. (2.25) into eq. (2.22) leads to

$$W_{t_0}\{\xi\} = \text{const.} \exp\left\{ -\sum_{i<0} \sigma(\Psi_i, \Psi_{i-1}, t_i, t_{i-1}) \right\} \tag{2.26}$$

where

$$\sigma(\Psi, \varphi, t, s) = \frac{1}{2}\eta(2\varepsilon D')^{-1}\eta - \varepsilon\lambda \frac{\partial f_\alpha(\Psi,t)}{\partial \Psi_\alpha} \tag{2.27}$$

and

$$-\eta_\mu = \Psi_\mu - \varphi_\mu - \varepsilon\lambda f_\mu(\Psi, t) - \varepsilon(1-\lambda) f_\mu(\varphi, s) \tag{2.28}$$

Going back to continuous time we obtain

$$W_\tau\{\Psi(t)\} = \text{const.} \exp\left\{ -\int_{-\infty}^{\tau} dt \; L(\Psi(t), \dot\Psi(t), t) \right\} \tag{2.29}$$

with the "Lagrangian"

$$L(\Psi, \dot\Psi, t) = \frac{1}{4}(\dot\Psi_\mu - f_\mu(\Psi,t)) D'^{-1}_{\mu\nu} (\dot\Psi_\nu - f_\nu(\Psi,t)) - \lambda \frac{\partial f_\alpha(\Psi,t)}{\partial \Psi_\alpha} \tag{2.30}$$

In terms of the functional (2.26) the probability $P(\Psi, t)$ may be expressed as

$$P(\Psi, t_i) = Z^{-1}_{t_i} \omega(\Psi, t_i) = \langle \delta(\Psi_i - \Psi) \rangle_{t_i} \tag{2.31}$$

where

$$\omega(\Psi_i, t_i) = \left(\prod_{k \lessgtr i} \int d\Psi_k\right) W_{t_i}\{\Psi\} \tag{2.32}$$

Going back to continuous time it is now evident that eq. (2.31) is the exact meaning of the average (2.2).

The unrenormalized one-time probability $\omega(\Psi_i, t_i)$ may be used to rederive the FP equation. Indeed, in view of (2.26) eq. (2.32) may be written in the form of a transfer matrix equation [5]

$$\omega(\Psi, t_{i+1}) = \int d\varphi \; e^{\sigma_0 - \sigma(\Psi, \varphi, t_{i+1}, t_i)} \, \omega(\varphi, t_i)$$

or, with eq. (2.31), as

$$P(\Psi, t+\varepsilon) = \int d\varphi \; e^{\sigma_0 - \sigma(\Psi, \varphi, t+\varepsilon, t)} \, P(\varphi, t) \qquad (2.33)$$

Making the substitution $\varphi_\mu \to \eta_\mu$ defined by eq. (2.28), which again introduces a functional determinant of the form (2.25) but for fixed time t_i, we obtain

$$\begin{aligned} P(\Psi, t+\varepsilon) &= \int d\eta \; e^{\sigma_0 - \frac{1}{2}\eta (2\varepsilon D')^{-1}\eta - \varepsilon \partial f_\alpha / \partial \Psi_\alpha} \\ &\cdot \left\{ 1 + (\eta_\mu - \varepsilon f_\mu)\, \partial_\mu + \tfrac{1}{2} \eta_\mu \eta_\nu \partial_\mu \partial_\nu + O(\partial^3) \right\} P(\Psi, t) \end{aligned} \qquad (2.34)$$

In order that this equality holds in the limit $\varepsilon \to 0$ we must have

$$\int d\eta \; e^{\sigma_0 - \frac{1}{2}\eta (2\varepsilon D')^{-1}\eta} = 1$$

With this condition eq. (2.34) immediately leads back to the form (2.15) of the FP operator, except that in this functional approach the matrix D' was assumed to be independent of Ψ's and of time.

2.4. Detailed Balance

We are now making explicit use of the Markovian character of the system by introducing the conditional probability

$$p(\Psi, t; \Psi'; t') = \frac{P(\Psi, t; \Psi', t'; \Psi'', t''; \ldots)}{P(\Psi', t'; \Psi'', t''; \ldots)} \qquad (2.35)$$

which depends only on the latest (or earliest) two times, as defined by the ordering $t > t' > t'' > \ldots$ (or $t < t' < t'' < \ldots$). It immediately follows from eq. (2.35) that

$$\begin{aligned} P(\Psi, t) &= \int d\Psi' \; P(\Psi, t; \Psi', t') \\ &= \int d\Psi' \; p(\Psi, t; \Psi', t') \, P(\Psi', t') \end{aligned} \qquad (2.36)$$

Hence the conditional probability satisfies the FP equation,

$$\partial_t\, p(\Psi,t;\Psi',t) = \mathcal{F}(\Psi,t)\, p(\Psi,t;\Psi',t') \tag{2.37}$$

which is called the forward equation [16,13,17].

Differentiating eq. (2.36) with respect to t' we find, making use of the FP equation for $P(\Psi',t')$,

$$0 = \int d\Psi' \Big\{ p(\Psi,t;\Psi',t')\, \mathcal{F}(\Psi',t')\, P(\Psi',t') + P(\Psi',t')\, \partial_{t'}\, p(\Psi,t;\Psi',t') \Big\}$$

By partial integrations we may shift the differentiations implied in $\mathcal{F}(\Psi',t')$ to $p(\Psi,t;\Psi',t')$. This leads to the backward equation [16,13,17]

$$-\partial_{t'}\, p(\Psi,t;\Psi',t') = \mathcal{L}(\Psi',t')\, p(\Psi,t;\Psi',t') \tag{2.38}$$

Here

$$\mathcal{L}(\Psi,t) = f_\mu \partial_\mu + D'_{\mu\nu}\, \partial_\mu \partial_\nu + \mathcal{O}(\partial^3) \tag{2.39}$$

is the Langevin operator which is conjugate to the FP operator in the sense [18,19]

$$\int d\Psi \int d\Psi' \ldots \big(\mathcal{F}(\Psi,t)\, P(\Psi,t;\Psi',t',\ldots)\big)\, A(\Psi,\Psi',\ldots) = \int d\Psi \int d\Psi' \ldots P(\Psi,t;\Psi',t',\ldots)\, \big(\mathcal{L}(\Psi,t)\, A(\Psi,\Psi',\ldots)\big) \tag{2.40}$$

In particular, we have for steady states where f_μ, $D'_{\mu\nu}$, P, $\mathcal{F}$ and $\mathcal{L}$ contain no parametric time dependence

$$P(\Psi,t) = P^s(\Psi) = \text{const.}\; e^{-\phi(\Psi)} \tag{2.41}$$

and

$$p(\Psi,t;\Psi',t') = p^s(\Psi,\Psi';t-t') \tag{2.42}$$

and eqs. (2.37), (2.38) become

$$\partial_t\, p^s(\Psi,\Psi';t) = \mathcal{F}(\Psi)\, p^s(\Psi,\Psi';t) = \mathcal{L}(\Psi')\, p^s(\Psi,\Psi';t) \tag{2.43}$$

Detailed balance is the condition

$$P^s(\Psi, t; \Psi', 0) = P^s(\widetilde{\Psi}', 0; \widetilde{\Psi}, -t) \tag{2.44}$$

in which initial and final arguments are exchanged and time reversal $\mathcal{T}$ has been applied as follows

$$\Psi_\mu(t) \xrightarrow{\mathcal{T}} \widetilde{\Psi}_\mu(t) = \sigma_\mu \Psi_\mu(-t)$$

$$f_\mu(\Psi(t),t) \xrightarrow{\mathcal{T}} \widetilde{f}_\mu(\Psi(t),t) = \sigma_\mu f_\mu(\widetilde{\Psi}(t),-t)$$

$$D'_{\mu\nu}(\Psi(t),t) \xrightarrow{\mathcal{T}} \widetilde{D}'_{\mu\nu}(\Psi(t),t) = \sigma_\mu \sigma_\nu D'_{\mu\nu}(\widetilde{\Psi}(t),-t) \tag{2.45}$$

$\sigma_\mu = \pm 1$ is the $\mathcal{T}$-signature of the variable Ψ_μ and the arguments in eq. (2.44) are to be understood as values at $t=0$.

In terms of the conditional probabilites $p^s \equiv p^s(\Psi, \Psi'; t)$ and $\widetilde{p}^s \equiv p^s(\widetilde{\Psi}', \widetilde{\Psi}; t)$ defined by eq. (2.35) detailed balance (2.44) reads

$$p^s P'^s = \widetilde{p}^s P^s \tag{2.46}$$

where $P^s \equiv P^s(\widetilde{\Psi}) = P^s(\Psi)$, which follows by integrating eq. (2.44) over Ψ', and $P'^s \equiv P^s(\Psi')$. Taking the time derivative of eq. (2.46) and inserting eq. (2.43) we obtain

$$\mathcal{F}(\widetilde{p}^s P^s) = (\widetilde{\mathcal{L}}\, \widetilde{p}^s)\, P^s \tag{2.47}$$

where

$$\mathcal{F} \equiv \mathcal{F}(\Psi) = -\partial_\mu f_\mu + \partial_\mu \partial_\nu D'_{\mu\nu}$$

$$\widetilde{\mathcal{L}} \equiv \mathcal{L}(\widetilde{\Psi}) = f_\mu(\widetilde{\Psi})\, \widetilde{\partial}_\mu + D'_{\mu\nu}(\widetilde{\Psi})\, \widetilde{\partial}_\mu \widetilde{\partial}_\nu$$

$$= \widetilde{f}_\mu \partial_\mu + \widetilde{D}'_{\mu\nu} \partial_\mu \partial_\nu$$

neglecting higher order derivatives. Making use of eq. (2.41) and of the fact that $\mathcal{F} P^s = 0$ eq. (2.47) leads to the relation

$$\left\{\widetilde{f}_\mu + f_\mu - 2(\partial_\mu D'_{\mu\nu}) + 2D'_{\mu\nu}(\partial_\nu \phi)\right\}(\partial_\mu \widetilde{p}^s)$$

$$+ (\widetilde{D}'_{\mu\nu} - D'_{\mu\nu})(\partial_\mu \partial_\nu \widetilde{p}^s) = 0 \tag{2.48}$$

Here $(\partial_\mu \widetilde{p}^s)$ and $(\partial_\mu \partial_\nu \widetilde{p}^s)$ are independent functions. To see this it is sufficient to put $t \doteq 0$ so that $\widetilde{p}^s = \delta(\widetilde{\Psi} - \widetilde{\Psi}')$, multiply eq. (2.48) with $\widetilde{\Psi}'$ and integrate over $\widetilde{\Psi}'$. Then the function $(\partial_\mu \partial_\nu \widetilde{p}^s)$ produces zero and eq. (2.48) splits into the two conditions

$$\frac{1}{2}(f_\mu + \widetilde{f}_\mu) - (\partial_\nu D'_{\mu\nu}) + D'_{\mu\nu}(\partial_\nu \phi) = 0 \tag{2.49}$$

and

$$\tilde{D}'_{\mu\nu} - D'_{\mu\nu} = 0 \tag{2.50}$$

which are the potential conditions of Graham and Haken [13,17] .

2.5. Equilibrium

For closed systems as considered in these lectures the first potential condition, which combines steady state and detailed balance, determines the equilibrium distribution. In order to show this we need the relations [14]

$$f^{\circ}_{\mu} \equiv + (\partial_{\nu} H) D^{\circ}_{\nu\mu} = \frac{1}{2}(f_{\mu} - \tilde{f}_{\mu}) \tag{2.51}$$

and

$$f'_{\mu} \equiv - (\partial_{\nu} H) D'_{\nu\mu} = \frac{1}{2}(f_{\mu} + \tilde{f}_{\mu}) \tag{2.52}$$

which are of course in accord with eqs. (1.22) and (1.23). From eqs. (2.49) and (2.52) and the symmetry of the matrix D' we obtain

$$\partial_{\nu} D'_{\mu\nu} = D'_{\mu\nu} \partial_{\nu} (\phi - H) \tag{2.53}$$

On the other hand, we find with the help of eqs. (2.49), (2.51) and (2.41)

$$0 = \mathcal{F} P^{s} = \partial_{\mu} (f^{\circ}_{\mu} P^{s}) = P^{s} \{ \partial_{\mu} f^{\circ}_{\mu} - f^{\circ}_{\mu} \partial_{\mu} \phi \} \tag{2.54}$$

and from the definition (2.51)

$$\begin{aligned} \partial_{\mu} f^{\circ}_{\mu} &= (\partial_{\nu} H)(\partial_{\mu} D^{\circ}_{\nu\mu}) \\ f^{\circ}_{\mu} (\partial_{\mu} H) &= 0 \end{aligned} \tag{2.55}$$

Eqs. (2.54) and (2.55) may be combined into

$$(\partial_{\nu} H)(\partial_{\mu} D^{\circ}_{\nu\mu}) = f^{\circ}_{\mu} \partial_{\mu} (\phi - H) \tag{2.56}$$

The consequence of eqs. (2.53) and (2.56) is important enough to be formulated as [9]

Theorem 1 : For the class of systems described by eqs. (1.21) and (1.22) the equilibrium probability distribution is given by

$$P^{eq}(\Psi) = Z^{-1} e^{-H(\Psi)} \tag{2.57}$$

provided that detailed balance holds and that

$$\partial_{\mu} D_{\mu\nu} = \partial_{\nu} D_{\mu\nu} = 0 \tag{2.58}$$

Indeed eqs. (2.58) imply the vanishing of the left hand sides of eqs. (2.53) and (2.56) so that $\phi = H + \text{const.}$ in eq. (2.41).

2.6. Time Evolution and Averages of Observables

In analogy to eq. (2.5) we introduce a time evolution operator U for observables by

$$A(\Psi(t),t) = U(\Psi,t)\, A(\Psi,t)\, U^{-1}(\Psi,t) \tag{2.59}$$

where Ψ_μ are the initial values (2.3) and we have allowed for a parametric time dependence of the one-time observables. U is determined by the Langevinian $\mathcal{L}$ defined through eq. (2.40). Indeed, from eqs. (2.35) and (2.37) it follows that the multi-time probability distribution $P(\Psi,t;\Psi',t';\ldots)$ also satisfies the FP equation, so that we obtain from eqs. (2.1) and (2.40), for $t \neq t',\ldots,$

$$\begin{aligned} &\partial_t \langle A(\Psi(t))\, B(\Psi(t'))\ldots\rangle \\ &= \int d\Psi \int d\Psi' \ldots P(\Psi,t;\Psi',t';\ldots)\, \mathcal{L}(\Psi,t)\, A(\Psi)\, B(\Psi')\ldots \\ &= \langle \mathcal{L}(\Psi(t),t)\, A(\Psi(t))\, B(\Psi(t'))\ldots\rangle \end{aligned} \tag{2.60}$$

or with eq. (2.59) [22]

$$\dot{U}(\Psi,t) = U(\Psi,t)\,\mathcal{L}(\Psi,t) = \mathcal{L}(\Psi(t),t)\, U(\Psi,t) \tag{2.61}$$

Eq. (2.59) also allows us to write

$$\begin{aligned} &A(\Psi(t+\tau))\, B(\Psi(t'+\tau))\ldots \\ &= \langle U(\Psi(\tau),t)\, A(\Psi(\tau))\, U^{-1}(\Psi(\tau),t)\, U(\Psi(\tau),t')\cdot \\ &\quad\cdot B(\Psi(\tau))\, U^{-1}(\Psi(\tau),t')\ldots\rangle = \int d\Psi\, P(\Psi,\tau)\cdot \\ &\quad\cdot U(\Psi,t)\, A(\Psi)\, U^{-1}(\Psi,t)\, U(\Psi,t')\, B(\Psi)\, U^{-1}(\Psi,t')\ldots \end{aligned} \tag{2.62}$$

For $\tau = 0$ this is the <u>Heisenberg representation</u> of an average whereas eq. (2.1) is the <u>Schrödinger representation</u> [22]. For arbitrary τ eq. (2.62) exhibits <u>time translation invariance of steady states</u>, since P^s does not depend on time, so that

$$\begin{aligned} &\langle A(\Psi(t+\tau))\, B(\Psi(t'+\tau))\ldots\rangle^s \\ &= \langle A(\Psi(t))\, B(\Psi(t'))\ldots\rangle^s \end{aligned} \tag{2.63}$$

Another important relation is obtained from eq. (2.60) by making use on the left hand side of the Langevin eqs.(1.21) in the form

$$\partial_t A(\Psi(t)) = \left\{ f_\mu(\Psi(t),t) + \xi_\mu(t) \right\} \frac{\partial}{\partial \Psi_\mu(t)} A(\Psi(t))$$

and on the right hand side of eq. (2.39). The result is, for $t \neq t',\ldots$,

$$\left\langle \left\{ \xi_\mu(t) - D'_{\mu\nu}(\Psi(t),t) \frac{\partial}{\partial \Psi_\nu(t)} \right\} \frac{\partial}{\partial \Psi_\mu(t)} A(\Psi(t))\, B(\Psi(t'))\ldots \right\rangle = 0 \tag{2.64}$$

This shows that <u>in an average the fluctuating random forces ξ_μ are equivalent to linear combinations of derivatives with respect to the dynamical variables Ψ_μ</u>. An important special case of eq. (2.64) is obtained for $A(\Psi) = \Psi_\lambda$, namely

$$\langle \xi_\lambda(t)\, B(\Psi(t'))\ \ldots \rangle = 0 \tag{2.65}$$

valid for $t \neq t',\ldots$.

III. Perturbation Theory

3.1. Introduction

A general and simple diagram technique based on the work of Martin Siggia and Rose [23] has recently been devised for canonical dynamics [22]. It is evident from the Introduction to Section II that a transposition of this diagram technique to the case of Langevin equations of motion is not automatic. So we will examine in this Section the steps required for such a transposition to hold. Since the Green's functions which combine into a matrix Dyson equation are response and correlation functions we first discuss the general connection between the two.

3.2. Fluctuation-Dissipation Theorem

The response functions obtained by calculating the perturbation of an external field are [14,22]

$$R_{\mu\nu}(t,t') = \Theta(t - t') \langle \hat{\Psi}_\nu(t') \Psi_\mu(t) \rangle \tag{3.1}$$

Here

$$\hat{\Psi}_\mu = D_{\mu\nu} \partial_\nu \tag{3.2}$$

have to be considered as new dynamical variables since they are introduced by the time evolution which according to eq. (2.61) is determined by the Langevinian (2.39). For the derivation of the fluctuation-dissipation theorem (FDT) [9] connecting the response-functions and the correlation functions

$$G_{\mu\nu}(t,t') = \langle \Psi_\mu(t) \Psi_\nu(0) \rangle \tag{3.3}$$

we will assume equilibrium . This is a more restricted assumption than was used in the proof of ref. [22] which is valid for <u>local</u> equilibrium.

With eqs. (2.57) and (2.63) we may write eqs. (3.1), (3.2) as

$$R_{\mu\nu}(t) = \Theta(t) \int d\Psi \, Z^{-1} e^{-H(\Psi)} D_{\nu\lambda} \partial_\lambda \Psi_\mu(t)$$

Partial integration under the condition (2.58) leads to

$$R_{\mu\nu}(t) = \Theta(t) \left\langle \frac{\partial H}{\partial \Psi_\lambda} \Psi_\mu(t) \right\rangle D_{\nu\lambda}(t) \tag{3.4}$$

This is to be compared with the time derivative of the equilibrium form of eq. (3.3), making use of eq. (2.63)

$$\dot{G}_{\mu\nu}(t) = -\langle \dot{\Psi}_\nu(-t) \quad \Psi_\mu(0) \rangle$$

Inserting eqs. (1.21) and (1.22) one obtains, in view of eq. (2.65),

$$\dot{G}_{\mu\nu}(t) = - \left\langle \frac{\partial H}{\partial \Psi_\lambda} \Psi_\mu(t) \right\rangle D_{\lambda\nu}(-t) \tag{3.5}$$

Now a closer analysis of the unperturbed motion shows [14] that the matrix D cannot be the same for positive and negative times; the correct relation turns out to be [14]

$$D_{\mu\nu}(t) = - D_{\nu\mu}(-t) \tag{3.6}$$

This then leads to the FDT

$$R_{\mu\nu}(t) = \Theta(t) \dot{G}_{\mu\nu}(t) \tag{3.7}$$

3.3. Unperturbed Motion

We designate the unperturbed time evolution as determined by the free Hamiltonian H_0 by an upper index o [22]. Then the analogue of eq. (2.59) is

$$A(\Psi^0(t),t) = U_0(\Psi,t) \, A(\Psi,0) \, U_0^{-1}(\Psi,t) \tag{3.8}$$

starting from the same initial values (2.3) ,

$$\Psi^0_\mu(0) = \Psi_\mu \tag{3.9}$$

and U_0 is determined by the analogue of eq. (2.61) ,

$$\dot{U}_0(\Psi,t) = U_0(\Psi,t) \, \mathcal{L}_0(\Psi,t) \tag{3.10}$$

Since H_0 is a quadratic form [22]

$$H_0 = \frac{1}{2} \Psi_\mu E_{\mu\nu} \Psi_\nu \tag{3.11}$$

the free forces $f_{0\mu}$ as determined by eq. (1.22) are linear in the Ψ's, provided that the D-matrix is Ψ-independent,

$$f_{0\mu} = \frac{\partial H_0}{\partial \Psi_\nu} D_{\nu\mu} = \Psi_\nu \Lambda_{\nu\mu} \tag{3.12}$$

Then the free Langevinian is according to eq. (2.39)

$$\mathcal{L}_0 = \Psi_\nu \Lambda_{\nu\mu} \partial_\mu + D'_{\mu\nu} \partial_\mu \partial_\nu \tag{3.13}$$

Since it is time-independent eqs. (3.9), (3.10) immediately lead to

$$U_0(\Psi,t) = e^{\mathcal{L}_0(\Psi)t} \tag{3.14}$$

For the proof of Wick's theorem [22] the commutativity of the unperturbed new variables (3.2),

$$[\hat{\Psi}^0_\mu (t) , \hat{\Psi}^0_\nu (t')] = 0 \tag{3.15}$$

is of importance. To show the validity of eq. (3.15) in Fokker-Planck theory we calculate

$$\begin{aligned} \partial_\nu \, \partial^o_\mu (t) \, A(\Psi^0 (t)) &= \partial_\nu \, e^{\mathcal{L}_0(\Psi)t} \, \partial_\mu \, A(\Psi) \, e^{-\mathcal{L}_0(\Psi)t} \\ &= e^{\mathcal{L}_0(\Psi)t} \, (t \Lambda_{\nu\lambda}\partial_\lambda + \partial_\nu) \, \partial_\mu \, A(\Psi) \, e^{-\mathcal{L}_0(\Psi)t} \end{aligned} \tag{3.16}$$

On the other hand

$$\begin{aligned} \partial^o_\mu (t) \, \partial_\nu \, A(\Psi^0(t)) &= \partial^0_\mu (t) \, \partial_\nu \, e^{\mathcal{L}_0(\Psi)t} A(\Psi) \, e^{-\mathcal{L}_0(\Psi)t} \\ &= \partial^0_\mu (t) \, e^{\mathcal{L}_0(\Psi)t} (t \Lambda_{\nu\lambda} \partial_\lambda + \partial_\nu) \, A(\Psi) e^{-\mathcal{L}_0(\Psi)t} \\ &= e^{\mathcal{L}_0(\Psi)t} \, \partial_\mu (t \Lambda_{\nu\lambda}\partial_\lambda + \partial_\nu) \, A(\Psi) \, e^{-\mathcal{L}_0(\Psi)t} \end{aligned} \tag{3.17}$$

and taking the difference of eqs. (3.16), (3.17) we find for any $A(\Psi)$

$$[\partial_\nu , \partial^0_\mu (t)] \, A(\Psi^0(t)) = 0 \tag{3.18}$$

which proves eq. (3.15).

3.4. Interaction Representation

It is defined with respect to the commom initial state (2.3), (3.9) as [22]

$$A(\Psi(t),\hat{\Psi}(t)) = S(0,t) \; A(\Psi^0(t), \hat{\Psi}^0(t)) \, S(t,0) \tag{3.19}$$

From eqs. (2.59), (3.8) then follows

$$S(0,t) = S^{-1}(t,0) = U(\Psi,t) \, U_0^{-1} (\Psi,t) \tag{3.20}$$

This operator may be extended to two arbitrary times and has the well-known solution [22]

$$S(t,t') = \bar{T} \exp\left\{\int_t^{t'} d\tau \, \mathcal{L}_1^0 \; (\tau)\right\} \tag{3.21}$$

where $\bar{T}$ is the anti-chronological operator which orders the factors of a product such that increasing time arguments run from left to right. This order rather than the chronological one is fixed by the requirement that the response functions may be written as time-ordered Green's functions [22].

In eq. (3.21) $\mathcal{L}_1^0 \; (\tau)$ is the interaction representation

$$\mathcal{L}_1^0 (\tau) = e^{\mathcal{L}_0 \tau} \mathcal{L}_1 \, e^{-\mathcal{L}_0 \tau} \tag{3.22}$$

of the interaction Langevinian $\mathcal{L}_1$ given as difference between eqs. (2.39) and (3.13) ,

$$\mathcal{L}_1 = \mathcal{L} - \mathcal{L}_0 = \frac{\partial H_1}{\partial \Psi_\mu} \, \hat{\Psi}_\mu \tag{3.23}$$

$\mathcal{L}_1$ has a simple diagrammatic interpretation in terms of vertices. In the case of the interaction Hamiltonian (1.28) for example $\mathcal{L}_1$ contains the three vertices of fig. 1 where the straight legs correspond to an $\vec{u}$ - factor and the wavy legs to ε -factors and each $\hat{\Psi}$-leg is distinguished by an arrow pointing into the vertex [22]

Fig. 1

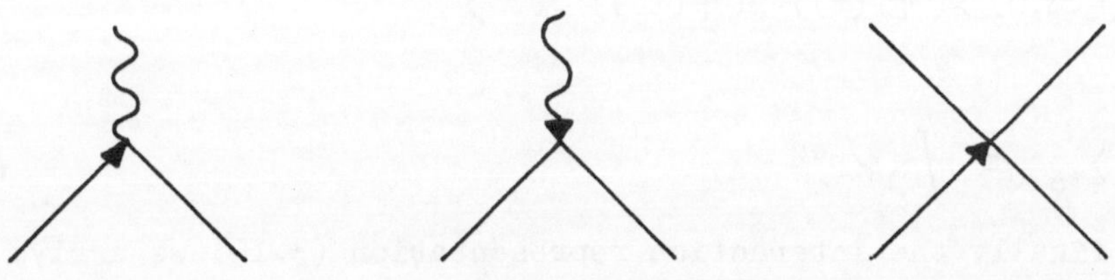

3.5. Anti-Chronological Products

For the proof of Wick's theorem in ref. [22] we finally need the reduction of $\bar{T}$-products from the Heisenberg to the interaction representation. As always in time-dependent perturbation theory this requires some kind of switch-on condition [22]. Thus we assume that in the remote past the system was asymptotically unperturbed

$$\lim_{t \to -\infty} P(\Psi,t) = P_0^{eq} (\Psi) \equiv Z_0^{-1} e^{-H_0(\Psi)} \tag{3.24}$$

According to eq. (2.62) at $\tau = 0$ the Heisenberg representation is determined by $P(\Psi,0)$ which we may now express with the help of eqs. (3.24), (2.5) and the analogous equation for the unperturbed time evolution

$$P_0(\Psi,t) = V_0(t)\, P_0(\Psi,0) \tag{3.25}$$

The result is

$$P(\Psi,0) = \lim_{t\to-\infty} V^{-1}(\Psi,t)\, V_0(\Psi,t)\, P_0^{eq}\ (\Psi,0) \tag{3.26}$$

From the equality between the Heisenberg representation (2.62) at $\tau = 0$ and the corresponding Schrödinger representation (2.1) we conclude in a chain of steps making use of eqs. (2.5), (2.8), (2.35) and (2.37) that

$$\begin{aligned} &\int d\Psi \ldots (V(\Psi,t)\ \ P(\Psi,0;\ \ldots))\ \ A(\Psi) \ldots \\ &= \int d\Psi P(\Psi,0)\quad U(\Psi,t)\, A(\Psi)\, U^{-1}(\Psi,t) \ldots \end{aligned} \tag{3.27}$$

This equality supplies the rule of how to transform a V-operator acting on the probability into a U-operator acting on the observables. Applying eqs. (3.26) and (3.27) and making use of eq. (3.20) we finally arrive at the basic expression

$$\begin{aligned} &\langle A(\Psi(t))\ B(\Psi(t'))\ \ldots \rangle = \\ &\langle S(-\infty,0)\ A(\Psi,(t))\ B(\Psi(t'))\ \ldots \rangle_o \end{aligned} \tag{3.28}$$

where

$$\langle A \rangle_0 \equiv \int d\Psi P_0^{eq}\ A \tag{3.29}$$

Applying finally the interaction representation (3.19) we arrive at the desired reduction of $\bar{T}$-products

$$\begin{aligned} &\langle \bar{T}(\phi_1(t_1)\ \ldots\ \phi_n(t_n)) \rangle = \\ &\langle \bar{T}(S(-\infty, +\infty)\ \phi_1^0\ (t_1)\ \ldots\ \phi_n^0\ (t_n)) \rangle_o \end{aligned} \tag{3.30}$$

from which the proof of Wick's theorem and hence the diagram technique of ref.[22] readily follow. In order to emphasize the importance of this result we formulate it as

<u>Theorem 2</u> : Replacing the Liouvillean by the Langevinian the diagram technique of Enz and Garrido [22] remains valid.

IV. Application to Structural Phase Transitions

4.1. Introduction

Structural phase transitions [24,25] may roughly be devided into the two broad groups of order-disorder and displacive types. An order-disorder transition is driven by a bistable bond (usually Hydrogen) in the crystal which may be described by an Ising model with a transverse part representing tunneling. A displacive transition is driven by a phonon mode which becomes soft at a high symmetry point q_0 in the Brillouin zone, usually the zone center $q_0 = 0$ (Nb_3Sn) or the R-corner $q_0 = (\pi/a)(1,1,1)$($SrTiO_3$). In the latter case a staggered representation in which the displacement field alternates between adjacent unit cells (as in an anti-ferromagnet) maps q_0 onto the zone center [15].

Ferro-electric transitions are generally of the first type, although there often is a displacive distortion coupled to it (KDP). The instability in a displacive transition may be due to electronic effects (electron-phonon coupling, Jahn-Teller effect) or to anharmonic interactions. Anharmonic phonon models actually are able to describe both the displacive and the order-disorder types [6].

In recent years a new feature in the excitation spectrum near structural phase transitions has been seen in inelastic neutron scattering near q_0 and at frequency $\omega = 0$ [25]. This central peak (CP) has recently received considerable publicity [26]. It has been shown earlier [15] that thermo-elastic coupling as defined by eqs. (1.26) to (1.29) may account for the central peak in $SrTiO_3$ (which is the most carefully studied case [26]) at least moderately close to T_c and q_0. More recently other explanations in terms of cluster dynamics [4,5,6] and of defect coupling [27] have been proposed. Leaving out the last possibility because it lies outside the scope of these lectures we will shortly summarize a recent analysis of the first two proposals [28]. This will be an opportunity to illustrate the diagram technique of ref. [22] and at the same time to give a description of the renormalization group (RG) analysis applied to critical dynamics.

4.2. Self-Energies

The unperturbed response functions obtained from the thermo-elastic equations (1.26) to (1.29) are respectively [14]

$$R^o(q,\omega) = \rho^{-1} (\omega^2 + i\,\omega\,\Gamma - \omega_q^2)^{-1} \tag{4.1}$$

and

$$G^o(q,\omega) = (i\omega\rho)^{-1} \left\{ (\omega^2 - i\,\omega\,\Gamma - \omega_q^2)^{-1} - (\omega^2 + i\,\omega\,\Gamma - \omega_q^2)^{-1} \right\} \tag{4.2}$$

for the order parameter $\vec{u}_q$,

$$S^o(q,\omega) = T\lambda q^2 (i\,\omega - D_T q^2)^{-1} \tag{4.3}$$

where $D_T = \lambda/\rho c_v$, and

$$D^o(q,\omega) = 2T\lambda q^2 \,(\omega^2 + D_T^2 q^4)^{-1} \tag{4.4}$$

for the energy density ε_q . It is easy to see that R^o and G^o on the one hand and S^o and D^o on the other are related by the FDT (3.7).

With these expressions and with the interaction Langevinian of eq. (3.23) and fig. 1 the self-energies defined by the Dyson equations [22,28]

$$\Pi = \Pi_\gamma + \Pi_v = R^{o-1} - R^{-1} \tag{4.5}$$

and

$$\Xi = S^{o-1} - S^{-1} \tag{4.6}$$

are obtained according to fig. 2 which also gives the diagrammatic meaning of the functions of eqs.(4.1) to (4.4). The result is

$$\Pi_\gamma(q,\omega) = 4\gamma^2 V^{-1} \sum_{q'} \int \frac{d\omega'}{2\pi} \left\{ R^o(q',\omega') \cdot D^o(q-q',\omega-\omega') + G^o(q',\omega')\, S^o(q-q',\omega-\omega') \right\} \tag{4.7}$$

$$\Pi_v(q,\omega) = 96\,(n+2)\, v^2\, V^{-2} \cdot \sum_{q'q''} \int \frac{d\omega'}{2\pi} \int \frac{d\omega''}{2\pi}\, G^o(q',\omega')\, G^o(q'',\omega'') \cdot R^o(q - q' - q'', \omega - \omega' - \omega'') \tag{4.8}$$

and

$$\Xi(q,\omega) = 4n\gamma^2 V^{-1} \sum_{q'} \int \frac{d\omega'}{2\pi} G^0(q',\omega') \cdot$$

$$\cdot R^0(q-q',\omega-\omega') \qquad (4.9)$$

Fig. 2

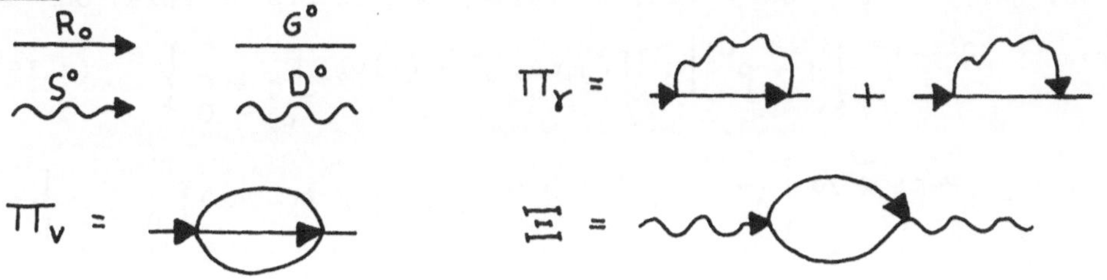

4.3. Renormalization Group Analysis

The frequency integrations in eqs (4.7) to (4.9) are easily done in the complex plane. On the other hand, for the RG recursion relations one has to evaluate the wave-number integrals in an interval $\Lambda b^{-1} < |q'| < \Lambda$ and at critical dimension d^*. In addition the following scale changes are involved

$$q' = bq \quad , \quad \omega' = b^z\omega$$

$$\vec{u}'_{q'}(t') = b^{-1+\eta/2}\,\vec{u}_q(t) \qquad (4.10)$$

$$\varepsilon'_{q'}(t') = b^{-\bar{\alpha}/2\nu}\,\varepsilon_q(t)$$

where $\bar{\alpha} = \max(\alpha, 0)$ [2]. With the definition (3.3) and the FDT (3.7) one then finds

$$R'^{-1}(q',\omega') = b^{2-\eta}\, R^{-1}(q,\omega) \qquad (4.11)$$

and

$$S'^{-1}(q',\omega') = b^{\bar{\alpha}/\nu}\, S^{-1}(q,\omega) \qquad (4.12)$$

Inserting here on the right hand side for R^{-1} and S^{-1} the expressions of eqs. (4.5) and (4.6) one obtains the RG recursion relations from which follow the fixed point values of the model parameters.

We are interested in the dynamical parameters Γ and λ which according to eqs. (4.1) and (4.3) may be expressed as

$$\Gamma = -\rho^{-1}\left[\partial R^{o^{-1}}(q,\omega)/\partial(-i\omega)\right]_{\substack{q=0\\ \omega=0}} \tag{4.13}$$

and

$$\lambda^{-1} = -T\left[q^2\,\partial S^{o^{-1}}(q,\omega)/\partial(-i\omega)\right]_{\substack{q=0\\ \omega=0}} \tag{4.14}$$

From eqs. (4.5), (4.6) and (4.10) to (4.14) the recursion relations

$$\Gamma' = b^{2-z-\eta}\left\{\Gamma + \rho^{-1}\left[\partial\Pi(q,\omega)/\partial(-i\omega)\right]_{\substack{q=0\\ \omega=0}}\right\} \tag{4.15}$$

and

$$\lambda'^{-1} = b^{2-z+\bar{\alpha}/\nu}\left\{\lambda^{-1} + T\left[q^2\,\partial\,\Xi(q,\omega)/\partial(-i\omega)\right]_{\substack{q=0\\ \omega=0}}\right\}$$

follow. (4.16)

The critical dimension d^* may be obtained by power counting [9]. By definition d^* is the limiting dimension above which the infrared ($q = 0$) divergences disappear. Therefore d^* is determined by the condition that the singular behaviour of any given diagram is not changed by arbitrary insertions. For this analysis we have to distinguish between an underdamped and an overdamped soft mode, distinguished respectively by $\Gamma \ll \omega_q$ and $\Gamma \gg \omega_q$ near T_c and $q = 0$. Writing the pole of R^o as $\omega \propto q^{z^o}$ we find from eqs. (4.1) and (4.3) for

$$\left.\begin{array}{ll}\Gamma \ll \omega_q : & z^o = 1\\ \Gamma \gg \omega_q : & z^o = 2\end{array}\right\}\quad R^o \propto q^{-2}\,,\; S^o \propto q^{2-z^o} \tag{4.17}$$

Since each closed loop introduces an integral over q and ω and since in our diagrams each vertex carries one arrow which according to the FDT (3.7) introduces a factor ω, eq. (4.17) leads to the following situation :

	Insertion	Power of q
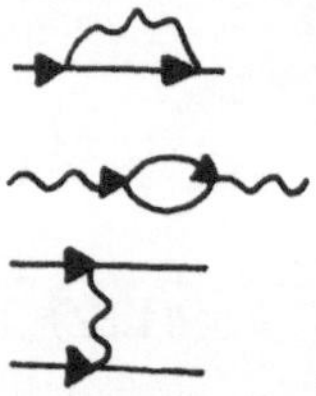	$d^dq\,d\omega\,R^{o^2}\,S^o/\omega$	$d - 2 - z^o$
	$d^dq\,d\omega\,R^{o^2}/\omega$	$d - 4$
	$(d^dq\,d\omega)^2\,R^{o^4}/\omega^2$	$2(d - 4)$

This shows that $d^* = 4$, as was to be expected.

The interesting fixed point of the recursion relations is $\Gamma^* = 0$ corresponding to critical slowing down and $0 < \lambda^* < \infty$ corresponding to a non-singular heat conductivity as found in $SrTiO_3$ [15] . These fixed points are indeed found to be stable [28] .

On the other hand, the fixed point of the coupling constant γ in eq. (1.28) is found from static RG recursion relations to be $\gamma^* \propto \bar{\alpha}$ [2]. Since in $SrTiO_3$ the soft mode is a transverse phonon we have $n = 2$. But for $d = 3$, $n = 2$ the exponent α found from RG calculations is slightly negative [29] which for $SrTiO_3$ is in agreement with the absence of a singularity in the specific heat [15]. But for α close to zero the fixed point value $\gamma^* = 0$ is reached very slowly so that transient effects are important in approaching T_c [28].

4.4. Central Peaks

The above conclusions from the RG analysis may now be applied to a more detailed evaluation of the self-energies in real dimension ($d = 3$) [9]. The quantity which may directly be compared with neutron scattering results is the dynamical structure factor

$$S(q,\omega) = \frac{1}{\omega} \operatorname{Im} \chi(q,\omega) \tag{4.18}$$

where χ is the correctly normalized and renormalized order parameter response function [28]

$$\chi^{-1}(q,\omega) = -\rho^{-1} R^{-1}(q,\omega) - \frac{n}{2} \delta^2 \tag{4.19}$$

where [15]

$$\delta^2 = 4\gamma^2 Q \rho^{-1} \int \frac{d^3q}{(2\pi)^3} \int \frac{d\omega}{2\pi} G^0(q,\omega) \tag{4.20}$$

It is found [28] that in an approximate evaluation of the self-energies (4.7) and (4.8) for the overdamped case the first diagram of Π_γ in fig. 2 reproduces exactly the CP obtained earlier [15] while Π_V produces a new CP. The first CP is "transient" in the sense that it is proportional to the δ^2 in eq. (4.20) which vanishes at the fixed point $\gamma^* = 0$ [30]. However its integrated intensity [15] does not vanish.

As to the new CP due to Π_V it is entirely due to the anharmonic interaction in eq. (1.28) which persists at T_c . This therefore is the CP attributed to cluster wall motion [4,5] and more generally to cluster dynamics [6]. It is important to note that this CP (as well as the first one) is a feature of critical dynamics since its width shrinks and its height diverges at T_c .

ACKNOWLEDGEMENT

During the Sitges School I benefited from many clarifying discussions with Mr. M. San Miguel (Barcelona) on the Fokker-Planck formalism presented in these lectures.

REFERENCES

1) T. SCHNEIDER, G. SRINIVASAN and C.P. ENZ, Phys. Rev. A 5, 1528 (1972).

2) B.I. HALPERIN, P.C. HOHENBERG and S.-K. MA, Phys. Rev. B 10, 139 (1974).

3) V.L. GINZBURG and L.D. LANDAU, Zh. Eksy. Teor. Fiz. 20, 1064 (1950). See , e.g., A.L. FETTER and J.D. WALECKA, "Quantum Theory of Many Particle Systems" (McGraw-Hill, New York, 1971), Sect. 50 and 55.

4) S. AUBRY, J. Chem. Phys. 62, 3217 (1975).

5) J.A. KRUMHANSL and J.R. SCHRIEFFER, Phys. Rev. B 11, 3535 (1975).

6) T. SCHNEIDER and E. STOLL, Phys. Rev. B 13, 1216 (1976).

7) E.P. GROSS, Nuovo Cimento 20, 454 (1961).

8) L.P. PITAEVSKII, Soviet Physics - JETP 13, 451 (1961).

9) S.-K. MA and G.F. MAZENKO, Phys. Rev. B 11, 4077 (1975).

10) B.I. HALPERIN, P.C. HOHENBERG and E.D. SIGGIA, Phys. Rev. B 13, 1299 (1976).

11) P. SZEPFALUSY, Lecture Notes (this volume)

12) L.P. KADANOFF and J. SWIFT, Phys. Rev. 166, 89 (1968) ; K. KAWASAKI, Ann. Physics (New York) 61, 1 (1970).

13) R. GRAHAM, Springer Tracts in Modern Physics (Springer, Berlin, 1973), Vol. 66.

14) C.P. ENZ, to be published.

15) C.P. ENZ, Helv. Phys. Acta 47, 749 (1974).

16) R.L. STRATONOVITCH, Topics in the Theory of Random Noise (Gordon and Breach, New York, 1963), Vol. 1 .

17) H. HAKEN, Rev. Mod. Phys. 47, 67 (1975).

18) H. MORI, H. FUJISAKA and H. SHIGEMATSU, Progr. Theor. Phys. 51, 109 (1974).

19) Y. KURAMOTO, Progr. Theor. Phys. 51, 1712 (1974).

20) R. BAUSCH, H.K. JANSSEN and H. WAGNER, to be published.

21) L. ONSAGER and S. MACHLUP, Phys. Rev. 91, 1505, 1512 (1953).

22) C.P. ENZ and L. GARRIDO, to be published.

23) P.C. MARTIN, E.D. SIGGIA and H.A. ROSE, Phys. Rev. A 8, 423 (1973).

24) J.F. SCOTT, Rev. Mod. Phys. 46, 83 (1974).

25) See, e.g., Structural Phase Transitions and Soft Modes, ed. E.J. Samuelsen, E. Anderson and J. Feder (Universitetsforlaget, Oslo, 1971) ; Proc. Nato Advanced Study Institute, Geilo, Norway, 1973, to be published.

26) G.B. LUBKIN, Physics Today 29, p. 17 (1976).
27) B.I. HALPERIN and C.M. VARMA, to be published.
28) C.P. ENZ, to be published.
29) F. WEGNER, Lecture Notes (this volume).
30) E. SIMANEK and R. WARD, Solid State Commun. 18, 841 (1976).

DYNAMIC CRITICAL PHENOMENA AND THE RENORMALIZATION GROUP - APPLICATION TO A LATTICE DYNAMIC MODEL

P. SZÉPFALUSY

Institute for Theoretical Physics
Eötvös University, Budapest, Hungary

Dynamic Critical Phenomena and the Renormalization Group - Application to a Lattice Dynamic Model

P. SZÉPFALUSY

Institute for Theoretical Physics
Eötvös University, Budapest, Hungary

I. Introduction

1.1. General Outline

When one extends the study of systems near their critical points from their static properties also to time dependent ones the picture of critical behavior becomes considerably richer. The reason behind this is that when we approach the critical point besides having very large spatial correlations the characteristic time also exhibits critical enlargement, the phenomenon called critical slowing down of the fluctuations takes place, which manifests itself in the singular behaviour of dynamic parameters.

In parallelism with statics, several theoretical approaches have been worked out also in the field of dynamic critical phenomena to explain and calculate the critical properties. Let me start by listing them.

1. Conventional theory [1,2] .
2. Mode-mode coupling approaches [3,4,5] .
3. Dynamic scaling [6,7].
4. Polyakov's theory [8] .
5. Renormalization group [9,10,11,12] .

The renormalization group theory played a unifying role among different approaches, namely from the existence of a fixed point and from the behaviour in the vicinity of this fixed point dynamic scaling follows. Moreover the series expansion techniques have provided a systematic means to evaluate mode-mode coupling theory. Also it has given the range of validity of the conventional theory as the region where the trivial fixed point is stable.

A discussion of Polyakov's theory is beyond the scope of this lecture. We only mention that his microscopic analysis of the λ transition of liquid helium has led to conclusions which are different from those of dynamic scaling assumptions and of the renormalization group treatment of semi-macroscopic models.

Wilson's renormalization group theory has also been able to give a natural framework in which universality can be understood. Universality is one of the most spectacular facts in static critical phenomena and has been a major driving force of the development of the theory. Later on the idea of searching for universality classes for critical dynamics has arisen quite naturally. It has turned out that this is a more involved problem and the static universality classes split more from the point of view of dynamics. The quantities determining the static universality classes are, in case of short range forces: dimensionality d, symmetry of the order parameter [13].

In case of dynamic universality classes, however, further decisive factors appear: conservation laws and Poisson bracket relations [9 - 12].

On the other hand, to get further insight into the problem, the classification of systems in time dependent critical phenomena from some other point of view also proves to be useful. The systems can roughly be classified into two large groups. The first group comprises all those systems in which the dynamic critical exponents can be expressed in terms of the static ones while in the second group this is not possible (unless the conventional, mean field theory applies) [5, 12].

In the following we will only be interested in the first group. This group splits into two according to whether a) the order parameter is a conserved quantity; b) it is not. Examples of a) are superfluid helium and its generalizations to multicomponent Bose systems, planar ferro- and antiferromagnets, Heisenberg antiferromagnet and the lattice dynamic model for structural phase transitions to be discussed during these lectures. Into category b) belongs the Heisenberg ferromagnet.

There is a special group within a) to which all the systems listed above but the multicomponent Bose system belong. The common feature of these systems is that the conserved field to which the order parameter is coupled is the infinitesimal generator for the rotation of the order parameter. This group will be called the "$z = \frac{d}{2}$ category" in the following, for reasons to be explained soon. The real part of the frequency of the propagating mode below T_c in the long wavelength limit can be written in the following form [6,14,15]

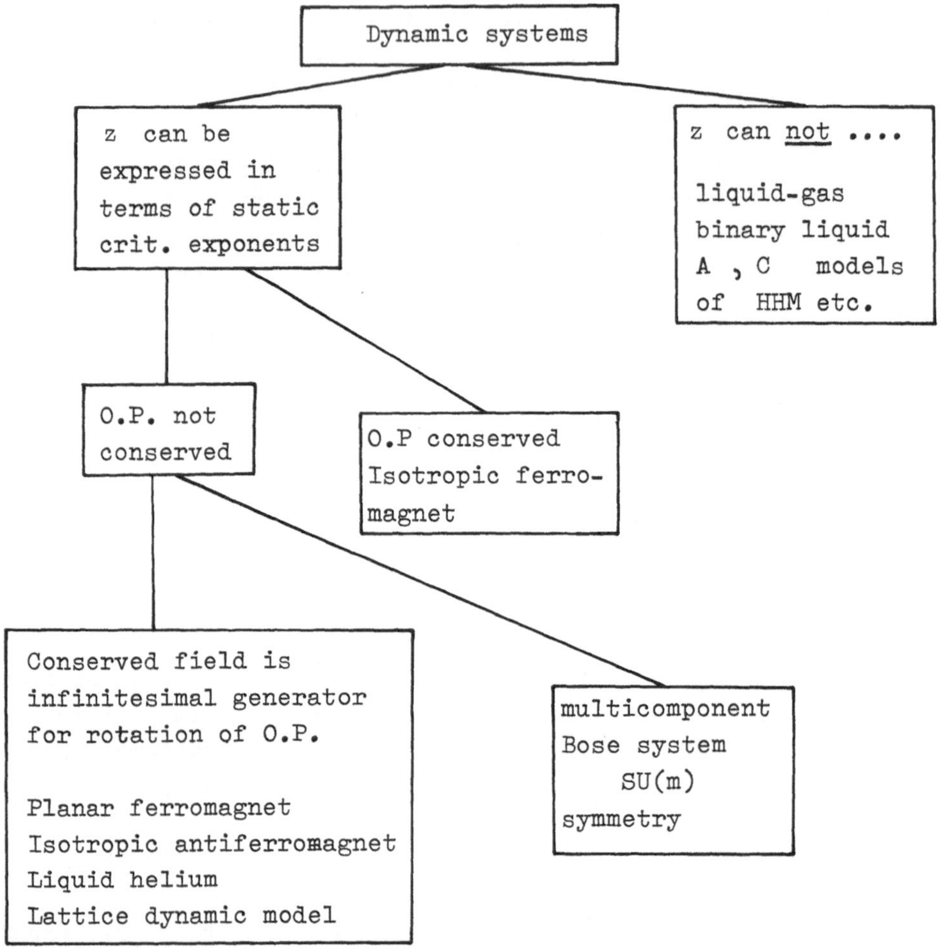

Fig. 1. Classification of dynamic systems.

for these systems

$$\omega_k = \frac{\Psi}{\sqrt{\chi_M \chi_T}} \tag{1.1}$$

where Ψ denotes the order parameter, χ_T is the transverse susceptibility of the order parameter and χ_M is the susceptibility of the generator-field which rotates the order parameter in the plane determined by its equilibrium direction and the direction of the transverse field. χ_T takes the very simple form in the long wavelength limit, namely [6,7,14]

$$\chi_T = \frac{\Psi^2}{\rho_s k^2} \tag{1.2}$$

which is the definition of ρ_s, a quantity playing in general a role similar to the superfluid density in liquid He II. The coherence length for the transverse order parameter correlation function is

$$\xi \sim \rho_s^{\frac{1}{2-d}} \tag{1.3}$$

For these systems this Goldstone-mode frequency plays the role of the characteristic frequency for critical behaviour. This assumption leads to definite consequences if we apply the dynamic scaling hypothesis [6,7]. The dynamic scaling hypothesis asserts that the characteristic frequency for the order parameter in the hydrodynamic region ($k\xi \ll 1$) and in the critical region ($k\xi \gg 1$) can be matched for $k = \xi^{-1}$. This matching condition is equivalent to the requirement that the characteristic frequency obeys, if no logarithms are present,

$$\omega(k, \xi) = k^z \Omega(k, \xi) \tag{1.4}$$

where z is the dynamic critical exponent.

To evaluate z from eq. (1.1) we still have to know the behaviour of χ_M. It can be well behaved or exhibit singularities in k or in the temperature. In the first case the dynamic critical exponent is

$$z = d/2 \; . \tag{1.5}$$

Examples of such behaviour are the lattice dynamical model [15], the symmetric planar spin model [14] and the isotropic antiferromagnet [14].

There are cases in which [6,7,14]:

$$\chi_M \sim (T_c - T)^{-\alpha} \qquad \alpha > 0 \qquad (1.6)$$

which leads to

$$z = \tfrac{1}{2}(d + \frac{\alpha}{\nu}) \qquad (1.7)$$

where α and ν are the critical exponents of the heat capacity and coherence length, respectively. We define the $z = \frac{d}{2}$ category to include also these cases.

It is worth mentioning that the formula (1.5) can be applied also for the isotropic ferromagnet [14]. In this system the generator field for the rotation of the order parameter is the order parameter itself. If the order parameter points in the z direction we can choose

$$\chi_T = \chi_{M_x}, \qquad \chi_M = \chi_{M_y} .$$

Since $\chi_{M_x} = \chi_{M_y}$ we obtain by using eq. (1.1) and (1.2)

$$\omega_k = \frac{\rho_s}{\psi} k^2$$

which yields a dynamic critical exponent [7, 11]

$$z = \tfrac{1}{2}(d + 2 - \eta) .$$

Furthermore the m-component Bose system [16] also shows similar behavior despite the essential difference that the symmetry group is U(m) in that case. In the following we will restrict our considerations to the systems belonging to the $z = \frac{d}{2}$ category. This category is represented by a single model, a lattice dynamic model (LDM in the following) [17,15,18] in which the order parameter has n components. Namely it reproduces all the properties appearing on a semimacroscopic level of all the systems in the $z = \frac{d}{2}$ category: for n = 2 its critical properties are equivalent to those of the planar systems while for n = 3 it has the same critical properties as the isotropic antiferromagnet.

In addition, further aspects of dynamic critical phenomena can be discussed in the framework of the LDM, these will be summarized in the next subsection where our model will be introduced explicitly.

1.2. The Model and the Constants of Motion

The quantum mechanical Hamiltonian of the model is as follows

[17]:

$$H = \frac{1}{2M} \sum_{\alpha,\ell} p^2_{\alpha\ell} + V(\sum_{\alpha} x_{\alpha\ell} \; x_{\alpha\ell'}) \qquad (1.8)$$

At each site ℓ of the d-dimensional lattice there are n displacive degrees of freedom $x_{\alpha\ell}$, their conjugate momenta $p_{\alpha\ell}$ with $\alpha = 1,2, \ldots, n$. (See Figure 2).

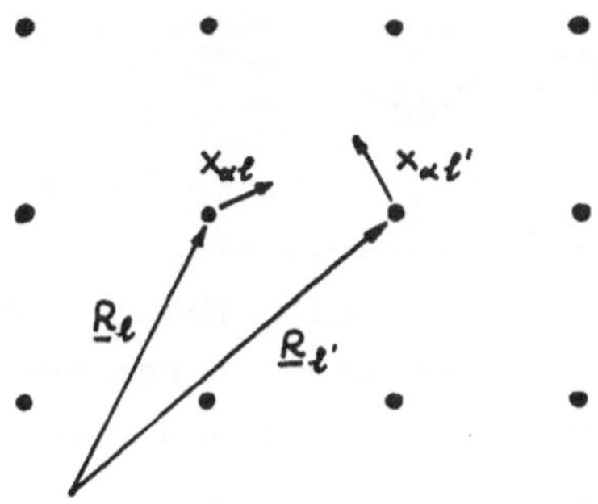

Figure 2.

These degrees of freedom are characterised by a common mass M and hence the kinetic energy as well as the interaction potential V are rotationally invariant in the "intrinsic space" of the displacive degrees of freedom. The symmetry group of the Hamiltonian (1.8) is the group O(n). Frequently it will be convenient to use Fourier transformed operators defined by

$$x_{\alpha q} = \frac{1}{N^{\frac{1}{2}}} \sum_{\ell} e^{iq\ell} x_{\alpha\ell} \quad , \quad p_{\alpha q} = \frac{1}{N^{\frac{1}{2}}} \sum_{\ell} e^{iq\ell} p_{\alpha\ell} \qquad (1.9)$$

where N denotes the number of unit cells. The non-vanishing commutators are ($\hbar = k_B = 1$)

$$[x_{\alpha q}, p_{\alpha' q'}] = i\, \delta_{\alpha\alpha'} \Delta(q + q') \qquad (1.10)$$

$$\Delta(p) = \begin{cases} 1, & \text{if } p = 0, \; p = \text{a reciprocal lattice vector} \\ 0, & \text{otherwise} \end{cases}$$

i.e. the generalized Kronecker symbol $\Delta(p)$ is defined to vanish except when its argument is zero or a reciprocal lattice vector, in which case it equals unity. The most important special case for the interaction potential V is

$$V((\sum_\alpha x_{\alpha\ell} \quad x_{\alpha\ell'})) = \frac{1}{2}\sum_{\ell\ell'} v_{\ell\ell'} \sum_\alpha x_{\alpha\ell} \, x_{\alpha\ell'} + \\ + \frac{u}{4!} \sum_\ell (\sum_\alpha x^2_{\alpha\ell})^2 \qquad (1.11)$$

where the constants $v_{\ell\ell'}$ and u characterise the strength of the interaction.

The system undergoes a ferrodistortive structural phase transition at some temperature T_c provided the Fourier-transformed interaction potential

$$v(k) = \sum_\ell e^{-ik(\ell - \ell')} \, v_{\ell\ell'} \qquad (1.12)$$

is minimal at $k = 0$ and $v(0) < 0$. It is hoped that the LDM incorporates some of the essential features of a structural phase transition driven by a zone-centred soft mode, though some relevant features like anisotropy are not included in the LDM. The model for an antiferrodistortive transition can be transformed by a suitable redefinition of the variables into the ferrodistortive model. For the sake of definiteness, we will consider here the ferrodistortive transition, and we will use its terminology throughout the lectures.

The $\frac{1}{2}n(n-1)$ generators $\Lambda_{\alpha\beta}$ of the group $O(n)$ are constants of motion of the Hamiltonian (1.8). They can be expressed in the form

$$\Lambda_{\alpha\beta} = \sum_\ell L_{\alpha\beta\ell} = N^{\frac{1}{2}} L_{\alpha\beta\, q=0} \qquad (1.13)$$

by the local angular momenta

$$L_{\alpha\beta\ell} = x_{\alpha\ell} \, p_{\beta\ell} - x_{\beta\ell} \, p_{\alpha\ell} \quad . \qquad (1.14)$$

In the second identity of eq. (1.13) we introduced

$$L_{\alpha\beta q} = N^{-\frac{1}{2}} \sum_k (x_{\alpha k} \, p_{\beta -k+q} - x_{\beta k} \, p_{\alpha -k+q}) \qquad (1.15)$$

Of course the total energy H is also a constant of motion and for the Fourier transformed energy density H_q we find the usual hydrodynamic equation.

The model defined above provides a good theoretical laboratory for studying several aspects of critical phenomena. If we restrict ourselves to static properties and to the classical limit it is equivalent to the Ginzburg-Landau-Wilson [13,19] n-vector model for the classical spin system. The order parameter is interpreted now as the n-component displacement of the atoms in the d-dimensional crystal lattice as in the work by Larkin and Khmelnitskii [20] who calculated the logarithmic corrections to the static mean field critical behavior at four dimensions.

The fact that the LDM was defined on a quantum mechanical level makes it possible to study the quantum corrections which can explicitly be demonstrated to be negligible near T_c in the long wavelength region provided T_c is finite. With a suitable choice of the parameters, however, $T_c = 0$ can be reached which is the so-called displacive limit [21]. In the vicinity of this limit and low enough temperatures quantum effects start playing an important role.

Concerning the dynamical aspects of the LDM we have already mentioned its similarity to other models for special values of n in the semi-macroscopic description. The fact that the LDM is defined for arbitrary n proves to be useful and leads to interesting cross-over behaviors for small and large n's. Moreover the model has a well-defined dynamics on the microscopic level due to the kinetic energy of the atoms which differs basically from the dynamics of spins and can be more easily dealt with by a Feynman diagram technique. In this respect the LDM is similar to the multicomponent Bose system [22,23].

It is of special interest that a model is obtainable in the $n \to \infty$ limit which is exactly solvable and nontrivial at the same time. The next section will be devoted to the treatment of this case.

For any finite n, however, the low-frequency, long-wavelength properties are determined by the conserved quantities, by their couplings among themselves and to the order parameter. The study of this behavior starting from the microscopic description encounters enormous difficulties. (We will return to this question later on.) Therefore we adopt here another way of attacking the problem which has been the basic idea of mode coupling approaches and which is the only one used in renormalization group calculations until now. Namely, we construct a semimacroscopic stochastic model to describe the critical behavior of the system. This will be given in Section IV. Preceding this, however, in Section III, the hydrodynamic properties of the system together with the consequences of the use of the dynamic scaling hypothesis will be described. Section V is devoted to the renormalization group analysis of the LDM.

II. The Dynamics of the Model in the $n \to \infty$ limit

Let us start by discussing some general features of the model and especially its dynamics in the $n \to \infty$ limit [17]. With the aim of a perturbational expansion we split the hamiltonian into a free part

$$H_o = \frac{1}{2M} \sum_{\ell\alpha} p_{\ell\alpha}^2 + \tfrac{1}{2} \sum_{\ell\ell'} \left\{ v_{\ell\ell'} + \left[r_T - v(0) \right] \right\} \delta_{\ell\ell'} \sum_{\alpha} x_{\ell\alpha}\, x_{\ell'\alpha} \tag{2.1}$$

and a perturbation

$$H_1 = -\tfrac{1}{2} \sum_{\ell\alpha} (r_T - v(0))\, x_{\ell\alpha}^2 + \frac{u}{4!} \sum_{\substack{\ell \\ \alpha,\beta}} x_{\ell\alpha}^2\, x_{\ell\beta}^2$$

$$H = H_o + H_1 \tag{2.2}$$

where r_T has been introduced to enable a mass renormalization to be performed.

In the symmetry breaking phase or in the presence of the external field the average value of the displacement is non zero, i.e.

$$\langle x_{\ell\alpha} \rangle = P\, \delta_{\alpha 1} \tag{2.3}$$

where it has been taken to point into the direction of the first axis. The displacement propagator represents the order parameter response function. In momentum space the longitudinal and transverse propagators are defined as

$$G_L(q,t) = -i\theta(t) \left\langle \left[x_{q,1}(t),\; x_{-q,1}(0) \right] \right\rangle + P^2 \delta_{q,0} \tag{2.4}$$

$$G_T(q,t) = -i\theta(t) \left\langle \left[x_{q,\alpha}(t),\; x_{-q,\alpha}(0) \right] \right\rangle, \quad n \geq \alpha \geq 2 \tag{2.5}$$

where $\theta(t)$ is the step function, $\langle\ \rangle$ denotes thermodynamic averaging, and the square bracket denotes the commutator. Rotational invariance makes non-diagonal propagators vanish.

The Fourier components of the displacement propagators satisfy Dyson's equation

$$G = G_o + G_o \Sigma G \tag{2.6}$$

where

$$G_o = \left[(i\,\omega_m)^2 - r_T - q^2 \right] \tag{2.7}$$

with

$$\omega_m = 2m\pi T, \quad m = 0,1,2,\ \ldots \quad \text{(in units } k_B = 1\text{)} \quad \text{and}$$

$$v(q) - v(0) = q^2 ,$$

in appropriate units and in the small q limit.

r_T is chosen to be the inverse of the transverse isothermal susceptibility:

$$r_T = G_T^{-1}\,(q \rightarrow 0,\ \omega_m = 0) \quad . \tag{2.8}$$

For $T > T_c$ there is no difference between longitudinal and transversal susceptibilities, $r_T = r$, and eq. (2.8) is equivalent to

$$\lim_{q \rightarrow 0} \Sigma(q,0) = 0 \tag{2.9}$$

according to eqs. (2.6) and (2.7).

In the broken symmetry phase, i.e. for $T < T_c$ with vanishing external field we have

$$r_T = 0 \tag{2.10}$$

which is a manifestation of the Goldstone Theorem. The equality

$$\lim_{q \rightarrow 0} \Sigma_T\,(q,0) = 0 \tag{2.11}$$

describes the equilibrium value of the order parameter.

It is worth noting here that the replacement of $v(0)$ by r_T was necessary to ensure the appropriate analytic behavior of G_o.

Now let us consider the limit $n \rightarrow \infty$. Following Wilson's $1/n$ expansion method we take u to be of order $1/n$ and P of order $n^{1/2}$. Moreover we have to take into account the fact that every closed loop gives a factor of n, [24,25,23].

For $T > T_c$ the diagrams contributing to (2.9) in leading order are

$\Sigma(0,0) =$  $= 0$

whence

$$r \sim (T - T_c)^{\gamma}$$

$$\gamma = \frac{2}{d - 2} \quad ; \qquad 2 < d < 4$$

is obtained. In this section it will be assumed that the dimensionality is $2 < d < 4$ without further indication. The coherence length is given by

$$\xi \sim r^{-\frac{1}{2}} \sim (T - T_c)^{-\nu}$$

with

$$\nu = 1/(d - 2) \quad .$$

These are the well-known critical exponents of the spherical model to which the statics of our model is equivalent in the $n \to \infty$ limit, a fact which is anticipated from Stanley's proof for the Heisenberg n-vector model [26]. From eqs. (2.6) and (2.7)

$$G = \frac{1}{q^2} \frac{1}{\Omega^2 - [1 + (q\xi)^{-2}]} \tag{2.12}$$

is obtained, where $\Omega^2 = \omega^2/q^2$.

The pole of eq. (3.12) yields the eigenfrequency

$$\omega = \sqrt{\frac{1}{\xi^2} + q^2}$$

which gives, for the dynamic critical exponent z, defined in eq. (1.4), $z = 1$. Here the expected soft mode behavior can be seen explicitly.

Below T_c, in the limit $n \to \infty$, the contributions to Σ_T are given by

$\Sigma_T(0,0) \;=$ 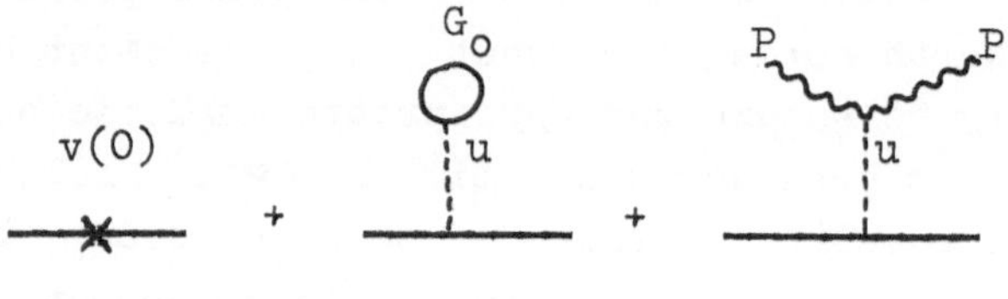 $= \; 0$.

G_T coincides with the propagator G_o

$$G_T = \frac{1}{q^2} \frac{1}{\Omega^2 - 1} \tag{2.13}$$

which has a gapless spectrum in accordance with the Goldstone theorem.

To determine the longitudinal propagator let us calculate Σ_L to leading order in n. It is given schematically by

$$\Sigma_L = \Sigma_T + $$  $$= \frac{1}{3} u_{eff} P^2$$

where ===== denotes the sum of bubble diagrams

and u_{eff} is defined by

$$u_{eff} = \frac{u}{1 - \frac{1}{6} u \, n \Pi(q,\omega)}$$

where $\Pi(q,\omega)$ is the contribution of the bubble diagram. In the classical approximation we have

$$\Pi(q,\omega) = \left(\frac{q_c}{q}\right)^{4-d} L(\Omega) \tag{2.14}$$

with $q_c^{4-d} = nuT/3$, taking the lattice constant as unity. $L(\Omega)$ is a dimensionless function of its variable.

By introducing two coherence lengths

$$\xi' = \left(\frac{nT}{8\pi P^2}\right)^{\frac{1}{d-2}} \tag{2.15}$$

and

$$\xi_{MF} = \left(\frac{3}{uP^2}\right)^{\frac{1}{2}} . \tag{2.16}$$

We can write

$$\frac{1}{6} u \, n\left(\frac{q_c}{q}\right)^{4-d} = \frac{(\xi' q)^{d-2}}{(\xi_{MF} q)^2}$$

and G_L can be written as

$$G_L = \frac{1}{q^2} \frac{1 + \frac{(\xi' q)^{d-2}}{(\xi_{MF} q)^2} L(\Omega)}{(\Omega^2 - 1)\left[1 + \frac{(\xi' q)^{d-2}}{(\xi_{MF} q)^2} L(\Omega)\right] - (\xi_{MF} q)^{-2}} . \tag{2.17}$$

The poles of G_L can be obtained from the equation

$$(\Omega^2 - 1)\left[1 + \frac{(q\xi')^{d-2}}{(q\xi_{MF})^2} L(\Omega)\right] - \frac{1}{(\xi_{MF}q)^2} = 0 \quad . \qquad (2.18)$$

Close to T_c ξ_{MF} becomes very large, i.e.

$$\xi_{MF} \longrightarrow \infty \qquad \text{if} \quad T \longrightarrow T_c$$

and the zeros are given by the first term in eq. (3.18). $\Omega \longrightarrow \pm 1$ corresponds to the critical mode while the vanishing of the quantity in the square brackets yields a noncritical mode. The $\Omega^2 = 1$ solutions will remain poles of the propagator also above T_c while the noncritical poles of G_L go over to the poles of the Fourier transform of the propagator

$$F = - i\theta(t) \left\langle \left[\sum_\alpha x^2_{\alpha\ell}(t), \; \sum_\alpha x^2_{\alpha\ell'}(0) \right] \right\rangle .$$

In general, if $P \neq 0$, G_L and F have common excitation spectra. For details see ref. [17].

G_L is scaled by two characteristic lengths, namely ξ_{MF} and ξ'. Both rise to infinity at the transition point but the divergence of ξ' is the stronger. However, there are regions of q and Ω where only one of these lengths prevails, while the other approximately drops out of G_L. In the mean field region the term containing $L(\Omega)$ can be neglected and G_L takes the form of G above T_c but with ξ replaced by ξ_{MF}. The static correlation function has the standard Ornstein-Zernike form in this region with the correlation length $\xi_{MF} \sim (|T - T_c|)^{-1/2}$.

By fixing a sufficiently small value of q and ω we necessarily leave this region by increasing the temperature and enter the critical region characterized by the dominance of the second term in the square brackets in eq. (2.18). Thus we have

$$(\Omega^2 - 1)(q\xi')^{d-2} L(\Omega) - 1 = 0 \quad . \qquad (2.19)$$

It is worth noting that ξ_{MF}, which contains, according to eq. (2.16), the interaction u, has dropped out from eq. (2.19) and thus G_L becomes independent of the anharmonic coupling constant just as required by universality.

Moreover the solution of (2.19) can clearly be written in the the form

$$\Omega = f(q\,\xi')$$

and remembering the definition $\Omega \equiv \omega/q$ we get

$$\omega = q\, f(q\,\xi')$$

which, according to eq. (1.5) yields

$$z = 1$$

for the dynamical critical exponent.

For $d = 3$ $L(\Omega)$ is given by

$$L(\Omega) = \frac{i}{2} \ln \frac{\Omega-1}{\Omega+1} \quad .$$

It is interesting to note that in this special case $d = 3$ the contribution of the bubble graph expressed in terms of Ω is the same as for the multicomponent Bose gas [22,23], while they are quite different in other dimensions. An essential difference occurs even at $d = 3$, namely that $\Omega = \omega/q$ in the LDM while $\Omega = \omega/q^2$ for the Bose gas.

The migration of the poles of eq. (2.17) with increasing temperature is shown on Fig. (3). Inside the dotted circle is the scaling region. Starting from low enough temperatures, first the $\xi_{MF} \gg \xi'$ region is reached. Here

$$\Omega_{12} = \pm \sqrt{\frac{1}{(q\,\xi_{MF})^2} - \Gamma_q^2} - i\,\Gamma_q$$

with

$$\Gamma_q = \frac{q\,\xi'}{2(q\,\xi_{MF})^2}$$

i.e. $\omega_{12} = \pm 1/\xi_{MF} + \ldots$ exhibits a gap.

With increasing temperature these eigenfrequencies get overdamped, the poles collide on the imaginary axis. After this collision one of the poles starts moving upwards while the other moves downwards on the imaginary axis. (The latter behaves noncritically around T_c while the former reaches the scaling region. For $\xi' \gg \xi_{MF}$ and $q\,\xi' \ll 1$ it is given by $\omega = -i\,\frac{1}{\xi'}$.) Here it meets a third pole, Ω_3, emerging from under a cut of the correlation function, then they leave the imaginary axis and arrive at $\Omega = \pm 1$ for $T = T_c$.

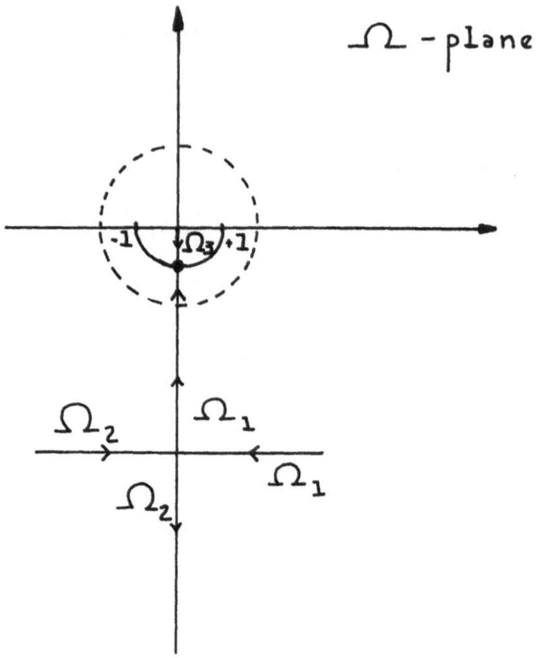

Figure 3

III. Hydrodynamics and Dynamic Scaling in the Lattice Dynamical Model

For small wavenumber q and frequency ω the response of a many-particle system at $T \neq 0$ is described by hydrodynamic equations. The enormous simplification in this region is due to the fact that the conserved densities oscillate with a characteristic time which is a power of the wavenumber and therefore is slow, whereas the nonconserved variables relax to their local-equilibrium values within a short time, leading to damping processes only and thus disappearing from the dynamic description. Thus hydrodynamic equations are determined solely by the conservation laws and the symmetry of the state. They constitute an exact description in the low-frequency and small-wavenumber limit. The velocities and the mass terms in the hydrodynamic equations can be related exactly to thermodynamic derivatives, while definite power laws in q for the damping terms are predicted. The LDM considered by us has rotational symmetry, i.e. the symmetry group of the Hamiltonian, as we already mentioned, is the group $O(n)$. The conserved quantities are the $\frac{1}{2}n(n-1)$ generators $\Lambda_{\alpha\beta}$ of the group $O(n)$ and the total energy [15]. Let us denote by H_q the Fourier transformed energy density.

In the high temperature phase we get the following diffusion-type equations for the conserved densities

$$\frac{d}{dt}(\delta H_q) = - D_E\, q^2 (\delta H_q) \tag{3.1}$$

$$\frac{d}{dt}\,\delta L_{\alpha\beta\, q} = - D q^2\, \delta L_{\alpha\beta\, q} \tag{3.2}$$

where $D_E = \lambda_E / C$, λ_E is the thermal conductivity and C is the specific heat. $L_{\alpha\beta\, q}$ was defined in eq. (1.15). Eq. (3.2) comprises $\frac{n(n-1)}{2}$ equations.

Before proceeding with our discussion of the hydrodynamic equations below T_c let us first examine the behavior of static susceptibilities in this region. Let us define the susceptibilities as

$$\chi(A,B;\,q) = i\int_0^{\infty} dt\, \langle [A_q(t),\; B_q^{+}(0)] \rangle \tag{3.3}$$

For identical operators we will use the notation

$$\chi(A;q) \equiv \chi\,(A,A;\,q).$$

It is easy to see that the angular momentum susceptibility can be related, in the classical limit, to the local fluctuations $\langle x^2_{\beta\ell} \rangle$, as

$$\chi(L_{\alpha\beta}) = \frac{1}{k_B T}\,\frac{\int d\Gamma\, e^{-\beta H}(p_\alpha x_\beta - x_\alpha p_\beta)^2}{\int d\Gamma\, e^{-\beta H}} =$$

$$= \frac{1}{k_B T}\left\{\langle x_\alpha^2\rangle\langle p_\beta^2\rangle + \langle x_\beta^2\rangle\langle p_\alpha^2\rangle\right\} = M\left\{\langle x_\alpha^2\rangle + \langle x_\beta^2\rangle\right\} \tag{3.4}$$

For $T < T_c$ the expectation value of the order parameter is different from zero and is chosen to point in the direction of the first axis (eq. (2.3)). When examining the susceptibilities of the angular momenta, we have to distinguish between the operators $L_{1\beta}$ which generate rotations of the order parameter and $L_{\alpha\beta}$ $(2 \leq \alpha, \beta \leq n)$ which leave it unchanged. The corresponding susceptibilities are

$$\chi_1(L;\,q) \equiv \chi(L_{1\alpha};q) \qquad 2 \leq \alpha$$

$$\chi_2(L;\,q) \equiv \chi(L_{\beta\gamma};q) \qquad 2 \leq \beta < \gamma \leq n\,.$$

According to eq. (3.4) we can write

$$\chi_1(L;q) = M(\langle x^2_{1\ell}\rangle + \langle x^2_{2\ell}\rangle) \tag{3.5}$$

and

$$\chi_2(L;q) = 2M\,\langle x^2_{2\ell}\rangle \tag{3.6}$$

Introducing the quantity Δ, defined by

$$\Delta = \langle x_1^2\rangle - \langle x_2^2\rangle\;, \tag{3.7}$$

$\chi_1(L;q)$ can be rewritten in the form

$$\chi_1(L;q) = \chi_2(L;q) + M\Delta \tag{3.8}$$

Here $\chi_2(L;q)$ is finite at T_c while Δ tends to zero approaching the critical point as

$$\Delta \sim \tau^{\beta_x} \quad , \qquad \tau \equiv (T - T_c)/T_c \tag{3.9}$$

where β_x can be related to the crossover exponent ϕ [28] as

$$\beta_x \ . = \ d\nu - \phi \ . \tag{3.10}$$

Also we have $\Delta \sim n$ which, together with (3.9) means that the $M\Delta$ term in (3.8) behaves differently depending upon the order of the limits $\tau \rightarrow 0$ and $n \rightarrow \infty$.

Let us carry out a $^1/n$ expansion calculation of Δ and β_x in order to confirm the behavior sketched above.

In terms of the Fourier transforms of the propagators G_L and G_T defined in eqs. (2.4) and (2.5), respectively Δ can be written as

$$P^2 - k_B T \sum_m \int \frac{d^d k}{(2\pi)^d} \left[G_L(k,\omega_m) - G_T(k,\omega_m) \right]$$

$$\approx P^2 - k_B T \int \frac{d^d k}{(2\pi)^d} \left[G_L(k,0) - G_T(k,0) \right] = \Delta$$

We evaluate the integral in the leading order, $^1/n = 0$, and can thus insert (2.13), (2.17) with $\omega = 0$ for the susceptibilities. The integration yields

$$\Delta = P^2 - \frac{4S_d}{d-2} \frac{P^2}{n} \ln \frac{P^2}{n} = P^{2 + \frac{4}{d-2} S_d} + O(^1/n^2) \tag{3.11}$$

where

$$S_d = \frac{\pi(\frac{d}{2} - 1) B(\frac{d}{2} - 1; \frac{d}{2} - 1)}{\sin \pi(\frac{d}{2} - 1)} \ .$$

Comparing equations (3.9) and (3.11) we get by using β to $O(\frac{1}{n})$ [30]

$$\beta_x = 2\beta + \frac{4}{d-2} \frac{S_d}{n} + O(\frac{1}{n^2}) = 1 - 8 \frac{d-3}{d-2} \frac{S_d}{n} + O(\frac{1}{n^2}) \ . \tag{3.12}$$

The coefficient in front of the temperature dependent factor in Δ is given by P^2 which is indeed proportional to n for $n \rightarrow \infty$.

Let us now return to hydrodynamics and proceed by discussing propagating hydrodynamic modes.

Because of the broken continuous symmetry, the transverse displacements $x_{\beta q}$ ($\beta = 2, \ldots, n$) also belong to the set of hydrodynamic variables. They are coupled to the angular momenta $L_{1\beta q}$, so the hydrodynamic equations consist of coupled equations determining the above quantities below T_c. These equations will not be given here, they can be found in ref. [15].

As a result we get $2(n-1)$ propagating modes with eigenfrequencies in the form given by eq. (1.1), characteristic of systems belonging to the category we are interested in, namely in which the conserved field is an infinitesimal generator for the rotation of the order parameter, i.e.

$$\mathrm{Re}\,\omega(q) = \frac{P}{\sqrt{\chi_1(L;q)\,\chi_T(q)}} . \tag{3.13}$$

The damping term of the eigenfrequencies is quadratic in the wave vector. The expression one obtains by taking into account eq. (1.2) for the transverse susceptibility can be simplified to the form

$$\omega(q) = \pm cq - \frac{1}{2}\bar{D}\,q^2 \tag{3.14}$$

with the sound velocity c given by

$$c = \left[\frac{\rho_s}{\chi_1(L;0)}\right]^{1/2} \tag{3.15}$$

There are also nonpropagating modes in our system below T_c; these are connected with the angular momenta $L_{\alpha\beta q}$ with $2 \leqslant \alpha < \beta \leqslant n$. They obey equations of the type

$$\frac{d}{dt} L_{\alpha\beta q} = -\Gamma^{<}(q)\, L_{\alpha\beta q} \qquad \alpha,\beta \geqslant 2.$$

For dimensionality d one finds

$$\Gamma^{<}(q) \sim (q^{d/2} + O(q^2))$$

which shows that the hydrodynamic damping applies for $d > 4$. There are $(n-1)(n-2)/2$ relaxing modes.

Now we will consider the critical behavior first for a fixed finite value of n and consider the large n limit afterwards. Using equations (1.2) and (1.3) and the property that $\chi_1(L;0)$ is finite at T_c we find from eqs. (3.14) and (3.15)

$$\mathrm{Re}\ \omega(q) \sim \xi^{\frac{2-d}{2}}\, q, \qquad q\xi \ll 1 ,$$

which implies

$$z = \frac{d}{2} .$$

It is now a simple matter, by using dynamic scaling assumption (1.4), to predict the frequency in the nonhydrodynamic critical region

$$\omega(q,\xi) \sim q^{d/2}, \quad (q\xi \gg 1)$$

and also the critical dependence of the damping term as

$$\bar{D} \sim |T - T_c|^{-\nu'(4-d)/2} .$$

Above T_c a mode-mode coupling analysis can be carried out which gives $z = d/2$.

In the limit of large n the structure of the hydrodynamic equations is unchanged but the angular momentum susceptibility has a different characteristic form.

χ_1 was given in eq. (4.9), the second term on the right hand side of which was $M\Delta$. We have also seen before that $\Delta \sim n\tau^{\beta_x}$ with β_x given by (4.11). When $\tau \ll \tau_c$, where τ_c is a characteristic reduced temperature $\tau_c \sim n^{1/(d\nu - \phi)}$, Δ can be neglected, χ_1 = const, and the previous results hold for the characteristic frequency. On the other hand for $\tau \gg \tau_c$

$$\chi_1 \sim \tau^{d\nu - \phi}$$

and consequently

$$\mathrm{Re}\,\omega \sim \xi^{1 - \frac{\phi}{2\nu}} q .$$

It means that outside the first asymptotic region mentioned above we get from eq. (1.4) an effective value for the dynamical critical exponent

$$\tilde{z} = \frac{\phi}{2\nu} = 1 - \frac{2}{4-d}\eta + O\left(\frac{1}{n^2}\right) \tag{3.16}$$

where in the last equality we used the results of the $1/n$ expansion [25,29,30] and η is to be expressed by its expansion to $O(1/n)$. Thus we have seen a crossover of the frequency of the hydrodynamic Goldstone-type modes in the limit of large n, which is a new feature of the model investigated here: for any finite n in the immediate vicinity of the critical point $z = \frac{d}{2}$, while for large n there is a second nonasymptotic region characterized by a critical exponent $\tilde{z} = \phi / 2\nu$.

Thus far we have investigated our model system in two opposite limits. The $n = \infty$ limit has some special features because of which the long wavelength and low frequency behavior of the system is different in that limit from that for any finite n. Namely, the hydrodynamic region shrinks to zero when $n \to \infty$. To see this we can argue in the same way as Ma and Senbetu [31] did in the case of the multicomponent Bose system. As a consequence of the fact that $u \sim 1/n$ which involves that the collision rate (cross section x number of modes) is of order $1/n$, $\omega < O(1/n)$ in the hydrodynamic region. It can be easily seen that the wave numbers are also restricted to values $O(1/n)$ for large n in the hydrodynamic regime.

For any finite n we necessarily enter the hydrodynamic region in the limit $\omega, k \to 0$. If we generate a perturbation expansion starting from a microscopic Hamiltonian we would meet singularities reflecting the existence of hydrodynamic modes. The calculation by Ma and Senbetu of the spectrum of the total density fluctuation in the case of the n-component Bose fluid shows explicitly such a behavior. These authors carried out the calculation to leading order in $1/n$ by taking the frequency and wavenumber of the modes to be of $O(1/n)$. (This limit is different from what we have discussed in Section II; there the number of components of the field went to infinity while keeping ω and k at a finite value. In this way we necessarily leave the hydrodynamic regime.) Though Ma and Senbetu have found the hydrodynamic modes in their calculation, they could not obtain the expected true critical behavior [16].

It is worth noting that in the limit $n \to \infty$ there is a smooth change-over into the nonhydrodynamic region .
Namely, taking into account that for $1/n = 0$ $\chi_1(L;q) = MP^2$, $\chi_T(q) = 1/q^2$ the frequency given by eq. (3.13) agrees precisely with the temperature-independent excitatioh spectrum of the transverse order parameter fluctuations below T_c obtained in Section II.

In the next section instead of starting from a microscopic theory we construct a semimacroscopic stochastic model which is hoped to describe the behavior of the system in the long wavelength low frequency region for a finite n value.

IV. Semimacroscopic Description of the System

4.1. Definition of the Stochastic Model

To describe the long wavelength, low frequency behavior of the system for a finite n (which includes the hydrodynamic region) we proceed as follows [18]. In this region dynamics are determined solely by variables which vary slowly, namely the long wavelength components of the order parameter and of the fields of the conserved quantities. As it has been discussed previously, besides the energy, the $\frac{1}{2}n(n-1)$ generators, $\Lambda_{\alpha\beta}$ of the group O(n) are constants of motion of the Hamiltonian (2.1). The $\Lambda_{\alpha\beta}$ are expressed in terms of the local angular momenta as given in eq. (1.13).

We describe the long wavelength fluctuations of the order parameter and angular momentum density by an n-component vector field $\varphi_\alpha(x,t)$ and an antisymmetric tensor field $L_{\alpha\beta}(x,t)$, respectively:

$$\varphi_\alpha(x,t) = \sum_{k<\Lambda} x_{\alpha k}(t)\, e^{ikx}$$

$$L_{\alpha\beta}(x,t) = \sum_{k<\Lambda} L_{\alpha\beta}(k,t) e^{ikx} \quad .$$

Here Λ is the cut-off wavenumber which is chosen to be much smaller than the reciprocal of the lattice constant a.

These fields are obtained by eliminating those quantities which vary appreciably over a length Λ^{-1} i.e. $\varphi_\alpha(x,t)$ and $L_{\alpha\beta}(x,t)$ should be regarded as the average values of the order parameter and the angular momentum density, respectively, in the volume $1/\Lambda^d$, i.e. fluctuations over short distances have been smeared out.

First, for the sake of simplicity, we do not take also the energy as a further conserved quantity into account.

The phenomenological description corresponds to decreasing the degrees of freedom. The system originally had $n/(a\Lambda)^d$ degrees of freedom in the volume $1/\Lambda^d$ while we introduced $n(n-1)/2$ angular momentum variables and n order parameter variables in the phenomenological description. The remaining degrees of freedom are incorporated into the parameters of the phenomenological model in the course of the averaging procedure. Clearly

$$\frac{n(n+1)}{2} < \frac{n}{(a\Lambda)^d}$$

should be fulfilled in order to avoid overcounting of the modes. This requirement becomes important when n is large and yields a restriction for the applicable cut-off

$$(a\Lambda)^d < \frac{n}{n(n+1)} = O(^1/n) .$$

The phenomenological model can give an adequate description of the microscopic system (1.8) only in a region of the wave-number space of $O(^1/n)$. Consequently, this semimacroscopic description is valid only for finite n. The cut-off will be chosen in such a way that $1/\Lambda$ be smaller than the correlation length of the fluctuations of the order parameter which makes possible to treat also the critical region. The construction of the equations of motion follows the usual lines of mode coupling theory and our model belongs to the family of models used in the dynamic renormalization group analysis [9,11] . The time variations of the fields φ_α and $L_{\alpha\beta}$ are governed by the following stochastic equations of motion

$$\frac{\partial \varphi_\alpha}{\partial t} = g \sum_\beta \frac{\delta F'}{\delta L_{\alpha\beta}} \varphi_\beta - \Gamma \frac{\delta F'}{\delta \varphi_\alpha} + \zeta_\alpha \tag{4.1}$$

$$\frac{\partial L_{\alpha\beta}}{\partial t} = g\left(\varphi_\alpha \frac{\delta F'}{\delta \varphi_\beta} - \frac{\delta F'}{\delta \varphi_\alpha}\varphi_\beta\right) + j \sum_{\gamma \neq \alpha, \beta} \left(\frac{\delta F'}{\delta L_{\alpha\gamma}} L_{\beta\gamma} - \frac{\delta F'}{\delta L_{\beta\gamma}} L_{\alpha\gamma}\right)$$
$$+ \lambda \nabla^2 \frac{\delta F'}{\delta L_{\alpha\beta}} + \theta_{\alpha\beta} \tag{4.2}$$

where $F'[\varphi_\alpha(x,t),\ L_{\alpha\beta}(x,t)]$ is the free energy functional; λ, Γ, j and g are constants and $\zeta_\alpha(x,t)$ and $\theta_{\alpha\beta}(x,t) = -\theta_{\beta\alpha}(x,t)$ are random fields with a Gaussian distribution and a white spectrum

$$\langle \zeta_\alpha(x,t) \rangle = 0$$
$$\langle \zeta_\alpha(x,t)\, \zeta_\beta(x',t') \rangle = \delta_{\alpha\beta}\, \delta(x-x')\, \delta(t-t')\, 2\Gamma$$
$$\langle \theta_{\alpha\beta}(x,t) \rangle = 0 \tag{4.3}$$
$$\langle \theta_{\alpha\beta}(x,t)\theta_{\gamma\delta}(x',t')\rangle = -2\lambda\nabla^2 \delta(x-x')\, \delta(t-t') \times$$
$$\times \left[\delta_{\alpha\gamma}\delta_{\beta\delta} - \delta_{\alpha\delta}\delta_{\beta\gamma}\right] .$$

F' is the free energy functional in the presence of the external fields $h_\alpha(x,t)$ and $h_{\alpha\beta}(x,t)$:

$$F' = F - \int d^dx \left\{ \sum_\alpha h_\alpha(x,t)\, \varphi_\alpha(x) + \sum_{\alpha<\beta} h_{\alpha\beta}(x,t) . L_{\alpha\beta}(x) \right\} \tag{4.4}$$

$$F = \int d^dx \left\{ \tfrac{1}{2} r_0 \varphi^2 + \tfrac{1}{2}(\nabla\varphi)^2 + \tfrac{1}{4} u \varphi^4 + \tfrac{1}{2} L^2 \right\}$$

where

$$\varphi^2 \equiv \sum_\alpha \varphi_\alpha^2 \quad ; \quad \varphi^4 \equiv (\varphi^2)^2$$

$$L^2 \equiv \sum_{\alpha<\beta} L^2_{\alpha\beta} \quad ; \quad (\nabla\varphi)^2 \equiv \sum_\alpha (\nabla\varphi_\alpha)^2 \ .$$

4.2. Relationship to Other Models

As it was already mentioned in Section I the LDM has close relationship to planar systems and to the isotropic antiferromagnet. This will be shown now explicitly.

Let us first take $n = 2$ and introduce the following notation

$$\Psi = \varphi_1 + i\varphi_2$$

$$\zeta = \zeta_1 + i\zeta_2$$

$$m = L_{12} \ , \quad \theta = \theta_{12}$$

then the free energy and the equations of motion, eqs. (4.4), (4.1) and (4.2), can be cast into the following form by simple algebraic manipulations

$$F = \frac{1}{2} \int d^dx \left\{ r_0 |\Psi|^2 + |\nabla\Psi|^2 + \frac{u}{2}(|\Psi|^2)^2 + m^2 \right\}$$

$$\frac{\partial\Psi}{\partial t} = -ig \frac{\delta F}{\delta m}\Psi - 2\Gamma \frac{\delta F}{\delta\Psi^*} + \zeta \tag{4.5}$$

$$\frac{\partial m}{\partial t} = 2g\, \mathrm{Im}\left(\Psi^* \frac{\delta F}{\delta\Psi^*}\right) + \lambda \nabla^2 \frac{\delta F}{\delta m} + \theta \tag{4.6}$$

where the terms representing the couplings to the external fields have been omitted, for the sake of simplicity. $\delta F/\delta\Psi^*$ is defined as

$$\frac{\delta F}{\delta\Psi^*} \equiv \frac{1}{2}\left[\frac{\delta F}{\delta \mathrm{Re}\,\Psi} + i \frac{\delta F}{\delta \mathrm{Im}\,\Psi} \right] .$$

The system of equations (4.5), (4.6) is the same as that of the symmetric planar model treated by Halperin, Hohenberg and Siggia [9].

The asymmetric model can be obtained if the external field h_{12} is kept finite. In the case of the planar ferromagnet [14] m is the magnetization perpendicular to the easy plane while in the case of liquid helium m corresponds to the entropy [6, 9]. In this latter case the first term on the right hand side of eq. (4.6) has the interpretation as the transport of mass by a supercurrent.

The equations of motion (4.5) and (4.6) are similar to those introduced by Kawasaki [3] in the mode-coupling theory.

Let us discuss now the case $n = 3$ and introduce the following notation

$$\vec{\varphi} = (\varphi_1, \varphi_2, \varphi_3) , \qquad \vec{\zeta} = (\zeta_1, \zeta_2, \zeta_3)$$

$$L_{\alpha\beta} = \begin{pmatrix} 0 & m_3 & -m_2 \\ -m_3 & 0 & m_1 \\ m_2 & -m_1 & 0 \end{pmatrix} \equiv \varepsilon_{\alpha\beta\gamma}\, m_\gamma$$

$$\theta_{\alpha\beta} = \varepsilon_{\alpha\beta\gamma}\, \theta_\gamma .$$

It is again an easy task to rearrange the equations of motion (4.1) and (4.2) into the form

$$\frac{\partial\vec{\varphi}}{\partial t} = g\vec{\varphi} \times \frac{\delta F}{\delta\vec{m}} - \Gamma \frac{\delta F}{\delta\vec{\varphi}} + \vec{\zeta} \tag{4.7}$$

$$\frac{\partial\vec{m}}{\partial t} = g\vec{\varphi} \times \frac{\delta F}{\delta\vec{\varphi}} - j\, \vec{m} \times \frac{\delta F}{\delta\vec{m}} + \lambda \nabla^2 \frac{\delta F}{\delta\vec{m}} + \vec{\theta} . \tag{4.8}$$

By taking $\vec{\varphi}$ as the staggered magnetization, $\vec{m}$ as the total magnetization and $g = j$, these equations are equivalent to those given by Halperin, Hohenberg and Siggia [9] for the isotropic antiferromagnet. The first term on the right hand side of eq. (4.7) and the second term on the right hand side of eq. (4.8) correspond to Larmor precession of the staggered magnetization and total magnetiza--tion in the effective magnetic field $\delta F/\delta\vec{m}$, respectively (which suggests the choice $g = j$).

It is worth mentioning that there is an additional term in the equations of motion, treated by Freedman and Mazenko [27], for the isotropic antiferromagnet. This term originates from the fact that the components of the staggered magnetization do not commute, in contrast to the displacement field. It does not play a role, however, in critical dynamics, at least to $O(\varepsilon)$ [27].

4.3. Perturbation Expansion

Following the procedure developed by Ma and Mazenko [11] the perturbational series will be obtained by iterating the equations of motion. We will restrict our discussion to the case $T > T_c$, for the treatment of the ordered phase, see ref. [18]. Let us apply a subtraction in the free energy similarly, and for the similar reason, as was done in the Hamiltonian in Section II.

$$F = \frac{1}{2}\int d^dx\left(r\varphi^2 + (\nabla\varphi)^2 + \frac{1}{2}u(\varphi^2)^2 + (r_0-r)\varphi^2 + \sum_{\alpha<\beta} L^2_{\alpha\beta}\right)$$

where r is the inverse of the order parameter susceptibility. For iterating we start from the solution of the linearized equations of motion; the non-linear couplings and the counter term coming from the subtraction are considered as perturbations.

The equations of motion in the Fourier-transformed form are represented graphically in Fig. 4.

The solutions of the linearized equations are

$$\varphi^o_\alpha \; (k,\omega) = G_o(k,\omega)\left(h_\alpha + \frac{\zeta_\alpha}{\Gamma}\right) \tag{4.9}$$

$$G_o(k,\omega) = 1/(r + k^2 - i\,\omega/\Gamma) \tag{4.10}$$

$$L^o_{\alpha\beta}(k,\omega) = -L^o_{\beta\alpha}(k,\omega) = K_o(k,\omega)\left(h_{\alpha\beta} + \frac{\theta_{\alpha\beta}}{\lambda k^2}\right) \tag{4.11}$$

$$K_o(k,\omega) = 1/(1 - i\,\omega/\lambda k^2) \quad . \tag{4.12}$$

The iterative forms of the fields $\varphi_\alpha(k,\omega)$ and $L_{\alpha\beta}(k,\omega)$ can be represented by diagrams which start from a single point and then branch off. In a given diagram the internal lines directed in compliance with the order of iterations represent the zeroth order response functions $G_o(k,\omega)$ and $K_o(k,\omega)$, while the branch-points denote the vertices corresponding to the non-linear couplings. The external fields and the counter term, coming from the subtraction, are represented by two-leg vertices. At all vertices momentum and energy conservation have to be fulfilled; the remaining momentum and frequency variables have to be integrated over. The lines outgoing from a given diagram represent the zeroth order solutions of the equation of motion.

As the random fields have Gaussian distributions, the averaging can be performed by simply pairing the outgoing lines. The new lines (see Fig. 5) obtained in this way correspond to the zeroth order correlation functions,

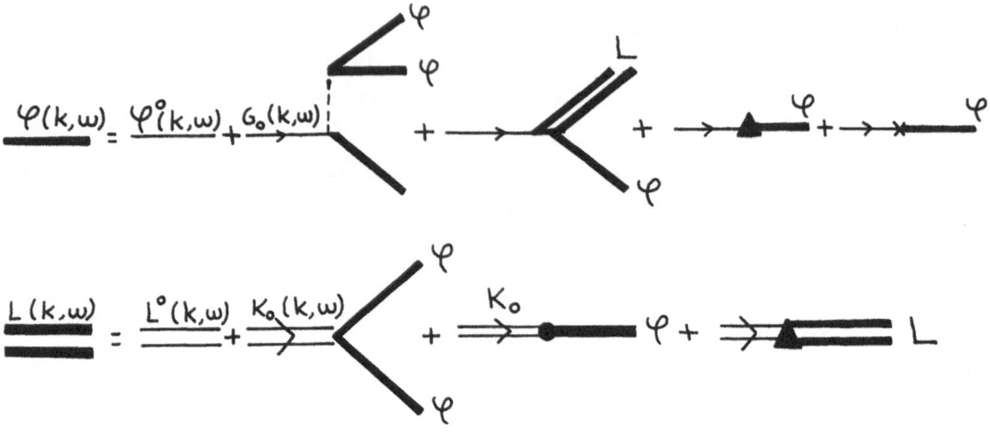

a) Equations of motion.

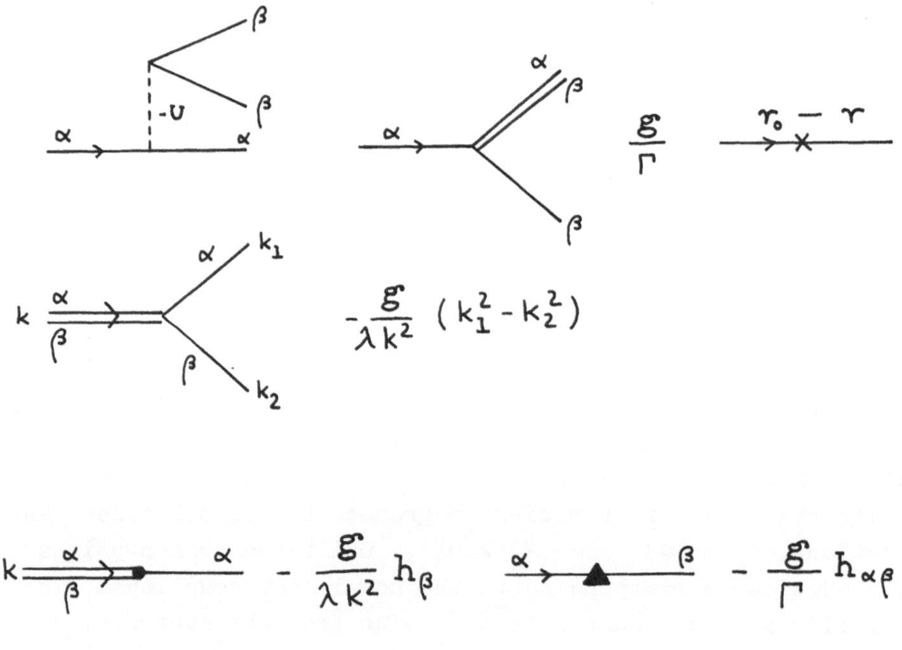

b) Contributions of the vertices.

Figure 4

$$\langle \varphi_\alpha^o(k,\omega)\varphi_\beta^o(k'\omega')\rangle_{h_\alpha,h_{\alpha\beta}=0} = \frac{2}{\omega}\,\mathrm{Im}\,G_o(k,\omega)\delta_{\alpha\beta}\delta_{k,-k'}2\pi\delta(\omega+\omega') \tag{4.13}$$

and

$$\langle L^o_{\alpha\beta}(k,\omega)\,L^o_{\gamma\delta}(k'\omega')\rangle_{h_\alpha,h_{\alpha\beta}=0} = \frac{2}{\omega}\,\mathrm{Im}\,K_o(k,\omega)(\delta_{\alpha\gamma}\delta_{\beta\delta}-\delta_{\alpha\delta}\delta_{\beta\gamma})\cdot \delta_{k,-k'}\,2\pi\,\delta(\omega+\omega'). \tag{4.14}$$

For the remaining outgoing lines $G_o h_\alpha$ and $K_o h_{\alpha\beta}$ are to be written after the averaging.

For determining the order parameter and the linear response functions it suffices to take into account the external fields to linear order, i.e. only those diagrams have to be drawn which contain at least one external-field-type vertex or outgoing line.

We define the response functions in the usual way

$$\langle \varphi_\alpha(k,\omega)\rangle = G(k,\omega)h_\alpha \tag{4.15}$$

$$\langle L_{\alpha\beta}(k,\omega)\rangle = K(k,\omega)h_{\alpha\beta}(k,\omega) \quad . \tag{4.16}$$

They obey Dyson's equations

$$G = G_o + G_o\Sigma G + G_o\Sigma' \tag{4.17}$$

$$K = K_o + K_oMK + K_oM' , \tag{4.18}$$

which are represented graphically on Fig. 6. Σ , M and Σ', M' are the sums of all irreducible diagrams in which the external field appears on an outgoing line and on a vertex, respectively. In the course of the perturbational procedure Σ , Σ', M and M' can be calculated. The "traditional" self-energy which is the difference between the inverses of the zeroth order and exact response functions can be obtained as a linear combination of the above self-energy parts. E.g. in the case of the order parameter response function we have

$$G_o^{-1} - G^{-1} = \frac{\Sigma + G_o^{-1}\Sigma'}{1+\Sigma'} \quad . \tag{4.19}$$

The excitation spectrum of the system is given by the poles of the response functions thus, e.g. by the solution of the equation

$$G^{-1}(k,\omega) = 0 \; .$$

If $\Sigma'(k,\omega)$ has no poles we can use the simpler equation,

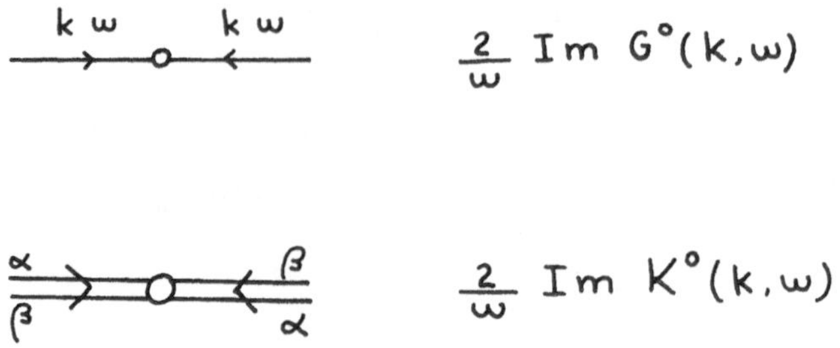

Figure 5

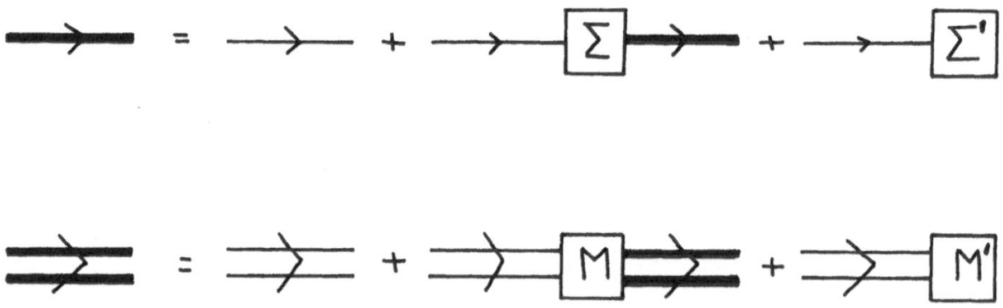

Figure 6

$$G_0^{-1} - \Sigma(k,\omega) = 0 \tag{4.20}$$

while in the opposite case one has to be more careful.

r should be determined from the condition that the right hand side of eq. (4.19) vanish in the limit $k \to 0$, $\omega \to 0$, i.e.

$$\Sigma(0, 0) + r\,\Sigma'(0,0) = 0 \tag{4.21}$$

V. Renormalization Group Analysis

5.1 The Dynamic Renormalization Group

The renormalization group analysis of our model follows the usual lines.

The dynamical renormalization group transformation is defined on the equations of motion and consists of two steps [9,10,11] .

a) In the first step an integration over momentum shells is carried out. The equations of motion are to be solved for the variables φ_q and L_q with $\Lambda/b < q < \Lambda$. Substituting the solutions in the remaining equations of motion and averaging over the random forces ζ and θ we arrive at new equations of motion for the remaining variables.

b) In the second step the following scale transformation is carried out

$$\begin{aligned} k &\longrightarrow bk \\ t &\longrightarrow b^{-z}t \\ \varphi_\alpha(k,t) &\longrightarrow b^{1-\frac{\eta}{2}}\ \varphi_\alpha(bk,\ tb^{-z}) \\ L_{\alpha\beta}(k,t) &\longrightarrow b^{\ell}\ L_{\alpha\beta}(bk,tb^{-z}) \end{aligned} \tag{5.1}$$

It is also necessary to take into account that the volume of the system shrinks by a factor b^{-d} under the renormalization group transformation.

After bringing the remaining equations of motion into the same form as they had before the above two manipulations, we can read off the transformation rules for the parameters. The renormalization group is defined as the transformation connecting the different sets of parameters. Symbolically

$$\mu' = R_b\mu, \tag{5.2}$$

where μ denotes the set of independent parameters. In our case these are

$$\mu(\tilde{\lambda},\ \tilde{g},\ r_0,\ u,\ h_\alpha,\ h_{\alpha\beta})$$

where

$$\tilde{\lambda} = \frac{\lambda}{\Gamma} , \qquad \tilde{g} = \frac{g}{\Gamma} .$$

We note that the forms of the correlation functions of the random forces should also be preserved for the transformed equations of motion, i.e. eq. (4.3) should also be valid if we insert transformed quantities on both sides of them. This is necessary to ensure that the free energy (4.4) describes the equilibrium state.

Let us examine next how the correlation functions and response functions transform under the renormalization group transformation. It is easy to see that the former ones should transform, when (5.1) is carried out, as

$$\begin{aligned} \langle \varphi_\alpha(k,t)\, \varphi_\alpha(0,0) \rangle_\mu &= b^{2-\eta} \langle \varphi_\alpha(bk, b^{-z}t)\, \varphi_\alpha(0,0) \rangle_{\mu'} \\ \langle L_{\alpha\beta}(k,t)\, L_{\alpha\beta}(0,0) \rangle_\mu &= b^{2\ell} \langle L_{\alpha\beta}(bk, b^{-z}t) L_{\alpha\beta}(0,0) \rangle_{\mu'} \\ k &< \Lambda/b . \end{aligned} \tag{5.3}$$

Now we invoke the fluctuation-dissipation theorem (see ref. [11] for the proof for such type of nonlinear stochastic models)which, in our case, is as follows

$$\begin{aligned} \langle \varphi_\alpha(k,t)\, \varphi_\alpha(0,0) \rangle &= \int \frac{d\omega}{2\pi}\, e^{-i\omega t}\, \frac{2\,\mathrm{Im}\, G(k,\omega)}{\omega} \\ \langle L_{\alpha\beta}(k,t) L_{\alpha\beta}(0,0) \rangle &= \int \frac{d\omega}{2\pi}\, e^{-i\omega t}\, \frac{2\,\mathrm{Im}\, K(k,\omega)}{\omega} . \end{aligned} \tag{5.4}$$

Comparing eqs. (5.3) and (5.4) we arrive at the transformation rules for the response functions

$$\begin{aligned} G(k,\omega;\mu) &= b^{2-\eta}\, G(bk, b^z\omega\, ; R_b\mu) \\ K(k,\omega;\mu) &= b^{2\ell}\, K(bk, b^z\omega\, ; R_b\mu) . \end{aligned} \tag{5.5}$$

The basic hypothesis is, that applying the renormalization group transformation many times at the critical point ($T = T_c$, $h_\alpha = h_{\alpha\beta} = 0$) we reach the fixed point μ^* defined by

$$R_b\, \mu^* = \mu^* . \tag{5.6}$$

By choosing $b \sim 1/k$ we get from eqs. (5.5) and (5.6)

$$\begin{aligned} G(k,\omega) &= k^{-2+\eta}\, \mathcal{G}_c\, (\omega/k^z) \\ K(k,\omega) &= k^{-2\ell}\, \mathcal{K}_c\, (\omega/k^z) , \end{aligned} \tag{5.7}$$

which are in accord with the dynamic scaling hypothesis when applied at the critical point [6,7] .

When moving away from the critical point by shifting the temperature, the coherence length, $\xi \sim (|T - T_c|)^{-\nu}$, can be chosen as b. When T is close enough to T_c the deviations of the irrelevant parameters from their fixed point values can be neglected and we arrive at forms for the response functions as required by dynamic scaling [6,7]

$$\begin{aligned} G(k,\omega) &= \xi^{2-\eta}\, \mathcal{G}(k\xi, \omega\xi^z) \\ K(k,\omega) &= \xi^{2\ell}\, \mathcal{K}(k\xi, \omega\xi^z) \end{aligned} \qquad (5.8)$$

Especially it can be seen that the characteristic frequency has the same form as in eq. (1.4). In this way we have demonstrated how the fixed point hypothesis involves dynamic scaling behavior.

5.2 Analysis of the LDM to O(ε)

It can be shown by a simple power counting that the critical dimension for the model is 4 around which an ε expansion can be generated, where ε , as usual, is defined as $\varepsilon = 4 - d$. The calculation has been carried out to first order in ε [18]. In order to determine the fixed point let us first perform the initial step of the renormalization group transformation. The graphs contributing to Γ and λ are given in Figs. 7 and 8, respectively.

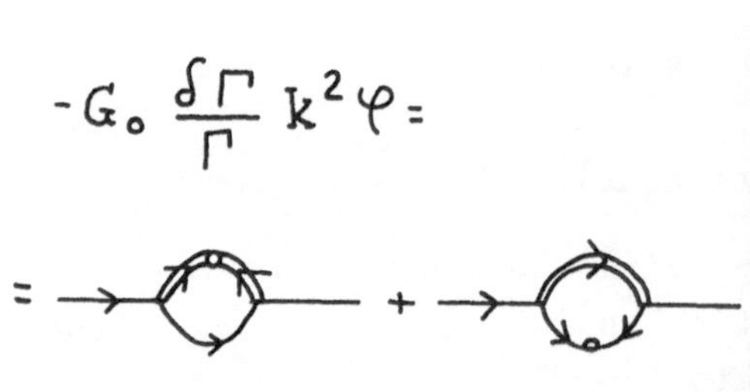

Fig. 7

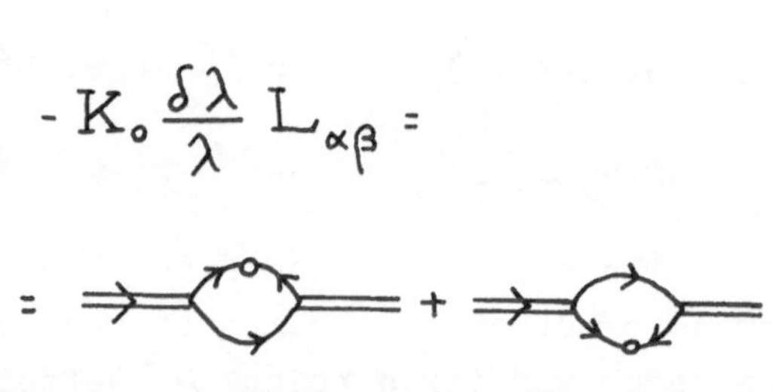

Fig. 8

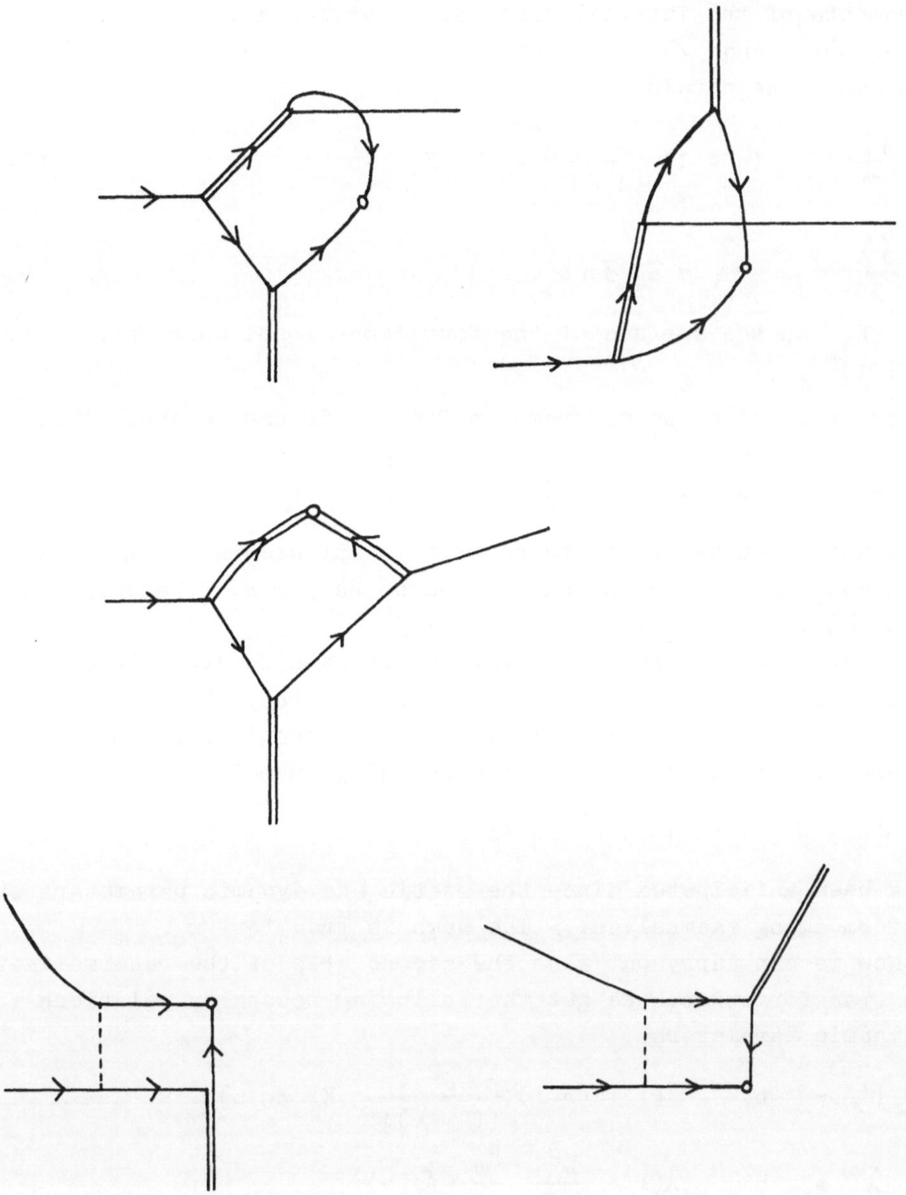

Fig. 9

The momenta of the internal lines are restricted to lie in the shell between Λ/b and Λ. The diagrams are to be evaluated at four dimensions. We obtain

$$\frac{\delta\Gamma}{\Gamma} = (n-1)\,\frac{g^2}{\Gamma(\Gamma+\lambda)}\,K_4\,\ln b \tag{5.9}$$

and

$$\frac{\delta\lambda}{\lambda} = \frac{g^2}{\lambda\Gamma}\,\tfrac{1}{2}\,K_4\,\ln b \tag{5.10}$$

where K_4 is the surface of the four-dimensional unit sphere divided by $(2\pi)^4$.

By evaluating the diagrams in Fig. 9, it can be shown that

$$\frac{\delta g}{\Gamma} = 0 \quad , \tag{5.11}$$

a fact which can be proved to be true to any orders in ε in a similar way as was done for the planar system by Halperin, Hohenberg and Siggia [9].

To get the transformation for u we have to evaluate diagrams with four point φ- vertices. It turns out that the contribution of the sum of the diagrams containing L lines is zero and we get the same result as in the static case [13], namely

$$\delta u = -(u+8)\,u^2\,K_4\,\ln b \tag{5.12}$$

as has been anticipated since the static and dynamic parameters should transform among themselves. The same is true for r_0.

Now we can carry out also the second step of the renormalization group transformation. We get the following recursion relations for the dynamic parameters

$$\Gamma' = b^{z-2}\,\Gamma\left(1 + (n-1)\,\frac{g^2}{(\Gamma+\lambda)\Gamma}\,K_4\,\ln b\right)$$

$$\lambda' = b^{z-2}\,\lambda\left(1 + \frac{g^2}{\lambda\Gamma}\,\frac{K_4}{2}\,\ln b\right)$$

$$g' = b^{z-d/2}\,g \quad .$$

Here we used the known result for static critical phenomena that $\eta = 0$ to $O(\varepsilon)$ and the fact that $\ell = 0$ (see eq. (5.1) for the definition of ℓ) which can be proven easily.

It is worth mentioning that there are two ways of getting these results for the wave function renormalization factors. One can look at the transformation of the free energy or alternatively one should use the relationships between the transformed random forces and the transformed kinetic coefficients, eq. (4.3).

We have two independent parameters, the following combinations

$$f = \frac{\tilde{g}^2}{\tilde{\lambda}} \quad \text{and} \quad \tilde{\lambda} = \frac{\lambda}{\Gamma}$$

where $\tilde{g} = g/\Gamma$.

The recursion relations for them can be cast in the following form

$$\begin{aligned} f' &= f + f \ln b \left\{ \varepsilon - f K_4 \left[\tfrac{1}{2} + (n-1) \frac{\tilde{\lambda}}{1+\tilde{\lambda}} \right] \right\} \\ \tilde{\lambda}' &= \tilde{\lambda} + \tilde{\lambda} f K_4 \ln b \left[\tfrac{1}{2} - (n-1) \frac{\tilde{\lambda}}{1+\tilde{\lambda}} \right] \end{aligned} \qquad (5.13)$$

which lead to the fixed points as follows

1) trivial fixed point

$$f^* = 0, \quad \tilde{\lambda}^* \text{ arbitrary}, \quad (\tilde{g}^* = 0)$$
$$z = 2 .$$

2) $$f^* = \frac{2\epsilon}{K_4}, \quad \tilde{\lambda}^* = 0, \quad (\tilde{g}^* = 0)$$
$$z = 2.$$

3) $$f^* = \frac{\varepsilon}{K_4}, \quad \tilde{\lambda}^* = \frac{1}{2n-3}, \quad (\tilde{g}^* \neq 0)$$
$$z = d/2 .$$

By linearizing the recursion relations we find that f and $\tilde{\lambda}$ are scaling fields. Writing the eigenvalues as b^{y_f} and $b^{y_{\tilde{\lambda}}}$, respectively, we get

1) $y_f = \epsilon$, $y_{\tilde{\lambda}} = 0$

2) $y_f = -\epsilon$, $y_{\tilde{\lambda}} = \varepsilon$

3) $y_f = -\varepsilon$, $y_{\tilde{\lambda}} = -\frac{2n-3}{4(n-1)} \varepsilon$.

One can see that for $d > 4$ the trivial fixed point is the stable one while for $d < 4$ provided $n > \frac{3}{2}$ the stable fixed point is the one which leads to the dynamic critical exponent

$$z = d/2 .$$

This result, similarly to the one $\delta g = 0$, can be proven to any order in ε .

We note that we have two slow transients around this stable fixed point. The trajectories are sketched in Fig. 10.

The physically relevant values of n are of course the positive integer ones. For $n = 2, 3, \ldots$ we found the dynamic critical

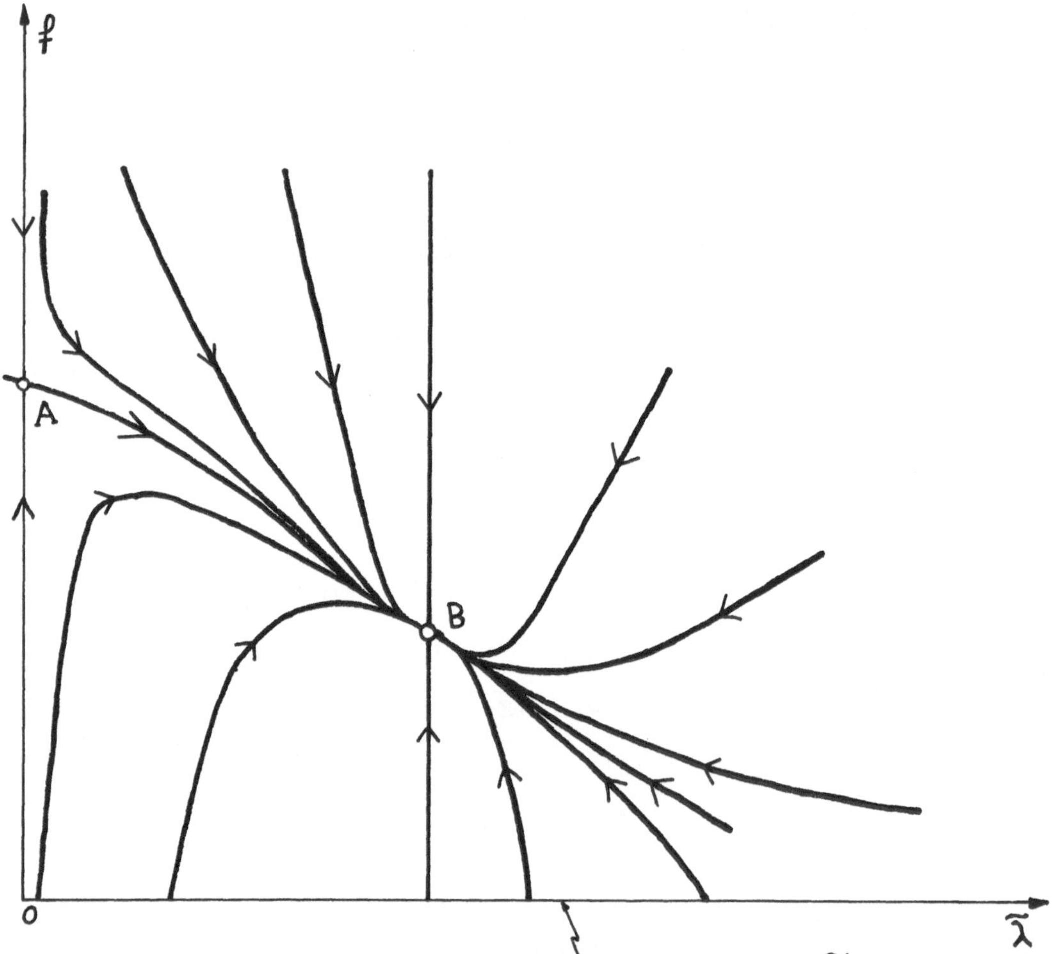

fixed line : $f^* = 0$; $\tilde{\lambda}^*$ = arbitrary
(stable for $d > 4$) $z = 2$.

A : unstable fixed point :
$f^* = \frac{2\varepsilon}{K_4}$, $\tilde{\lambda}^* = 0$, $z = 2$.

B : fixed point :
$f^* = \varepsilon/K_4$, $\tilde{\lambda}^* = \frac{1}{2n-3}$ (stable for $d < 4$)

$z = 2 - \varepsilon/2 = d/2$.

Fig. 10

exponent $z = d/2$ for $d < 4$ while for $n = 1$ our model should reduce to model A of Halperin, Hohenberg and Ma, [10], which has $z = 2$ to $O(\varepsilon)$. It is interesting to follow the behavior of the system when n varies from 2 to 1, which is, of course, an unphysical region. At $n = \frac{3}{2}$ the stable fixed point goes to infinity; for $1 < n < \frac{3}{2}$ λ, the diffusion constant for the L field, becomes negative.

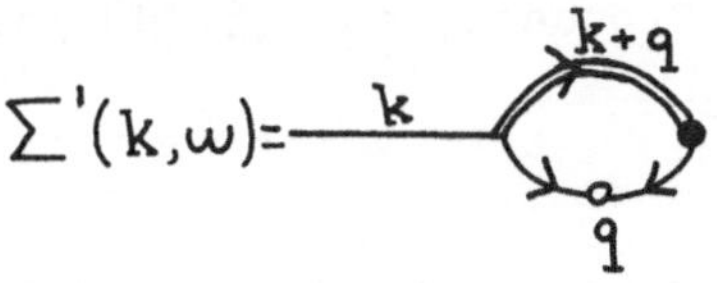

Fig. 11

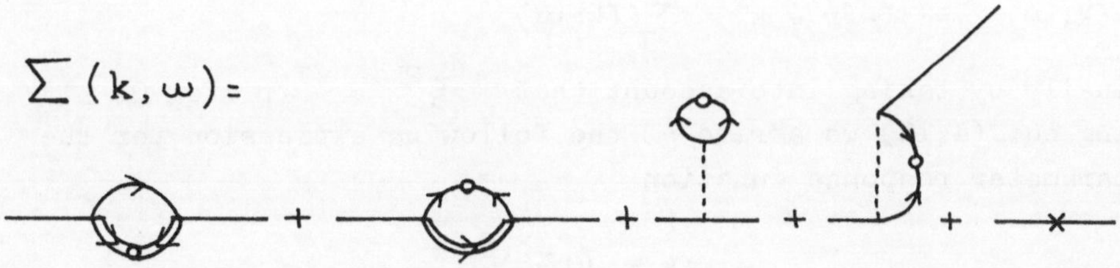

Fig. 12

As an example of calculating a response function we determine $G(k,\omega)$, the order parameter response function, to $O(\varepsilon)$ above T_c. For u, $\tilde{\lambda}$ and $\tilde{g}$ their fixed point values should be substituted to avoid difficulties connected with slow transients. The diagrams giving the leading contributions to Σ' and Σ are shown on Figs. 11 and 12, respectively. Σ' is given as

$$\Sigma'(k,\omega) = (n-1)\frac{g^2}{\Gamma}\int\frac{d^4q}{(2\pi)^4}\,\frac{d\nu}{2\pi}\,\frac{1}{\lambda(k+q)^2}\,\cdot$$
$$\cdot\, K_0(k+q,\,\omega+\nu)\,\frac{2}{\nu}\,\mathrm{Im}\,G_0(q,\nu)\quad. \tag{5.14}$$

For zero external momentum and frequency eq. (5.14) yields

$$\Sigma'(0,0) = \frac{\varepsilon}{4}\,\ln\frac{\Lambda^2}{r}$$

with the help of which we can write to the given order in ϵ

$$1+\Sigma'(k,\omega) = \left(\frac{\Lambda^2}{r}\right)^{\epsilon/4}\left(1+\Sigma'(k,\omega)-\Sigma(0,0)\right) =$$
$$= \left(\frac{\Lambda^2}{r}\right)^{\epsilon/4}\sigma\left(\frac{k^2}{r}\,,\,\frac{\omega}{r}\right) \tag{5.15}$$

As for the self energy, Σ, it can be easily seen, that the sum of the last three diagrams of Fig. 12 gives zero, which just means the condition determining r (see eq. 4.21). Then we obtain, by a simple calculation

$$\Sigma(k,\omega) = -(r+k^2)\,\Sigma'(k,\omega)\quad.$$

Finally by taking into account that $\xi^2 = 1/r$ to $O(\varepsilon^2)$ and using eq. (4.19) we arrive at the following expression for the order parameter response function

$$G(k,\omega) = \frac{\sigma(k\xi,\,\omega\xi^2)}{(k^2+\xi^{-2})\sigma(k\xi,\omega\xi^2) - \frac{i\omega}{\Gamma}(\Lambda\xi)^{-\epsilon/2}}\quad.$$

$G(k,\omega)$ has the form of eq. (5.8), since it is consistent with our approximations that the dynamical critical exponent has its zero order value in the numerator and in the first term of the denominator. The characteristic frequency is determined as the pole of the response function. The results in two limiting cases are

$$\omega = -i\Gamma k^z \quad , \qquad k\xi >> 1 \, ,$$

$$\omega = -i\Gamma \xi^{-z} \quad , \qquad k\xi << 1 \, ,$$

$$z = 2 - \frac{\varepsilon}{2} \; .$$

In the small wavelength limit the solution shows the expected soft mode behavior.

Other response functions can be calculated in a similar way but will not be considered here. (See ref. [18]). We only note that the non-hydrodynamic behavior of the relaxation frequency for $L_{\alpha\beta}$ ($\alpha, \beta \geqslant 2$) below T_c for $k\xi << 1$, mentioned in Section III, can be verified to $O(\varepsilon)$ in this way.

5.3 Consequences of the Energy Conservation

I want to discuss briefly the influence of energy conservation on the critical dynamics of the LDM [18].

In the region of linearized hydrodynamics the energy field does not couple to other fields characterizing our system either above or below T_c and we have a diffusion equation for the energy density (see Section III). A coupling can arise, however, in the non-linear regime by additional terms in the free energy. These additional terms are the same as those appearing in model C of Halperin, Hohenberg and Ma [10]

$$\delta F = \int d^d x \left\{ \gamma \varphi^2 \mathcal{E} + \beta \mathcal{E} + \tfrac{1}{2} c \mathcal{E}^2 \right\} \; .$$

Consider first the equilibrium situation; we note that the $L_{\alpha\beta}$ field can be transformed out of the free energy and consequently the results of ref. [10] apply also here for the parameter $v \equiv \gamma^2/c$. Namely the fixed point value of v is zero when $n \geqslant 4$, while it is of $O(\varepsilon)$ if $n < 4$. Consequently for $n \geqslant 4$ the energy becomes uncoupled from the other fields and our recursion relations, eq. (5.13) remain unchanged near the fixed point.

In cases when $2 \leqslant n < 4$ we arrive at similar conclusions as Halperin, Hohenberg and Siggia [9] for the planar system which corresponds to $n = 2$ in our model. Namely, to order ε the dynamic critical exponent for the stable fixed point remains $z = d/2$. This result applies for example to the isotropic antiferromagnet by choosing $n = 3$.

ACKNOWLEDGEMENTS

I would like to thank Dr. L. Sasvari, co-author in all my work done on the lattice dynamic model, for his valuable remarks on the subject matter of this talk. I am also highly indebted to my wife Dr. N. Menyhárd-Szépfalusy for useful comments and for her great help during the preparation of the manuscript at Sitges.

REFERENCES

(1) L. VAN HOVE, Phys. Rev. 93, 1374 (1954).

(2) L.D. LANDAU and I.M. KHALATNIKOV, Dokl. Akad. Nauk SSSR 96, 469 (1954) (English translation in Collected Papers of L.D. LANDAU, edited by D. ter Haar (Pergamon, London, 1965).

(3) K. KAWASAKI, Ann. Physics 61, 1 (1970).

(4) L.P. KADANOFF and J. SWIFT, Phys. Rev. 166, 89 (1968).

(5) See for a recent review and further references: J.D. GUNTON, K. KAWASAKI in Progress in Liquid Physics, edited by C.A. CROXTON (Wiley, 1976).

(6) R.A. FERRELL, N. MENYHÁRD, H. SCHMIDT, F. SCHWABL and P. SZEPFALUSY, Ann. Phys. (NY), 565 (1968).

(7) B.I. HALPERIN and P.C. HOHENBERG, Phys. Rev. 177, 952 (1969).

(8) A.M. POLYAKOV, Zh. Eksp. Teor. Fiz. 57, 2144 (1969). (English translation Soviet Physics JETP 30, 1164 (1970)).

(9) B.I. HALPERIN, P.C. HOHENBERG and E.D. SIGGIA, Phys. Rev. B13, 299 (1976).

(10) B.I. HALPERIN, P.C. HOHENBERG and S.K. MA, Phys. Rev. B10, 139 (1974).

(11) S. MA and G.F. MAZENKO, Phys. Rev. B11, 4077 (1975).

(12) B.I. HALPERIN, in Collective Properties of Physical Systems, Nobel Symposium 24, ed. by B. LUNDQUIST and S. LUNDQUIST (Academic, N.Y., (1973), p. 54.).

(13) K.G. WILSON and J. KOGUT, Phys. Reports C12, 95 (1974). K.G. WILSON, Rev. Mod. Phys. 47, 773 (1975).

(14) B.I. HALPERIN and P.C. HOHENBERG, Phys. Rev. 188, 898 (1969).

(15) L. SASVÁRI, F. SCHWABL and P. SZÉPFALUSY, Physica 81A, 108 (1975).

(16) B.I. HALPERIN, Phys. Rev. B11, 178 (1975).

(17) L. SASVÁRI and P. SZÉPFALUSY, J. Phys. C 7, 1061 (1974).

(18) L. SASVÁRI and P. SZÉPFALUSY, Lecture at the UPAP International Conference on Statistical Physics, Budapest, August 1975. L. SASVÁRI and P. SZÉPFALUSY, to be published.

(19) S. MA, Rev. Mod. Phys. 45, 589 (1973).

(20) A.I. LARKIN and D.E. KHMELNITSKII, Zh. eksp. teor. Fiz. 64, 2087 (1969). (Soviet Physics - JETP 29, 1123 (1969)).

(21) H. BECK, T. SCHNEIDER, E. STOLL, Phys. Rev. B11, 5198 (1975).

(22) P. SZÉPFALUSY and I. KONDOR, Annals of Physics 82, 1 (1974).

(23) P. SZÉPFALUSY and I. KONDOR, Local Properties at Phase Transitions: Proceedings of the International School of Physics "Enrico Fermi", Varenna, LIX Corso, p. 806 (Editrice Compositori-Bologna-Italy, 1976).

(24) E. BREZIN, D.J. WALLACE and K.G. WILSON, Phys. Rev. B7, 232 (1973).

(25) S. MA, Phys. Rev. A7, 2172 (1973).

(26) H.E. STANLEY, Phys. Rev. 176, 718 (1968).

(27) R. FREEDMAN, G.F. MAZENKO, Phys. Rev. Letters 34, 1575 (1975).

(28) E. RIEDEL and F. WEGNER, Z. Physik 225, 195 (1969); M.E. FISHER and D. JASNOW, in Phase Transitions and Critical Phenomena, Vol. 4, C. DOMB and M.S. GREEN, eds., 1975.

(29) R. ABE, Progr. Theor. Phys. 48, 1414. Ibid., 49, 442 (1973).

(30) E. BREZIN, D.J. WALLACE, Phys. Rev. B7, 1967 (1973).

(31) S. MA and L. SENBETU, Phys. Rev. A10, 2401 (1974).

THE ALGEBRAIC APPROACH TO QUANTUM STATISTICAL MECHANICS. EQUILIBRIUM STATES AND HIERARCHY OF STABILITY

R. HAAG

II. Institut für Theoretische Physik
der Universität Hamburg, Deutschland

The Algebraic Approach to Quantum Statistical Mechanics. Equilibrium States and Hierarchy of Stability

R. HAAG

II. Institut für Theoretische Physik
der Universität Hamburg, Deutschland

I. Introduction

Before going into any details of the subject matter I have to face two questions

i) What does the so-called "algebraic approach" help in the understanding of critical phenomena ?

ii) Is this series of talks understandable to a physicist who is not specially trained in a sophisticated branch of functional analysis, namely the theory of C^*-algebras and von Neumann algebras ?

Marc Kac used to refer (quite benevolently but a little sarcastically) to the "C^*-boys", which expresses at the same time the feeling that this approach is carried by a closed shop of specialists and that they work somewhere on the fringes, producing rarely results of interest to the main stream of the development in statistical mechanics.

The reason why I dare to give you this series of lectures here is because I believe that the essential ideas, concepts and results of the algebraic approach are easily understandable to every theoretical physicist and that they should be understood because they often provide a simplification of the language, a clarification of the issue and also some useful tools. I shall try therefore, to make these talks understandable. On the other hand I must admit at the outset that very little has been done so far with these methods in the area of critical phenomena. So the relation of my talks to the main topic of the conference lies mainly in hopes for the future.

The organization of the lectures is as follows : In Section II, the basic mathematical notions and their correspondence to some physical concepts are described. In Section III, I discuss equilibrium states with special emphasis on a direct characterization of these states by three simple properties, among them a stability condition. In Section IV we illustrate the use of this stability condition in the simple examples of the free Bose- and Fermi gas. We compute from this condition directly the equilibrium states without using the Gibbs Ansatz or the KMS-condition or extremality of entropy and find that temperature, chemical potential and - in the Bose case in addition the superfluid density - appear as the only parameters labelling the equilibrium states of these simple systems. In Section V, if time permits, we shall discuss the stability further; of particular interest will be the different degree of stability of states in relation to phase transitions.

One word of warning : the mathematical precision and reliability of these notes is not on a high level. This is partly because of lack of mathematical education on my side but also because I want to keep these talk as free as possible from technical details which may distract from the central concepts and may hamper the intuitive understanding.

II. Basic Concepts

2.1. Quantum Physics

In traditional courses one learns to consider Hilbert space as the mathematical playground of Quantum Physics. Physical states are then mathematically represented by vectors in this space, or, in the case of more general ensembles which correspond to statistical mixtures (imperfect knowledge) by "density matrices", i.e. positive Hermitean operators with trace 1. Observables are described by linear operators acting in this space.

Consider the set of all (continuous) linear operators acting on the Hilbert space $\mathcal{H}$. This set is usually denoted by $\mathcal{B}(\mathcal{H})$. Within it the following algebraic operations are natural

i) linear combination $\alpha A + \beta B$
   $A, B \in \mathcal{B}(\mathcal{H})$; α, β complex numbers

ii) product $A B$

iii) transition to the adjoint (Hermitean conjugate)
   $A \rightarrow A^*$

One may summarize this by saying that $\mathcal{B}(\mathcal{H})$ is a (non commutative) algebra with a $*$-operation, in short a $*$-algebra.

The algebraic approach to Quantum Physics arises from the observation that one may turn the emphasis around in the consideration of what is basic and what is secondary, i.e. one may consider an abstract $*$ - algebra as the basic mathematical object and Hilbert space as a secondary notion.

Historically this point of view goes back indeed to the origins of quantum mechanics. In "matrix mechanics" the algebraic relations (in particular commutation relations) between position, momentum, angular momentum, energy, etc. are the starting point, not their realization by operators on a Hilbert space. Compare also Dirac's notion of "q-numbers". This point of view is natural and has precisely the same advantages which are familiar in group theory : the group as such, the abstract group, is defined by its multiplication law. It may be realized as a transformation group of some space but frequently in physical problems many such realizations have to be considered side by side and the unifying link between all of them is the abstract group.

The first reasonably complete and mathematically precise formulation of Quantum Mechanics which uses a $*$-algebra as the primary mathematical object was given by I.E. Segal [1]. The following steps were necessary. First, for an abstract algebra with infinitely many linearly independent elements one needs a notion of convergence, i.e. one has to introduce a topology. Although I want to stay away from mathematical technicalities I have to say a few words about this. The simplest topology is a norm topology and this is what we shall use. However, if one considers the Heisenberg commutation relations between position and momentum to start with then one cannot directly introduce a norm topology but needs a more subtle procedure. This is ultimately because the spectrum of position or momentum runs from $-\infty$ to $+\infty$ and so one cannot introduce a finite norm for these quantities. Thus, if one wants a simple topology one should consider only bounded observables (which cannot have infinitely large measured values) e.g. instead of the position coordinate the yes-no-questions : "is the position within a certain specified interval ?", etc. This being understood we may assume the algebra to be normed, i.e. to each element A we have a positive number $\|A\|$ satisfying the requirements for a norm :

$$\|\alpha A\| = |\alpha|\,\|A\| \;;\; \|A + B\| \leq \|A\| + \|B\| \;;\; \|AB\| \leq \|A\|\,\|B\| \;;\; \|A^*\| = \|A\| \tag{2.1}$$

A $*$-algebra equipped with a norm (and completed) is called a Banach$*$-algebra. It turns out that in a certain sense the algebraic structure itself determines the natural way to introduce this norm. Roughly speaking this comes from the fact that the spectrum of an element is determined algebraically and the natural norm of the element is just the supremum of the absolute values which lie in the spectrum of the element. This natural norm satisfies in addition to (2.1) the relation

$$\|A^* A\| = \|A\|^2 \tag{2.2}$$

A Banach $*$-algebra equipped with a norm satisfying (2.2) is called a C^*-algebra.

We consider now an abstract C^*-algebra $\mathfrak{A}$ as the primary mathematical object. How does one make the connection to physical interpretation ? The algebra corresponds to a physical system. What is an observable, what is a physical state ?

A physical observable is obviously represented by a selfadjoint element of $\mathcal{A}$ i.e. an element for which $A^* = A$. The possible measured values of A correspond to the spectral values of A and, as mentioned before, the latter are algebraically determined. A "state" will be understood, on the side of physics, to mean a statistical ensemble of copies of the system under consideration. It is mathematically represented by an expectation function ω on the algebra. Specifically, a state ω assigns to each $A \in \mathcal{A}$ a complex number $\omega(A)$ respecting the three properties

$$\begin{aligned} &\text{linearity :} && \omega(\alpha A + \beta B) = \alpha\, \omega(A) + \beta\, \omega(B) \\ &\text{positivity :} && \omega(A^* A) \geq 0 \quad \text{for any } A \in \mathcal{A} \\ &\text{normalization :} && \omega(1) = 1 \quad \text{(where 1 denotes the unit element of } \mathcal{A}\text{)} \end{aligned} \tag{2.3}$$

It follows that $\omega(A^*) = \overline{\omega(A)}$ (the bar denoting the complex conjugate of a number). Thus, if A is a self adjoint element (observable) $\omega(A)$ is real and can be interpreted as the expectation value of the observable A in the state ω.

Any numerical function ω on $\mathcal{A}$ satisfying (2.3) is a state. So a C^*-algebra has very many states. Too many in fact. We shall be interested only in states having some purity properties. This arises from the following consideration. If ω_1 and ω_2 are states then ω defined by

$$\omega(A) = \lambda\, \omega_1(A) + (1-\lambda)\, \omega_2(A)$$

is again a state, provided $0 \leq \lambda \leq 1$. It is immediate to see that ω will indeed satisfy the three requirements (2.3). We write $\omega = \lambda \omega_1 + (1-\lambda)\, \omega_2$ and interpret it as the mixture of ω_1 and ω_2 with respective weights λ, $(1-\lambda)$. The converse question would be : given a state ω can we find two different states ω_1, ω_2 so that

$$\omega = \lambda \omega_1 + (1-\lambda)\, \omega_2$$

for some $0 < \lambda < 1$?

If this is not possible then ω is called a pure state. In "cold physics" (in contrast to Thermodynamics) we usually deal only with pure states. In statistical mechanics we have to consider mixtures, but not arbitrarily impure one. We are for instance not interested in mixtures between states of different temperatures. More generally, we shall be interested in states which are not absolutely pure but "relatively pure", i.e. ensembles in which at least every macroscopic quantity has a sharp value. If we overidealize or oversimplify this a little we may say that we restrict attention to the so-called "primary states"[1]. For the purposes of this series of lectures this will be adequate.

2.2. Comparison between Hilbert Space and Algebraic Approach

Given a C^*-algebra $\mathfrak{A}$ we may consider its (Hilbert space) representations. A representation π assigns to each algebraic element A a (bounded, linear) operator $\pi(A)$ acting in a Hilbert space $\mathcal{H}$ and this assignment has to carry the algebraic structure over i.e.

$$\pi(AB) = \pi(A)\,\pi(B)\ ; \quad \pi(\alpha A + \beta B) = \alpha\,\pi(A) + \beta\,\pi(B) \tag{2.4}$$

$$\pi(A^*) = (\pi(A))^* \tag{2.5}$$

where the asterisk on the right hand side of (2.5) means the Hermitean adjoint operator in the Hilbert space. If, up to equivalence, $\mathfrak{A}$ has only one irreducible representation then there is no advantage in considering the abstract algebra. We might as well use $\pi(\mathfrak{A})$ and thereby return to the customary description. This situation prevails typically in Quantum Mechanics, where one deals with a finite number of coordinates of the system.

In the case of a system with infinitely many degrees of freedom (Quantum Field Theory, Quantum Statistical Mechanics in the thermodynamic limit) the algebra $\mathfrak{A}$ is such that it has many inequivalent irreducible representations and, moreover, the choice of a representation in which we are interested depends on the physical aspect we want to discuss. The essential point here is the correspondence between a representation and a particular family of states. This goes in both directions. If a representation π is given (acting in the Hilbert space $\mathcal{H}$) then every density matrix ρ (pos. definite self adjoint operator in $\mathcal{H}$ with trace 1) defines a state ω_ρ of $\mathfrak{A}$ by

$$\omega_\rho(A) = \mathrm{Tr}(\rho\,\pi(A)) \tag{2.6}$$

The set of all these states, as ρ runs through all the density matrices of $\mathcal{H}$, is the family of states associated with the representation π. We shall denote it by $\mathcal{S}_\pi$. Conversely, if a state ω is given then there is a canonical construction of a representation π_ω and thereby of the whole family of states to which ω belongs. This construction, the Gelfand-Naimark-Segal construction (GNS-construction) will be briefly

[1] For a mathematical definition see appendix.

described in the appendix. To tighten things let us restrict attention to primary states and representations arising from such states (the so-called primary or "factorial" representations). Then we have the simple fact : the different families of primary states are completely disjoint. In fact, if ω_1 and ω_2 are primary and belong to different families then [2]

$$\|\omega_1 - \omega_2\| = 2 \tag{2.7}$$

The parameters which distinguish different families of primary states may be called "superselection" parameters (in the terminology of [2], where the need for "superselection rules" was first pointed out in connection with integer vs. half integer spin values). In Quantum Field Theory the significant superselection parameters are the charges (electric, baryonic ...) ; in statistical mechanics they are the equilibrium parameters (temperature, chemical potential ...).

2.3. Infinitely Extended Systems. Locality. Nets of Algebras and Physical Interpretation

In the treatment of an infinitely extended quantum system the distinction between local and global aspects of states becomes important. The theory is formulated in terms of local quantities. To bring this local structure into evidence we have to focus on a collection of subalgebras of $\mathcal{A}$. We shall consider to each space region $\mathcal{V}$ and each time t a C^*-algebra $\mathcal{A}(\mathcal{V},t)$. It shall be the algebra generated from those observables which can be measured at the time t within the space region $\mathcal{V}$. The correspondence between space-time regions and subalgebras of $\mathcal{A}$

$$\mathcal{V},\, t \longrightarrow \mathcal{A}(\mathcal{V},t) \tag{2.8}$$

we call a <u>net of C^*-algebras</u>. We claim that the formulation of a specific physical theory, the characterization of a specific physical system consists precisely in the assignment (2.8). A physical theory, (even including the interpretation of quantities as far as one has to know it) is synonymous with giving (or implicitly defining) the net of local algebras $\mathcal{A}(\mathcal{V},t)$ [3]. Of course, this assignement of a net

---

2 $\varphi = \omega_1 - \omega_2$ is a linear form over $\mathcal{A}$. Its norm is defined as

$$\|\varphi\| = \frac{\sup |\varphi(A)|}{\|A\|}$$

has to respect some general physical principles. Let us denote the algebra generated by all quasilocal observables at time t by $\mathcal{A}(t)$. Mathematically it shall be the (norm closure of) the union of all $\mathcal{A}(\mathcal{V},t)$ where $\mathcal{V}$ runs through all space-regions of finite size, say all spheres of finite radius :

$$\mathcal{A}(t) = \overline{\bigcup_{\mathcal{V}} \mathcal{A}(\mathcal{V},t)} \tag{2.9}$$

Similarly, the algebra $\mathcal{A}$ of all quasilocal observables $\mathcal{A}$ shall be the norm closure of the union of the $\mathcal{A}(t)$ for all t :

$$\mathcal{A} = \overline{\bigcup_{t} \mathcal{A}(t)} \tag{2.10}$$

In the case of the non relativistic many body problem and statistical mechanics we may consider the creation and destruction operators of a particle at position $\vec{x}$ time t, denoted by $a^*(\vec{x},t)$, $a(\vec{x},t)$, as the building stones from which the net $\mathcal{A}(\mathcal{V},t)$ is constructed. Essentially $\mathcal{A}(\mathcal{V},t)$ consists of bounded functions of objects like

$$a^*(f,t) = \int a^*(\vec{x},t)\, f(\vec{x},t)\, d^3x$$

and their adjoints $a(f,t)$ where the wave function f has its support in the space region $\mathcal{V}$.

Causality is expressed by the Heisenberg equations of motion which relate $a^*(\vec{x},t)$, $a(\vec{x},t)$ at an arbitrary time t to the quantities at another time t_o. In the algebraic language this means that [3]

$$\mathcal{A}(t_1) = \mathcal{A}(t_2) = \mathcal{A} \tag{2.11}$$

Time translation (the dynamical law) is then described by a 1-parameter automorphism group α_t acting on the "quasilocal algebra". Specifically, if A is some observable, then $\alpha_t(A)$ denotes the same quantity measured a time t later.

---

[3] In lattice systems (and in a relativistic theory due to its limited propagation velocity) the dynamical law is such that (2.11) results automatically for any reasonable definition of the local C^*-algebras. In a continuous non relativistic system there is a problem with (2.11) if local algebras are not very judiciously defined (or the definition (2.9) relaxed) because the delocalization of a strictly local quantity after a finite time interval may be so considerable that it makes an approximation by local quantities in the norm topology impossible. This difficulty should, however, not concern us here.

More generally, any symmetry group (invariance group) of the theory will be described by an automorphism group of $\mathcal{A}$. For our purposes, apart from time translations, we shall be only concerned with spatial translations $\alpha_{\vec{x}}$ and, occasionally with gauge transformations γ_φ. These latter, defined by their action on creation operators

$$\gamma_\varphi(a^*(\vec{x},t)) = e^{i\varphi} a^*(\vec{x},t) \tag{2.12}$$

are not really symmetries of the observable algebra since we assume that all observables are gauge invariant. The observables are formed from densities and currents which have an equal number of creation and destruction operators in each term and are thus unaffected by the application of γ_φ. Strictly speaking the a^*, a generate a net $\mathcal{F}(\mathcal{V},t)$ which is larger than the observable net $\mathcal{A}(\mathcal{V},t)$. The latter consists of all the gauge invariant elements of $\mathcal{F}$.

We have called $\mathcal{A}$ the algebra of quasilocal quantities because, for every element $A \in \mathcal{A}$ and arbitrarily prescribed precision ε we can find an element A_1 belonging to the algebra of some finitely extended region $\mathcal{V}_1$ (at time t) which has norm distance less than ε from A. If we increase the demands in precision i.e. make ε smaller the size of the region $\mathcal{V}_1$ has to be increased but as long as $\varepsilon > 0$ a finite region suffices. How do global quantities (e.g. averages of local quantities over the whole space) or asymptotic observables ("observables at infinity") enter ? An example of the latter would be to start from an arbitrary $B \in \mathcal{A}$, pick some sequence of points $\vec{x}_n$ moving to infinity as $n \to \infty$ and consider the sequence

$$B_n = \alpha_{\vec{x}_n}(B) \tag{2.13}$$

Obviously this sequence does not converge in the norm topology. It will not have a limit in $\mathcal{A}$. But $\mathcal{A}$ is naturally embedded in a larger algebra $\mathcal{A}^{**}$, the "double dual" of $\mathcal{A}$ or the "enveloping von Neumann algebra". This object is closed in the so-called "weak topology" ; it contains all the weak limit points of $\mathcal{A}$ and among these we find the global and the asymptotic observables. Let me not try to give a mathematically precise definition of the weak topology here but just say that a sequence $A_n \in \mathcal{A}$ is called weakly convergent if for every state ω over $\mathcal{A}$ the numerical sequence $\omega(A_n)$ converges. It is a fact that every infinite sequence A_n with uniformly bounded norm ($\|A_n\| < c$ for all n) has weak limit points. It usually has more than one i.e. the sequences need not be weakly convergent as such but may for instance oscillate between different limit points in $\mathcal{A}^{**}$ each of which can be obtained by picking from $\{A_n\}$ a suitable convergent subsequence.

One more mathematical comment and some standard notation : Every representation π of the algebra $\mathfrak{a}$ extends naturally to a representation of $\mathfrak{a}^{**}$. The set $\pi(\mathfrak{a}^{**})$ is a set of bounded operators acting on the representation space $\mathcal{H}$. It is thus a subset of $\mathcal{B}(\mathcal{H})$ and it coincides with the closure of the set $\pi(\mathfrak{a})$ in the ordinary weak topology of Hilbert space operators. A weakly closed operator $*$-algebra on a Hilbert space is called a von Neumann algebra. If $\mathfrak{M}$ is a von Neumann algebra on $\mathcal{H}$ then one denotes by $\mathfrak{M}'$ its "commutant" i.e. the set of all operators from $\mathcal{B}(\mathcal{H})$ which commute with every operator in $\mathfrak{M}$.

$\mathfrak{M}'$ is again a von Neumann algebra (i.e. a weakly closed $*$-algebra). By a theorem of von Neumann the weak closure of a $*$-algebra of operators, say $\mathcal{B}$, is precisely the "bicommutant" $\mathcal{B}''$ of $\mathcal{B}$ (the commutant of $\mathcal{B}'$). Thus in particular $\mathfrak{M}'' = \mathfrak{M}$ (since $\mathfrak{M}$ was already a weakly closed $*$-algebra). In our context we have then

$$\pi(\mathfrak{a}^{**}) = \{\pi(\mathfrak{a})\}'' \tag{2.14}$$

The intersection of $\mathfrak{M}$ with its commutant is called the center $\mathfrak{z}$ of $\mathfrak{M}$:

$$\mathfrak{z} = \mathfrak{M} \cap \mathfrak{M}' \tag{2.15}$$

It is obviously an Abelian von Neumann algebra.

A primary (factorial) representation π of $\mathfrak{a}$ has the property that the center of $\pi(\mathfrak{a})''$ is trivial, i.e. it consists only of the multiples of the identity operator :

$$\pi(\mathfrak{a})'' \cap \pi(\mathfrak{a})' = \{\lambda \cdot 1\} \quad ; \lambda \text{ a complex number} \tag{2.16}$$

if π is primary. Compare with Schur's lemma : if π is irreducible then already $\pi(\mathfrak{a})'$ is trivial.

2.4. Asymptotic Abelianness. Clustering

Observables belonging to disjoint space regions $\mathcal{V}_1$, $\mathcal{V}_2$ at one time t are compatible i.e.

$$[A_1 , A_2] = 0$$

if $A_1 \in \mathfrak{a}(\mathcal{V}_1,t)$, $A_2 \in \mathfrak{a}(\mathcal{V}_2,t)$ and $\mathcal{V}_1$ is disjoint from $\mathcal{V}_2$. By (2.9), (2.11) every element in $\mathfrak{a}$ can be approximated in the norm topology by elements from a finite region at an arbitrarily chosen time. On the other hand, by a sufficiently large spatial translation, we can shift any finite region so far that it becomes disjoint from an arbitrarily given (but fixed) other finite region. Therefore we have

$$\lim_{|\vec{x}|\to\infty} \| [A, \alpha_{\vec{x}}(B)] \| = 0 \tag{2.17}$$

for any pair $A,B \in \mathcal{A}$. The space translations shift any element of $\mathcal{A}$ so that it will commute ultimately with any other element in $\mathcal{A}$. One says that the algebra $\mathcal{A}$ is asymptotically Abelian (commutative) with respect to the space translation group.

Although local observables at different times do not in general commute a reasonable dynamical law will be such that a local observation at time t_0 does not greatly perturb the measurement of a local observable at a very much later time (due to the spreading of the perturbation due to the first measurement over a larger and larger volume as one waits longer and longer and thereby weakening the effect in any local neighborhood). Therefore we also expect that $\mathcal{A}$ is asymptotically Abelian with respect to time translations i.e.

$$\lim_{|t|\to\infty} \| [A, \alpha_t(B)] \| = 0 \quad \text{for any } A,B \in \mathcal{A} \tag{2.18}$$

The properties (2.17), (2.18) entail the important clustering properties of primary states (vanishing of correlation functions for large spatial or temporal distances). The connection is the following.

Consider e.g. the sequence (2.13). Since $\alpha_{\vec{x}}$ is an automorphism $\|\alpha_{\vec{x}}(B)\| = \|B\|$. So the sequence is uniformly bounded in norm and therefore has weak limit points in $\mathcal{A}^{**}$. Let $\hat{B}$ be one of them. Due to asymptotic Abelianness $\hat{B}$ will commute with every element of $\mathcal{A}$ (and, in fact, with every element of $\mathcal{A}^{**}$ since the commutativity is preserved if one takes weak closures). Thus $\hat{B}$ is an element of the center of $\mathcal{A}^{**}$. In any primary representation π of $\mathcal{A}$ therefore $\pi(\hat{B})$ will be a multiple of the unit operator. If we remember the connection between families of states and representations we may express this also in this way : if ω is a primary state then $\omega(\hat{B}A) = \lambda\,\omega(A)$ with $\lambda = \omega(\hat{B})$ or

$$\lim_{|\vec{x}|\to\infty} \left| \omega(\alpha_{\vec{x}}(B) \cdot A) - \omega(\alpha_{\vec{x}}(B))\, \omega(A) \right| = 0 \tag{2.19}$$

The quantity under the $|\;|$-sign is the spatial correlation function between A and B in the state ω. Obviously, if we have asymptotic Abelianness in time then a similar statement holds for the temporal correlation function

$$\lim_{|t|\to\infty} \left| \omega(\alpha_t(B)A) - \omega(\alpha_t(B))\, \omega(A) \right| = 0 \tag{2.20}$$

If ω is homogeneous (invariant under space translation) $\omega(\alpha_{\vec{x}}(B)) = \omega(B)$ so the left hand side of (2.19) simplifies to $\omega(\alpha_{\vec{x}}(B)A) - \omega(B)\omega(A)$. Similarly, if ω is stationary (invariant under time translation) α_t may be omitted in the second term of (2.20).

III. Equilibrium States. Stability Condition. KMS-condition

Having seen how an infinitely extended dynamical system may be mathematically described by a net of C^*-algebras $\mathcal{A}(\mathcal{V},t)$ we are faced with the question : there are myriads of families of states over $\mathcal{A}$; how do we distinguish within all these the thermodynamic equilibrium states, how do we define them and how can we compute them ?

The historic path would be to go back to a finite system and then use a Quantum Mechanical Gibbs ensemble to compute expectation values of local quantities (densities, products of densities, etc.). For a 1-component system the Gibbs ensemble is characterized by 2 parameters β, μ of which the first, as the expert knows, corresponds to the inverse temperature, the second to the chemical potential. If the size of this finite system is $\mathcal{V}$ and we restrict attention to expectation values of quantities in a much smaller subvolume $\mathcal{V}_0$ we may say that in this manner we compute a (stationary) state over $\mathcal{A}(\mathcal{V}_0)$ which we may denote by $\omega_{\beta,\mu,\mathcal{V}}$ because it depends on the three parameters $\beta, \mu, \mathcal{V}$. Then, keeping β, μ and $\mathcal{V}_0$ fixed we take the thermodynamic limit $\mathcal{V} \to \infty$ and we expect that if β, μ do not accidentally happen to lie on a phase coexistence line then the limit of the sequence of states $\omega_{\beta,\mu,\mathcal{V}}$ over $\mathcal{A}(\mathcal{V}_0)$ will exist. It will then define the restriction of the equilibrium state labeled by β, μ to the subalgebra $\mathcal{A}(\mathcal{V}_0)$ and since this procedure can be done for arbitrary finite $\mathcal{V}_0$ it will define the equilibrium state $\omega_{\beta,\mu}$ over the whole quasilocal algebra $\mathcal{A}$.

An honest disciple of the algebraic approach can, however, not be satisfied with this answer but wants a direct characterization of the set of equilibrium states which neither presupposes the notions of temperature and chemical potential, nor relies on the historical arguments which led to the consideration of quantum mechanical Gibbs ensembles. In order to show my honesty I shall therefore reverse the historical order of the development of equilibrium statistical mechanics.and report first on some work I did together with Daniel Kastler and Eva Trych-Pohlmeyer about two years ago [4] , then come to the Kubo-Martin-Schwinger condition and to Gibbs ensembles.

Our starting question was : what properties must a state over $\mathfrak{A}$ possess in order to qualify as an equilibrium state ? Two requirements are rather evident :

1) An equilibrium state is stationary

$$\omega(\alpha_t(A) = \omega(A)$$

2) As indicated in Section II we shall consider only primary states [4].

The third requirement is a certain stability property. It may happen (and at least for systems with a high degree of symmetry it does happen) that there are primary states which are very sensitive to slight perturbations of the dynamics. Since we will never know the dynamical laws with absolute precision and experimentally cannot guarantee the absolute absence of any contamination (a grain of dust in a bulk medium for instance) we can make meaningful theoretical statements which are experimentally reproducible only about such states which do not change drastically if the dynamical law is locally and infinitesimally changed. Of course, at a phase transition there will be less stability of the state than in a one-phase region of the parameters. If temperature and pressure lie on the liquid-gas coexistence curve then, in a large finite system a small change of the boundary conditions at the surface will suffice to change the state essentially from the liquid to the gas. But this perturbation,in the case of an infinite system, is an infinitely extended one (though of lower dimension). There is still even at a phase transition point stability under bounded, local perturbations. The only known example in which there may be some doubt about this is the Kondo effect. But if there should be a problem in that case with local stability then it will a forteriori overthrow the use of the Kubo-Martin-Schwinger condition and Gibbs ensembles. So it will be a deep problem. But I am not capable of judging whether this problem really exists.

Let me now define presicely what we mean by a local perturbation of the dynamics. If on the same kinematical algebra $\mathfrak{A}$ (union of local algebras at a fixed time t_0) we put two different dynamical laws, the automorphism group α_t and α'_t , then we may characterize the difference between them by a "derivation"

---

[4] If one wants to be completely honest one should not require that ω is primary but that it is extremal among the stationary states. For the concept of extremal invariant states see [5] . This is a priori a somewhat weaker requirement but it is not clear at this moment under what circumstances this weakening of the requirement is relevant.

$$d_h = -i\left(\frac{\partial}{\partial t} \alpha_t^{-1} \alpha_t'\right)_{t=0} \qquad (3.1)$$

We shall be interested in those derivations d_h whose action on the algebra is of the form

$$d_h(A) = [h,A] \qquad (3.2)$$

with $h \in \mathfrak{A}$ a self adjoint element of the algebra itself i.e. a bounded, quasilocal quantity. This h may be considered as the difference between the Hamiltonians which generate the group α_t respectively α_t'. To connect with the well known situation in the Hilbert space version : There we would write

$$\alpha_t(A) = e^{iHt} A e^{-iHt} \quad ; \quad \alpha_t' = e^{iH't} A e^{-iH't}$$

giving

$$-i \frac{\partial}{\partial t} (\alpha_t^{-1} \alpha_t'(A))\Big|_{t=0} = [(H'-H), A]$$

which agrees with (3.2) if we put $h = H' - H$. In the case of an infinite system neither H' nor H will be elements of the algebra $\mathfrak{A}$ since they are both global quantities. But we are interested in perturbations which are local so that the difference $H' - H$ is an element of $\mathfrak{A}$. Consider now a one parameter family of such perturbations, with h replaced by λh where λ is a coupling constant which we want to let tend towards zero. If ω is a primary state which is stationary with respect to the unperturbed dynamics α_t we call it "stable under the perturbations λh if for small enough values of λ there exist primary states $\omega^{\lambda h}$ stationary with respect to the perturbed dynamics $\alpha_t^{\lambda h}$ (for clarity we now write $\alpha_t^{\lambda h}$ instead of α_t') such that

$$\lim_{\lambda \to 0} \|\omega^{\lambda h} - \omega\| = 0 \qquad (3.3)$$

This means in particular that the perturbed state shall lie for sufficiently small λ, say for $\lambda < \lambda_0$, in the same primary family as ω.

With these definitions our third requirement for a state ω to be called an equilibrium state is

3) ω shall be stable under all quasilocal perturbations.

In other words, (3.3) shall hold for all self adjoint $h \in \mathfrak{A}$.
Comment : In the original spirit of the ergodic hypothesis one might

hope that if the dynamics α_t has no symmetries then the requirements 1) and 2) alone would suffice to characterize equilibrium states. Then 3) would just serve the purpose of making sure that all symmetries are destroyed if there happened to be any in the original dynamical law. If this should be true (which I do not know) then we might call the dynamics α_t "ergodic" if 3) is already automatically fulfilled for the primary states which are stationary under α_t.

Let me now proceed to combine the requirements 1), 2), 3) to get a useful equation which ω has to satisfy. The relation between $\alpha^{\lambda h}$, α and h, given differentially at $t = 0$ in (3.1), (3.2) can also be written as

$$\frac{d}{dt} \left(\alpha_t^{\lambda h}\right)^{-1} \alpha_t(A) = -i\lambda \left(\alpha_t^{\lambda h}\right)^{-1} \left([h, \alpha_t(A)] \right) \tag{3.4}$$

Since $\omega^{\lambda h}$ shall be stationary under $\alpha^{\lambda h}$ i.e. $\omega^{\lambda h}(\alpha_t^{\lambda h}(B)) = \omega^{\lambda h}(B)$ and since $\omega^{\lambda h}$ is a linear function on the algebra, so that $\omega^{\lambda h}(\frac{d}{dt} C(t)) = \frac{d}{dt}\omega^{\lambda h}(C(t))$ we get rid of the automorphisms $(\alpha_t^{\lambda h})^{-1}$ by taking the expectation value of (3.4) in the state $\omega^{\lambda h}$:

$$\frac{d}{dt} \omega^{\lambda h}(\alpha_t(A)) = -i\lambda\, \omega^{\lambda h}\left([h, \alpha_t(A)]\right) \tag{3.5}$$

Or, integrating from t_1 to t_2

$$\omega^{\lambda h}(\alpha_{t_2}(A)) - \omega^{\lambda h}(\alpha_{t_1}(A)) = -i\lambda \int_{t_1}^{t_2} \omega^{\lambda h}\left[h, \alpha_t(A)\right] dt \tag{3.6}$$

Now, since $\omega^{\lambda h}$ shall be primary, $\omega^{\lambda h}(\alpha_t(A))$ becomes asymptotically equal for $t \to \pm\infty$ to the expectation value of $\alpha_t(A)$ in any other state which belongs to the same primary family [5]. But this family contains ω when $\lambda < \lambda_0$ and $\omega(\alpha_t(A)) = \omega(A)$. So if we let $t_1 \to -\infty$, $t_2 \to +\infty$ in (3.6) both terms on the left converge to $\omega(A)$; their difference becomes zero. Thus

$$\lambda \int_{-\infty}^{+\infty} \omega^{\lambda h}\left([h, \alpha_t(A)] \right) dt = 0 \qquad \text{for any } \lambda < \lambda_0 \tag{3.7}$$

as long as $\lambda > 0$ we may divide by λ and, because the integrand shall, by (3.3), approach $\omega([h, \alpha_t(A)])$ for any t as $\lambda \to 0$, we would like to conclude

$$\int_{-\infty}^{+\infty} \omega\left([h, \alpha_t(A)] \right) dt = 0 \quad . \tag{3.8}$$

5 In the corresponding representation the weak limits of $\pi(\alpha_t(A))$ are multiples of the unit operator.

This condition, demanded for any pair h,A of elements from $\mathfrak{A}$ will be called the stability condition for the state ω and taken as an explicit expression of requirement 3).

The replacement of 3) by (3.8) needs some discussion. The step from (3.7) is rigorously justified only if the function $\omega^{\lambda h}([h,\alpha_t(A)])$, considered as a numerical function of time, is absolutely integrable. This seems to be the case under our assumption since in (3.6) the left hand side tends to zero no matter in which way $t_2 \to +\infty$, $t_1 \to -\infty$. Conversely, one may think that (3.7) contains more information than (3.8) and this is probably true if no further structure is known. However, it turns out that if the temporal correlation functions in the state ω decrease rapidly enough to be absolutely integrable then (3.8) suffices to show that also 3) is satisfied.

Let me turn now to the Kubo-Martin-Schwinger (KMS)-condition [6], [7] . In the case of a finite system this property is easily derived for a state which corresponds to a (quantum mechanical) Gibbs ensemble, and under suitable circumstances, it survives in the thermodynamic limit. So it may be considered as an adaptation of the Gibbs prescription to an infinitely extended system. In our notation it can be stated as follows [8] : Consider, for any pair of quasilocal quantities $A,B \in \mathfrak{A}$ the two temporal correlation functions (in an equilibrium state ω_β of inverse temperature β)

$$\begin{aligned} F_{A,B}(t) &= \omega_\beta(B\,\alpha_t(A)) - \omega_\beta(A)\;\omega_\beta(B) \\ G_{A,B}(t) &= \omega_\beta(\alpha_t(A)\,B) - \omega_\beta(A)\;\omega_\beta(B) \end{aligned} \tag{3.9}$$

Then their Fourier transforms $\tilde{F}_{A,B}(E)$, $\tilde{G}_{A,B}(E)$ are related by

$$\tilde{F}_{A,B}(E) = e^{\beta E}\,\tilde{G}_{A,B}(E) \tag{3.10}$$

Relations (3.9), (3.10) are the KMS-condition for the state ω_β . We see that the left hand side of the stability condition (3.8) is precisely

$$\int_{-\infty}^{\infty} (F_{h,A}(t) - G_{h,A}(t))dt = \tilde{F}_{h,A}(0) - \tilde{G}_{h,A}(0)$$

Thus (3.8) is just the KMS-condition at the special point $E = 0$, where the value of the parameter β drops out. It is rather amazing, however, that one can come from the stability condition (3.8) to the full KMS-condition provided the temporal correlation functions are absolutely integrable. In other words, a state satisfying the requirements 1), 2), 3) determines a parameter β such that (3.10) holds for all pairs $A,B \in \mathfrak{a}$. The significance of β is then that of a modul of stability : The first order change in state due to a perturbation is proportional to β. Defining

$$\omega_h^{(1)} = \frac{d\ \omega^{\lambda h}}{d\lambda}\Big|_{\lambda=0} \tag{3.11}$$

we have

$$|\omega_h^{(1)}(A)| < C_h \ \|A\| \ \beta \tag{3.12}$$

where C_h is a constant depending on h . At infinite temperature the state is insensitive to perturbations, at zero temperature it is most sensitive [6].

I shall only give a very brief sketch of the argument leading from (3.8) to (3.10) and refer for all details and rigour to [4]. Replace in (3.8) h by $h' = h_1 \alpha_s(h_2)$, A by $A' = A_1 \alpha_s(A_2)$ which is allowed if s is a finite time since h', A' are just two other elements of $\mathfrak{a}$. Then let $s \to \infty$, use clustering in time and interchange the two operators $\int_{-\infty}^{\infty} dt$ and $\lim_{s\to\infty}$ (performing the limit $s \to \infty$ first under the integral sign). This is permitted if the correlation functions are absolutely integrable. Then one obtains

$$\int_{-\infty}^{\infty} (F_{h_1,A_1}(t) \cdot F_{h_2,A_2}(t) - G_{h_1,A_1}(t)\ G_{h_2,A_2}(t))dt = 0 \tag{3.13}$$

or, Fourier transformed,

$$\int \tilde{F}_1(E)\ \tilde{F}_2(-E)\ dE = \int \tilde{G}_1(E)\ \tilde{G}_2(-E)\ dE \tag{3.14}$$

where we have for short written F_1 instead of F_{h_1,A_1} , etc.. The functions $\tilde{F}_i$, $\tilde{G}_i$ are continuous if the correlation functions are absolutely integrable. Now one can, by suitable choice of h_2, A_2 , make F_2 and simultaneously $\tilde{G}_2$ concentrated in an arbitrarily small interval around any chosen point, say $-E_o$. In other words, we can make a sequence of choices $h_2^{(n)}, A_2^{(n)}$ so that the functions $F_2^{(n)}\ \tilde{G}_2^{(n)}$ approach respectively $a\,\delta(E + E_o)$, $b\,\delta(E + E_o)$ and (3.12) can be simplified to

$$a\tilde{F}_1(E_o) = b\ \tilde{G}_1(E_o)$$

---

[6] One may invoke the picture of hell and heaven to understand this intuitively.

where a and b depend on the choice of the sequence $h_2^{(n)}, A_2^{(n)}$. Since the choice of h_1, A_1 is arbitrary and unaffected by the choice of the sequence $h_2^{(n)}, A_2^{(n)}$ the ratio $\frac{b}{a}$ is independent of h_1, A_1 (and therefore also of the sequence $h_2^{(n)}, A_2^{(n)}$) and can only depend on E_0.

Thus there is a function $\phi(E)$ associated with the state so that

$$F_{h_1 A_1}(E) = \phi(E)\, G_{h_1 A_1}(E) \tag{3.15}$$

where ϕ is independent of A_1, h_1. If one iterates the trick leading from (3.8) to (3.13) one finds that

$$\phi(E_1)\,\phi(E_2) = \phi(E_1 + E_2)$$

Also, ϕ is a continuous function of E, since the $\tilde{F}_i, \tilde{G}_i$ are continuous and it is real valued. Hence it must be of the form

$$\phi(E) = e^{\beta E}$$

The shortness of time forces me to limit the discussion of the KMS- and stability conditions to a few remarks. First, in what sense can the variable E in (3.10) be interpreted as an energy ? If we construct the representation π_ω which is obtained by the GNS-construction from the state ω then one can (uniquely) find a one parameter group of unitary operators $U(t)$ in the representation space which implement α_t and leave the vector Ω (corresponding to the state ω in the representation space) invariant :

$$\pi(\alpha_t(A)) = U(t)\,\pi(A)\,U^{-1}(t) \tag{3.16}$$

$$U(t)\,\Omega = \Omega \tag{3.17}$$

The generator of this group may be called the Hamiltonian H :

$$U(t) = e^{iHt} \tag{3.18}$$

and the E-values are indeed the spectral values of H. One finds that the spectrum of H is continuous, running from $-\infty$ to $+\infty$ with a single discrete point at $E=0$ in the middle of the continuum (corresponding to the normalized eigenvector Ω, the only (normalizable) eigenvector of H). Usually, if a discrete eigenvector sits in the middle of a continuum it will be dissolved by a small perturbation (a discrete level becomes a resonance with finite width in the continuum). That this shall not happen here is precisely guaranteed by the stability condition.

We may ask what is the relation between the Hamiltonian H and the energy density operator [7] u(x,t) ? One finds [8]

$$H = \lim_{\mathcal{V}\to\infty} \int_{\mathcal{V}} (u(\vec{x},t) - u'(\vec{x},t))\, d^3x \tag{3.19}$$

where u' is a "miror energy density" which is associated with the commutant of $\pi(\mathfrak{A})$. In computing equations of motion

$$\frac{\partial}{\partial t}\pi(A) = i\,[H, \pi(A)]$$

the u'- term in (3.19) does not contribute since it commutes with all $\pi(\mathfrak{A})$ but without the subtraction of u' the integral would not give any operator in $\mathcal{H}$. We do not only have to subtract an infinite constant to adjust the "energy" of Ω to the value zero but we have to subtract an infinite operator in the commutant; the fluctuations of the u-term alone increase proportional to $\sqrt{\mathcal{V}}$ and for $\mathcal{V}\to\infty$ this term alone would become meaningless [8].

Let me finally mention that the structure of the representation $\pi_\omega(\mathfrak{A})$ resulting from a KMS-state ω is closely connected with the recently developed beautiful mathematical theory of modular automorphisms. The basic mathematical ideas there were introduced by Tomita [9] without being aware of the KMS-condition in physics, see also [10]. This is one more example of the prestabilized harmony between canonical questions in mathematics and physics. Once the connection between KMS-states and Tomita-Takesaki theory became clear the interaction between mathematical and physical ideas in this area became fruitful for both disciplines. Of the interesting results I mention only one which is due to Araki [11] : If ω is a KMS-state, π_ω the corresponding representation, H the Hamiltonian and $h \in \pi(\mathfrak{A})$ a perturbation then $H + \lambda h$ has one discrete eigenvector and it can be calculated by a perturbation series in λ. This is the converse of our stability argument (and preceded it). It shows e.g. that if the Kondo effect should be interpreted to mean that a bounded local perturbation leads out of the family of the unperturbed state then the latter cannot be a KMS-state.

---

[7] $E(\mathcal{V},t) = \int_{\mathcal{V}} u(\vec{x},t)d^3x$ is an (unbounded) observable of the region $\mathcal{V}$ at time t , associated with $\mathfrak{A}(\mathcal{V},t)$, that is bounded functions of $E(\mathcal{V},t)$ belong to $\mathfrak{A}(\mathcal{V},t)$.

IV. Direct Computation of Equilibrium States from the Stability Condition for the Free Bose- and Fermi Gas

This is intended as an illustration of the discussion in Section III. It is instructive because the (infinite) free Bose- or Fermi gas has many stationary, primary states which are not stable. It is a highly non ergodic system (many symmetries) and hence the stability condition 3) is vital in extracting the equilibrium states. It will appear directly that besides β there appears the chemical potential μ as an equilibrium parameter as a result of the fact that all observables have to be gauge invariant (no perturbations are admitted which are sources of particles). In the Bose case, if $\mu = 0$, there appears another equilibrium parameter ρ_0 , the superfluid density. This is, of course, well known from standard treatments of the ideal quantum gases but it is gratifying to see how all of it comes out from condition (3.8) .

In constructing the net of algebras we start from the creation operator $a^*(f,t)$ of a particle with position space wave function $f(\vec{x})$ at time t . In terms of the creation operator of the particle at a point $\vec{x}$ at time t (which was denoted before by $a^*(\vec{x},t)$)

$$a^*(f,t) = \int a^*(\vec{x},t) \; f(\vec{x}) \, d^3x \qquad (4.1)'$$

The adjoint of this is the destruction operator

$$a(f,t) = \int a(\vec{x},t) \, \overline{f}(\vec{x}) \, d^3x \qquad (4.1)''$$

In the Fermi case the commutation relations at fixed time t are

$$\begin{aligned} \{a^*(f,t) , a^*(g,t)\} &= \{a(f,t), a(g,t)\} = 0 \\ \{a^*(f,t) , a(g,t)\} &= (g,f) = \int \overline{g}(\vec{x}) \, f(\vec{x}) \, d^3x \end{aligned} \qquad (4.2)$$

where $\{A,B\} = AB + BA$ denotes the anticommutator. The natural norm (C^*-norm) of these objects (in the Fermi case) is then

$$\| a^*(f,t) \| = \left(\int |f(\vec{x})|^2 \, d^3x \right)^{\frac{1}{2}} = \|f\|_2 \qquad (4.3)$$

Here $\|f\|_2$ is the norm of the wave function f considered as an $\mathcal{L}^{(2)}$-function in space. So we can admit all square integrable functions f if we want to get algebraic elements with finite norm. The $a^*(f,t)$, $a(g,t)$ generate a net of local algebras, which we shall denote by $\mathcal{F}(\mathcal{V},t)$, in the obvious way : Take all wave functions which have support in $\mathcal{V}$ i.e. which are zero outside of $\mathcal{V}$. Consider all linear combinations of all products with an arbitrary but finite number of factors $a^*(f_1,t)\, a^*(f_2,t) \ldots a(g_n,t)$ (all f_i, g_i having support in $\mathcal{V}$). If two such expressions which look different can be converted by the use of the commutation relations (4.2) to the same form then they are considered as the same algebraic element. Then take the completion of this set in the norm topology. This is $\mathcal{F}(\mathcal{V},t)$.

Again the kinematical algebra of all quasilocal objects at time t is

$$\mathcal{F}_t \equiv \bigcup_{\mathcal{V}} \mathcal{F}(\mathcal{V},t) \tag{4.4}$$

We shall now see that the dynamical law is such that the time translations are an automorphism group of $\mathcal{F}_t$ so that we may take the algebra of all quasilocal objects at any time to be just the kinematical algebra at a fixed time, say $t = 0$.

$$\mathcal{F} = \mathcal{F}_{t=0} \tag{4.5}$$

In the case of the non-interacting (free) Fermi gas the a^* satisfy linear equations of motion and

$$\alpha_t\,(a^*(f,t_o)) = a^*(f_t,\, t_o) \tag{4.6}$$

where f_t is the solution of the one-particle Schrödinger equation

$$\frac{\hbar}{i}\,\frac{\partial f_t}{\partial t} = -\,\frac{\hbar^2}{2m}\,\Delta f_t \tag{4.7}$$

with the initial condition $f_{t_o} = f$. Since the Schrödinger equation (4.8) conserves the scalar products of wave functions $(g_t,f_t) = (g,f)$, the commutation relations and the norms are conserved. So (4.6) defines indeed an automorphism of $\mathcal{F}_{t_o}$.

The dynamical law (4.6) looks simpler in momentum space. If we introduce the creation operator $a^*(\vec{k},t)$ of a particle with wave vector $\vec{k}$ (by taking for f the plane wave $e^{i\vec{k}\vec{x}}$) then (4.6) becomes

$$\alpha_t\,(a^*(\vec{k},t)) = e^{i\,\varepsilon_k t}\,a^*(\vec{k},t)\ ;\quad \varepsilon_k = \frac{\hbar^2 k^2}{2m} \tag{4.8}$$

Of course $a^*(\vec{k},t)$ is not an element of $\mathcal{F}$ since a plane wave is not a square integrable function but it is a computationally convenient object from which we can always return to true elements of $\mathcal{F}$ by integrating $a^*(\vec{k},t)$ with a square integrable function $\tilde{f}(\vec{k})$. In fact, it will be sufficient and convenient in the following to consider $a^*(\vec{k},t)$ and expectation values of products of such objects like $a^*(\vec{k}_1 t)\, a^*(\vec{k}_2,t) \dots$ as distributions in the Laurent Schwartz sense over test functions which are smooth. (e.g. infinitely often differentiable and fast decrease at infinity).

The net $\mathcal{F}$ is not the net of observables because of the latter we demand that they shall be "gauge invariant". Defining a gauge transformation as the automorphism transforming $a^*(f,t)$ into $e^{i\varphi}\, a^*(f,t)$:

$$\gamma_\varphi(a^*(f,t)) = e^{i\varphi}\, a^*(f,t) \; ; \quad \gamma_\varphi(a(f,t)) = e^{-i\varphi}\, a(f,t) \qquad (4.9)$$

the observable net $\mathcal{A}$ shall consist of all those objects of the net which are invariant under (4.9) :

$$A \in \mathcal{A}(\mathcal{V},t) \quad \text{if} \quad A \in \mathcal{F}(\mathcal{V},t) \quad \text{and} \quad \gamma_\varphi(A) = A \qquad (4.10)$$

In plain language an observable will always have an equal number of creation and destruction operators in each term. One easily checks that the observable algebras of disjoint regions at one time commute due to (4.2) so that the net $\mathcal{A}$ is asymptotically Abelian under space translations and one also checks that the dynamical law (4.6) or (4.8) is such that we have asymptotic Abelianness in time. In fact, if we take the norm dense subset in $\mathcal{A}$ which is generated by the $a^*(f,t)$ with smooth momentum space wave functions then for these

$$\left\| [A, \alpha_t(B)] \right\| < C \cdot t^{-\frac{3}{2}} \qquad (4.11)$$

so the commutator function is absolutely integrable in norm.

I discussed these essentially trivial matters at this extent just to illustrate how the well known quantities fit in detail into our general frame.

A state over $\mathcal{A}$ is completely determined if we know the set of expectation values of Wick ordered products with equal numbers of creators and annihilators at time $t = 0$.

$$W^{(n)}(k_1 \dots k_n \; ; \; k_1' \dots k_n') = \omega(a^*(k_1) \dots a^*(k_n)\; a(k_n') \dots a(k_1')) \qquad (4.12)$$

By the commutation relations they have to be antisymmetric in the undotted variables and in the dotted variables. By the linearity of ω the expectation value of any element of $\mathcal{A}$ is then obtained e.g.

$$\omega(a^*(f)\ a(g)) = \int \tilde{f}(k)\ \tilde{g}(k')\ \ W^{(1)}(k,k')\ d^3k\ d^3k'$$

where $\tilde{f},\tilde{g}$ are the momentum space wave functions i.e. the Fourier transforms of f and g.

Let us now turn the crank and apply the stability condition

$$\int \left[A,\ \alpha_t(B)\right]\ dt\ = 0 \tag{4.13}$$

in this case. First choose

$$B = a^*(g)\ a(g')\ ;\ A = a^*(f_1)\dots a^*(f_n)\ a(f_n')\dots a(f_1') \tag{4.14}$$

and denote the (momentum space) supports of $\tilde{g}$, $\tilde{g}'$ respectively by K, K' the support of $\tilde{f}_i$ by K_i, the support of $\tilde{f}_i'$ by K_i'.
For simplicity choose

$$K' \text{ disjoint from } K_2 \cup K_3 \dots \cup K_n \tag{4.15a}$$

and

$$K \text{ disjoint from } K_1' \cup K_2' \dots \cup K_n' \tag{4.15b}$$

then the only contributing term in the commutator will come from the interchange in the position of $a^*(f_1)$ and $a(g')$ and condition (4.13) gives[8]

$$0 = \int_{-\infty}^{\infty} dt\ \ W^{(n)}(q,\ k_2,\ \dots\ k_n\ ;\ k_n'\ \dots\ k_1')\ \times \tag{4.16}$$

$$\times\ F(q)f_2(k_2)\ \dots\ f_n(k_n)\ f_n'(k_n')\ \dots\ f_1'(k_1')\ d^3q\ \prod_2^n d^3k_i\ \prod_1^n d^3k_i'$$

where

$$F(q) = g(q) \int \delta(\ \varepsilon_q - \varepsilon_{q'})\ \bar{g}'(q')\ f_1(q')\ d^3q' \tag{4.17}$$

Consider now $2n$ points $p_1, \dots p_n, p_1', \dots p_n'$ such that $p_1 \neq p_j'$ for any j and $p_1 \neq 0$. Then there exists a neighborhood K of the point p_1 and neighborhoods $K_i (i = 2, \dots n)$, $K_j' (i = 1, \dots n)$ of the points p_i, p_j' respectively such that (4.15b) is satisfied and that

8 For convenience I omit now the $\sim$ over the wave functions. All wave functions from now on will be in momentum space.

an arbitrary test function F with support in K can be represented in the form (4.17) with an appropriate choice of test functions f_1, g, g' such that the support K' of g', satisfies (4.15a) and supp $g=K$. This is, because if g', f_1 are smooth functions and if

$$\varphi(q) = \int \delta(\varepsilon_q - \varepsilon_{q'}) \, \bar{g}'(q') \, f_1(q') \, d^3q' \tag{4.18}$$

does not vanish for any $q \in K$ then $\varphi(q)^{-1}$ is smooth within K and then we may, for any smooth F with support in K define a smooth g with the same support by $g = F \cdot \varphi^{-1}$. The origin $p_1 = 0$ has to be excluded because the function $\varphi(q)$ in (4.18) always vanishes at $q = 0$ (if the energy-momentum relation is (4.8) and if the space is more than 2-dimensional). The condition (4.16) therefore implies

$$W^{(2n)}(p_1 \dots p_n ; p_n' \dots p_1') = 0$$

unless $p_1 = p_j'$ for some j or $p_1 = 0$. In other words $W^{(2n)}$ with respect to the variable p_1 has support at most in the points $p_1 = 0$ and $p_1 = p_j'$ i.e. $W^{(2n)}$ contains a δ - function $\delta(p_1)$or $\delta(p_1-p_j)$or derivatives thereof. Since there is nothing special about the index 1 nor about the role of unprimed variables we find that $W^{(2n)}$ must be a linear combination of terms

$$\Delta_i(p_i) \dots \Delta_{kj}(p_k - p_j') \dots \Delta_l(p_l') \dots \tag{4.19}$$

where Δ_i, Δ_{jk} stand for distributions with point support.
Since $|\omega(A)| \leq \|A\|$ we must have

$$\int W^n(p_i ; p_i') \prod f(p_i) \, f'(p_j') \, d^3p_i \, d^3p_j' \leq \prod \|f_i\| \ \|f_j'\| \tag{4.20}$$

which excludes factors of the type $\Delta(p_i)$ or $\Delta(p_j')$ because for fixed $\mathcal{L}^{(2)}$-norm of f the values of f or its derivatives at $p = 0$ may be arbitrarily large. Similarly we see that $\Delta(p_i - p_j')$ must be ordinary δ - functions in order that (4.20) be satisfied. Finally, the clustering property of the state means that the so-called truncated functions $W_T^{(n)}$, (Ursell functions) which result from the $W^{(n)}$ by subtracting the uncorrelated part cannot have more than one δ-function factor because otherwise the correlation functions in x - space would not decrease if some difference $x_i - w_j$ gets large. Therefore

$$\begin{aligned} W_T^{(n)} &= 0 \quad \text{for } n > 1 \\ W_T^{(1)}(p,p') &= \rho(p) \, \delta^3(p - p') \end{aligned} \tag{4.21}$$

To determine the function ρ, the one particle momentum distribution function, we use the stability condition again, this time for

$$A = a^*(f_1)\ a^*(f_2)\ a(f_2')\ a(f_1')\ ; \quad h = A^* \tag{4.22}$$

choosing for simplicity K_1 disjoint from K_2', K_1' disjoint from K_2. Then (3.8) becomes

$$\int |(f_1(p_1)|^2\ |f_2(p_2)|^2\ |f_2'(p_2')|^2\ |f_1'(p_1')|^2\ \delta(\varepsilon_{p_1}+\varepsilon_{p_2}-\varepsilon_{p_1'}-\varepsilon_{p_2'})\ \cdot$$

$$\cdot\ B(p_i\ ;\ p_i')\ \prod d^3p_i\ d^3p_j' = 0 \tag{4.23}$$

with

$$B = \rho(p_1)\ \rho(p_2)\ (1-\rho(p_1')\ (1-\ \rho(p_2'))-(1-\rho(p_1)(1-\rho(p_2))\rho(p_1')\rho(p_2') \tag{4.24}$$

This implies

$$B = 0 \quad \text{whenever} \quad \varepsilon_{p_1} + \varepsilon_{p_2} - \varepsilon_{p_1'} - \varepsilon_{p_2'} = 0 \tag{4.25}$$

Introducing

$$R(p) = \log\left(\frac{1-\rho(p)}{\rho(p)}\right) \tag{4.26}$$

(4.24), (4.25) means that

$$R(p_1) + R(p_2) = R(p_1') + R(p_2') \quad \text{whenever} \quad \varepsilon_{p_1} + \varepsilon_{p_2} = \varepsilon_{p_1'} + \varepsilon_{p_2'}$$

i.e. R has to be a linear function of ε

$$R(p) = \alpha + \beta\,\varepsilon_p \tag{4.27}$$

or

$$\rho(p) = \frac{1}{1 + e^{\alpha+\beta\varepsilon_p}} \tag{4.28}$$

which is the Fermi distribution corresponding to an inverse temperature β and chemical potential $\mu = \beta^{-1}\alpha$.

In the Bose system, where we have in the commutation relations (4. 2) commutators instead of anticommutators, the $a^*(f)$ are not bounded elements (they have infinite norm). In order to form a C^*-algebra one would have to go over to the Weyl operators

$U(f) = e^{i(a^*(f) + a(f))}$ or use some other bounded functions of the a^*, a. For simplicity and because the argument up to (4.19) does not involve topological properties in any significant way we make the calculation with the unbounded elements $a^*(f)$ in the same way as for the Fermi case and come up to the form (4.19). Now, however, we cannot exclude anymore factors of the type $\Delta_i(p)$ or infer that $\Delta_{jk}(p-p')$ is a δ-function by the boundedness argument. For $W^{(1)}$ we find from the stationarity that it must be of the form

$$W^{(1)}(p,p') = \rho_0 \delta^3(p)\, \delta^3(p') + \rho(p)\, \delta^3(p-p') \tag{4.29}$$

since any derivatives of δ-functions would produce powers of the time t in the expectation value of $a^*(f_t)\, a(f_t')$. The clustering argument then determines the higher functions. For instance

$$\begin{aligned} W^{(2)}(p_1p_2,p_1'p_2') = {} & \rho(p_1)\, \rho(p_2)\, (\delta(p_1-p_1')\, \delta(p_2-p_2') + \delta(p_1-p_2') \cdot \\ & \cdot \delta(p_2-p_1')) \\ & + \rho_0\, \rho(p_1)\, \delta(p_1-p_1')\, \delta(p_2)\, \delta(p_2') + \text{permutations} \\ & + \rho_o^2\, \delta(p_1)\, \delta(p_2)\, \delta(p_1')\, \delta(p_2') \end{aligned} \tag{4.30}$$

To find the restrictions on the function $\rho(p)$ and the constant ρ_0 we use the (3.8) again with the choice (4.22) for A and h and get again (4.25) but now with

$$\begin{aligned} B \equiv {} & \rho(p_1)\, \rho(p_2)\, (1+ \rho(p_1'))\, (1+ \rho(p_2')) - (1+ \rho(p_1))\, (1+ \rho(p_2)) \cdot \\ & \cdot \rho(p_1')\, \rho(p_2') \end{aligned} \tag{4.31}$$

and in addition the requirement

$$\rho_0(\rho(p_2)\, (1+ \rho(p_1'))\, (1+ \rho(p_2')) - (1 + \rho(p_2))\, \rho(p_1')\, \rho(p_2')) = 0 \tag{4.32}$$

whenever $\varepsilon_{p_2} = \varepsilon_{p_1'} + \varepsilon_{p_2'}$

Eqs. (4.25), (4.31) give that $\rho(p)$ is a Bose distribution

$$\rho(p) = \frac{1}{e^{\alpha + \beta \varepsilon_p} - 1} \tag{4.33}$$

Equation (4.32) gives the alternative

either

$$\rho_0 = 0 \tag{4.34'}$$

or

$$R(p_2) = R(p_1') + R(p_2') \quad \text{whenever} \quad \varepsilon_{p_2} = \varepsilon_{p_1'} + \varepsilon_{p_2'}$$

with (4.34")

$$R(p) = \log \frac{1 + \rho(p)}{\rho(p)} = \alpha + \beta \varepsilon_p$$

(4.34") gives $\alpha = 0$.
So altogether one has (4.33) for the function ρ and the alternative

$$\text{either} \quad \rho_0 = 0 \ , \quad \alpha \text{ arbitrary} \tag{4.35'}$$

$$\text{or} \quad \rho_0 \text{ arbitrary} \ , \quad \alpha = 0 \tag{4.35"}$$

One more instructive exercise : Denote the state over the observable algebra $\mathcal{A}$ corresponding to (4.33), (4.35') by $\omega_{\alpha\beta}$, the state corresponding to (4.33), (4.35") by $\omega_{\rho_o\beta}$. We may extend these states to states over the larger algebra $\mathcal{F}$ (to non gauge invariant elements). In the case of $\omega_{\alpha\beta}$ the extension is unique and primary. In the case of $\omega_{\rho_o\beta}$ the extension is non unique. There is a unique gauge invariant extension $\hat{\omega}_{\rho_o\beta}$ .But this extension is not primary over $\mathcal{F}$.
The decomposition of $\hat{\omega}_{\rho_o\beta}$ into primary states leads to the appearance of a phase angle φ $(0 \leq \varphi < 2\pi)$ and non vanishing expectation values of non gauge invariant elements :

$$\omega(a^*(p)) = \sqrt{\rho_0}\, e^{i\varphi} \, \delta^3(p) \tag{4.36}$$

The parameter φ , unobservable in the true equilibrium state, will play a role in the analogue of the Josephson effect. It follows, however, from the above calculation that in the free Bose gas there are no inhomogeneous equilibrium states ; in particular we cannot have a stationary spatial phase coexistence, where in one half space we have the phase φ_1 in the other the different phase φ_2 even with $\beta_1 = \beta_2$, $\mu_1 = \mu_2 = 0$, $\rho_0^{(1)} = \rho_0^{(2)}$.

V. Hierarchy of Stability

As expected the time has run out. But in order to justify the title of this series of lectures let me add a few comments. One expects that an equilibrium state which is not just at a phase transition or critical point will be stable not only under local perturbations of the dynamics. We are therefore invited to study the degree of stability of equilibrium states.

In the simplest form, take $h \in \mathfrak{a}$ and consider the effect of a perturbation

$$h' = \int g(\bar{x}) \; \alpha_{\bar{x}}(h) \; d^3 x \tag{5.1}$$

where $g(\bar{x})$ is some continuous function of $\bar{x}$. If g is absolutely integrable then $h' \in \mathfrak{a}$ and we have still an essentially local perturbation. The interesting cases are therefore

$$\int_{v_R} |g(\bar{x})| \; d^3 x \leq c \cdot R^n \qquad n > 0 \tag{5.2}$$

Here v_R denotes a sphere of radius R around the origin. If the equilibrium state is such that

$$\int_{-\infty}^{+\infty} \omega([\alpha_t(A), h']) \; dt = 0 \tag{5.3}$$

For h' of the form (5.1) for all g satisfying (5.2) we may say that ω is stable of degree n.

We may illustrate this hierarchy of stability in the example of the free quantum gases discussed in the last section. For the Fermi gas and for the case $\mu \neq 0$, $\rho_0 = 0$ in the Bose gas, we have stability against global perturbations (degree 3), whereas in the superfluid case ($\mu = 0$, $\rho_0 \neq 0$) the state is instable already for n = 1. Typically, taking in (5.3) A and h of the form

$\int c(p_1 p_2, p_1' \; p_2') \; a^+(p_1) a^+(p_2) a(p_2') a(p_1') \prod d^3 p_i \; d^3 p_i'$ with smooth wave functions c in momentum space the ρ - term in the state ω gives contributions to the left hand side of (5.3) which are of the form

$$\int \varphi(p_i,p_0')B(p_i,p_j')\, e^{i(\bar{p}_1+\bar{p}_2-\bar{p}_1'-\bar{p}_2')\bar{x}-i(\varepsilon_{p_1}+\varepsilon_{p_2}-\varepsilon_{p_1'}-\varepsilon_{p_2'})t} \cdot$$

$$\cdot g(\bar{x})\, d^3\mathbf{x} \prod d^3p_i d^3p_j'\, dt \qquad (5.4)$$

whereas the ρ_0-term contributes, for instance, expressions of the form

$$\rho_0^3 \int \varphi(p)\, e^{i(px - \varepsilon_p t)}\, g(x)\, d^3\mathbf{x}\ d^3\mathbf{p}\, dt \qquad (5.5)$$

where $\varphi(p_i,p_i')$ and $\varphi_{|p|}$ are smooth functions, B is defined in (4.31) and the time integration should be performed as the last. Now, if $\mu \neq 0$, $\rho_0 = 0$ $B = (\varepsilon_{p_1} + \varepsilon_{p_2} - \varepsilon_{p_1'} - \varepsilon_{p_2'})\, b$
where b is still infinitely differentiable. Thus (5.4) can be written as

$$\int_{-\infty}^{\infty} dt\, \frac{\partial}{\partial t} \int \chi(p_i,p_i')\, g(\bar{x})\, e^{i(\bar{p}_1+\bar{p}_2-\bar{p}_1'-\bar{p}_2')\bar{x}-i(\varepsilon_{p_1}+\varepsilon_{p_2}-\varepsilon_{p_1'}-\varepsilon_{p_2'})t} \cdot$$

$\cdot \prod d^3p_i d^3p_i' d^3x$ with smooth χ . This will vanish if the inner integral vanishes in the limits $t \to \pm\infty$. This is the case even if $g(\bar{x})$ does not decrease at all for large $|\bar{x}|$. The expression (5.5) on the other hand will vanish only if $g(\bar{x})$ decreases faster than $\frac{1}{|\bar{x}|^2}$ for large $|\bar{x}|$. So, if $\rho_0 \neq 0$ we have instability when $n \gg 1$ in (5.2).

APPENDIX

A.1 The GNS-Construction

The algebra $\mathfrak{A}$ may be considered in a dual role. On one hand it is a vector space over the complex numbers if we only consider its linear structure. On the other hand we may consider each element as a linear operator acting in this vector space. If, in order to distinguish these two roles we attach a subscript v to an element when we want to consider it as a vector and the subscript op when we consider it as an operator we have

$$A_{op}B_v = (AB)_v \tag{A.1}$$

where on the right hand side the multiplication in the algebra is used. Then $A \rightarrow A_{op}$ is a representation of the algebra by operators acting in the linear space $\mathfrak{A}$ itself. In order to get a representation in a Hilbert space we have to introduce a scalar product in $\mathfrak{A}$. This can be done if a state ω over $\mathfrak{A}$ is given. We define

$$(A_v, B_v) = \omega(A^+B) \tag{A.2}$$

This is linear in the right member, complex conjugate linear in the left member and $\omega(A^+A) \geqslant 0$. So it is a semidefinite Hermitian form. In order to get a positive definite Hermitian form one singles out the subspace $\mathfrak{J} \subset \mathfrak{A}$ consisting of all elements $x \in \mathfrak{A}$ for which $\omega(x^+x) = 0$. Then one divides $\mathfrak{A}$ into classes modulo $\mathfrak{J}$; two elements A_1, A_2 belong to the same class if $A_1 - A_2 \in \mathfrak{J}$. Letting then the symbol A_v denote the whole class to which A belongs rather than an individual element one checks that (A.1) and (A.2) are still consistent i.e. the results do not depend on the choice of the element with a class. In the space of classes $\mathfrak{A}/\mathfrak{J}$ the right hand side of (A.2) defines a positive definite scalar product. This space can then be completed to become a Hilbert space and (A.1) defines A_{op} as a bounded linear operator acting in this Hilbert space.

A.2 Primary States and Factorial (Primary) Representations

If $\mathfrak{M}$ is a von Neumann algebra (weakly closed + -algebra of bounded operators acting on a Hilbert space $\mathcal{H}$) and $\mathfrak{M}'$ its "commutant" then the "center" of $\mathfrak{M}$

$$\mathfrak{Z} = \mathfrak{M} \cap \mathfrak{M}'$$

is a commutative von Neumann algebra. We can simultaneously diagonalize all elements of $\mathfrak{Z}$ and thereby decompose $\mathcal{H}$ into a direct sum (or direct integral) of spaces $\mathcal{H}_\lambda$ where λ stands for a set of simultaneous spectral values of the elements of $\mathfrak{Z}$. Since $\mathfrak{Z}$ commutes both with $\mathfrak{M}$ and with $\mathfrak{M}'$ we have thereby reduced $\mathfrak{M}$ and $\mathfrak{M}'$: in each $\mathcal{H}_\lambda$ we have the von Neumann algebras $\mathfrak{M}_\lambda$ and $\mathfrak{M}_\lambda'$ but now $\mathfrak{M}_\lambda \cap \mathfrak{M}_\lambda'$ is trivial i.e. $\mathfrak{Z}_\lambda$, consists only of multiples of the unit operator in $\mathcal{H}_\lambda$.

A von Neumann algebra is called a <u>factor</u>, if its center consists only of multiples of the unit operator. We have described above the decomposition of an arbitrary von Neumann algebra into factors (the central decomposition).

Given a state ω over the (abstract) c^+ - algebra $\mathfrak{a}$. By the GNS-construction we obtain a representation π_ω of $\mathfrak{a}$ by operators acting in a Hilbert space $\mathcal{H}$. $\pi_\omega(\mathfrak{a})''$ is a von Neumann algebra. If this is a factor, i.e. if its center is already trivial, then ω is called a primary state. If not, then corresponding to the central decomposition of $\pi_\omega(\mathfrak{a})''$ we have a decomposition of ω into primary states.

REFERENCES

1) I.E. SEGAL, Ann. Math. 48, 930 (1947)

2) G.C. WICK, A.S. WIGHTMAN and E.P. WIGNER, Phys. Rev. 88, 181 (1952)

3) R. HAAG and D. KASTLER, J. Math. Phys. 5, 848 (1964).

4) R. HAAG, D. KASTLER and E.B. TRYCH-POHLMEYER, Comm. Math. Phys. 38, 173 (1974).

5) DOPLICHER, D. KASTLER and ROBINSON

6) R. KUBO, J. Phys. Soc. Japan 12, 570 (1957).

7) P.C. MARTIN and J. SCHWINGER, Phys. Rev. 115, 1342 (1959).

8) R. HAAG, N. HUGENHOLTZ and M. WINNINK, Comm. Math. Phys. 5, 215 (1967).

9) M. TOMITA, Standard Forms of von Neumann Algebras, Vth Functional Analysis Symp. Math. Soc. of Japan 1967.

10) M. TAKESAKI, Tomita's Theory of Modular Hilbert Algebras and its Applications, Springer-Verlag, Berlin, 1970.

11) H. ARAKI, Ann. Sci. École Norm. Sup. 6, Nº 1 (1973).

THEOREMS ON PHASE TRANSITIONS WITH A TREATMENT FOR THE ISING MODEL

S. MIRACLE-SOLÉ

Facultad de Ciencias
Universidad de Zaragoza, Spain

Theorems on Phase Transitions with a Treatment for the Ising Model

S. MIRACLE-SOLE

Facultad de Ciencias
Universidad de Zaragoza, Spain

Introduction

In these lectures we would like to discuss some recent developments on phase transitions from the point of view of mathematical physics. Some of them concern the Ising model with nearest neighbour ferromagnetic interactions. Although not unique anymore as a non-trivial soluble problem this model continues to attract interest as a testing ground for general ideas. The classical solution of Onsager refers only to the macroscopic equation of state, in the case of magnetic field h=o, but there are other interesting questions which cannot be answered in such explicit terms. See for instance, in relation with our discussion, the review article by Gallavotti [1]. One of these questions is the description of the equilibrium states and the phase diagram of the system. In fact a complete description, from the theoretical point of view, was only known in the case of one dimensional systems with short range forces, and this justifies the interest in obtaining it in the case of a system exhibiting phase transitions. As we shall see, a number of results obtained in the last few years, have provided for the Ising ferromagnet in two dimensions a complete description of its phase diagram. They also give interesting information about the general behaviour of other lattice systems.

In presenting these results we shall have the opportunity of discussing a certain number of methods and problems which arise in the description of equilibrium states in statistical mechanics.

I. Notations and Definitions

We denote by L the set of lattice sites. In each site $x \in L$ we suppose that the system can be in two different states $\{+1,-1\}$ (spin up and spin down). A configuration σ^L of the system is given by specifying the state σ_x in each lattice site $x \in L$. Let Λ be a finite set of L which we identify as a box containing the system. For simplicity we will take Λ to be a parallelopiped. Once the interaction is given one can determine the energy $H_\Lambda(\sigma^L)$ of the system inside Λ for every configuration σ^L.

For instance in the Ising model, L is a cubic lattice $\mathbb{Z}^\nu$, and the interaction is such that only pairs of nearest neighbouring points have an interaction different from zero, which is the same for all pairs, and single points interact with a constant external magnetic field h . The energy has therefore the following expression

$$H_\Lambda(\sigma^L) = - \sum_{\substack{x \in \Lambda \\ y \in L}} J(x-y)\, \sigma_x \, \sigma_y + h \sum_{x \in \Lambda} \sigma_x$$

where $J(x-y) = J$ if $|x-y| = 1$ and $J(x-y) = 0$ otherwise. More general functions $J(x)$ on the lattice points will give us more general lattice systems with pair interactions. A certain condition on the decreasing to zero of $J(x)$ when $|x| \to \infty$, saying that points which are far apart have weak interaction, is then necessary to give sense to H_Λ , namely $\sum_{x \in L} |J(x)| < +\infty$. Such conditions are also needed in more fundamental grounds, in order to ensure a thermodynamic behaviour of the system. One can also consider many body interactions by including terms of the form $J(x_1,\dots,x_n)\ \sigma_{x_1} \dots \sigma_{x_n}$. The system is said to be ferromagnetic if all interaction potentials J are positive, as this has the effect that the spins tend to point in the same direction in order to decrease the interaction energy.

We consider the configuration $\sigma^L = \{\sigma_x\}$, $x \in L$, as the union $\sigma^\Lambda \sigma^{\Lambda'}$ of two restricted configurations: $\sigma^\Lambda = \{\sigma_a\}$, $a \in \Lambda$ which represents a configuration inside Λ, and $\sigma^{\Lambda'} = \{\sigma_a\}$, $a \in \Lambda'$ with $\Lambda' = L \setminus \Lambda$ which represents a configuration outside Λ , or in other terms a boundary condition.

Let $\beta = 1/kT$ where T is the temperature and k the Boltzmann constant. The Gibbs ensemble at a temperature T is defined as a probability measure on the configurations inside Λ given by

$$g_{\Lambda}(\sigma^{\Lambda}) = Z_{\Lambda}^{-1} \exp - \beta H_{\Lambda}(\sigma^{\Lambda}\sigma^{\Lambda'})$$

where

$$Z_{\Lambda} = \sum_{\sigma^{\Lambda}} \exp - \beta H_{\Lambda}(\sigma^{\Lambda}\sigma^{\Lambda'})$$

is the so called partition function.It is known that the thermodynamic limit

$$p = \lim_{\Lambda \to \infty} \frac{1}{|\Lambda|} \log Z_{\Lambda}$$

always exists and is independent of the boundary conditions,($|\Lambda|$ denotes the number of points of Λ). It corresponds to the free energy of the system.

In order to study the thermodynamic limit of the Gibbs ensemble one introduces the expectation values

$$\rho_{\Lambda,\sigma^{\Lambda'}}(\sigma_{x_1} \cdots \sigma_{x_n}) = \sum_{\sigma^{\Lambda}} \sigma_{x_1} \cdots \sigma_{x_n} \; g_{\Lambda}(\sigma^{\Lambda})$$

of products of spin variables contained in Λ . They are called the correlation functions of the system.

We shall be interested (following Dobrushin [2] and Lanford and Ruelle [3]) in the set of functions $\rho(\sigma_{x_1} \cdots \sigma_{x_n})$ which can be written, for a suitable choice of the sequence of boxes and boundary conditions, as

$$\rho(\sigma_{x_1} \cdots \sigma_{x_n}) = \lim_{\Lambda \to \infty} \rho_{\Lambda,\sigma^{\Lambda}}(\sigma_{x_1} \cdots \sigma_{x_n})$$

for all finite sets $\{x_1, \ldots, x_n\}$ of points of the lattice, or which are convex combinations of such limits. We also introduce the condition of translation invariance

$$\rho(\sigma_{x_1+a} \cdots \sigma_{x_n+a}) = \rho(\sigma_{x_1} \cdots \sigma_{x_n})$$

for all $a \in L$.

A set of correlation functions verifying these two conditions will be called an (invariant) equilibrium state of the system at temperature T.

A set of correlation functions verifying the first of the two conditions will be called a Gibbs state.

It is easy to show by a compactness argument, that there is at least one Gibbs state at any given β and at least one invariant equilibrium state for a system with translation invariant interactions.The latter coming from the fact that invariant equilibrium states appear also as limits of the averaged correlation functions

$$\rho_{\Lambda,\sigma^{\Lambda}}(\sigma_{x_1}\cdots\sigma_{x_n}) = \frac{1}{|\Lambda|}\sum_{a\in\Lambda_1}\rho_{\Lambda_1\sigma^{\Lambda}}(\sigma_{x_1+a}\cdots\sigma_{x_n+a})$$

where Λ_1 is the set of all $a\in L$ such that $\{x_1+a,\dots,x_n+a\}\subset\Lambda$.

The occurrence of a phase transition reveals itself by the existence of more than one equilibrium state; the averaged correlation functions can have different limits depending on the boundary conditions. The set of all possible equilibrium states, as well as the set of all Gibbs states, can be seen to form a convex and compact set (for the topology of point-wise convergence of the correlation functions). They form also a Choquet simplex and, hence, any state has a unique integral representation in terms of the extremal states (in the sense of convexity theory). The proof of these facts can be found for instance in Ruelle's book [4] for the equilibrium states and in ref. [3,5] for the Gibbs states. Because the equilibrium states appear also as limits of the averaged correlation functions they account for the mean properties of the system at equilibrium. The extremal equilibrium states can then be identified with the pure thermodynamic phases, that can coexist at a given temperature, since by linear convex combination, or mixtures, of them one obtains all the possible equilibrium states. On the other hand extremal equilibrium states have ergodic properties telling us that the mean properties of the system do not fluctuate, this is the situation which is expected in a pure phase. The significance of the Gibbs states will appear later in the lectures.

II. First Remarks on the Equilibrium States

Let us first point out that in a ferromagnetic system with two-body interaction potentials the following limits exist

$$\rho^+(\sigma_{x_1}\cdots\sigma_{x_n}) = \lim_{\Lambda\to\infty} \rho_\Lambda^+(\sigma_{x_1}\cdots\sigma_{x_n})$$

$$\rho^-(\sigma_{x_1}\cdots\sigma_{x_n}) = \lim_{\Lambda\to\infty} \rho_\Lambda^-(\sigma_{x_1}\cdots\sigma_{x_n})$$

where (+) denotes the boundary condition $\sigma_x=+1$ for all $x\in\Lambda'$, and (-) the boundary condition $\sigma_x=-1$ for all $x\in\Lambda'$. They define translation invariant correlation functions if the interaction is translation invariant.

The proof of this fact is a consequence of the monotonic dependence of ρ_Λ^+ and ρ_Λ^- on the box Λ which follows from the Fortuin, Kasteleyn and Ginibre (F.K.G.) inequalities [6] . Let us introduce the variables $n_x = \frac{1}{2}(1-\sigma_x)$ giving the occupation numbers at the lattice sites. Then it follows from F.K.G. that $\rho_\Lambda^+(n_{x_1}\dots n_{x_n})$ is a decreasing function of Λ while $\rho_\Lambda^-(n_{x_1}\dots n_{x_n})$ is an increasing function of Λ . The F.K.G. inequalities say that if $f(n_{x_1},\dots,n_{x_n})$ is a function of the n_x in a finite box Λ which is increasing separately in each variable then $\rho_\Lambda(f)$ is an increasing function of the interactions and the external fields at the lattice points. They hold if the pair interactions are ferromagnetic for positive or negative external fields, no translation invariance is required.

These correlation functions ρ^+ and ρ^- define then invariant equilibrium states. Moreover these states are extremal in the set of all Gibbs states. From this it follows that they correspond to pure thermodynamic phases and have short range correlations (in the sense of ref. [3]). This last property implies the following cluster property

$$\lim_{|a|\to\infty}\rho^\pm(\sigma_{x_1}\cdots\sigma_{x_n}\,\sigma_{y_1+a}\cdots\sigma_{y_m+a})=\rho^\pm(\sigma_{x_1}\cdots\sigma_{x_n})\,\rho^\pm(\sigma_{y_1}\cdots\sigma_{y_m})$$

The proof of the extremality of ρ^+ and ρ^- in the set of all Gibbs states is a consequence of the following inequalities

$$\rho^+(n_{x_1}\dots n_{x_n}) \geqslant \rho(n_{x_1}\dots n_{x_n}) \geqslant \rho^-(n_{x_1}\dots n_{x_n})$$

where $\rho(n_{x_1}\dots n_{x_n})$ represents the correlation functions of any Gibbs state of the system. These inequalities are again a consequence of F.K.G. inequalities, because the field acting on any spin in Λ caused by a fixed configuration outside Λ lies between that caused by the boundary condition(+) and by the boundary condition (-). They also tell us that the Gibbs' state is unique if $\rho^+(n_{x_1}\dots n_{x_n}) = \rho^-(n_{x_1}\dots n_{x_n})$.

Let us mention in passing, that for ρ^+, the decreasing in Λ and its extremality property can be seen from the Griffiths, Kelly and Sherman(G.K.S.) inequalities. These inequalities are the first kind of correlation inequalities which have been derived. They say that $\rho_\Lambda(\sigma_{x_1}\dots\sigma_{x_n})$ is an increasing function of the interaction potentials and external fields for any system (it can be non translation invariant and have many body potentials) with only ferromagnetic interactions and positive external fields.(see [7]).

In the case h=0 these states are simply related by

$$\rho^-(\sigma_{x_1}\dots\sigma_{x_n}) = (-1)^n \rho^+(\sigma_{x_1}\dots\sigma_{x_n})$$

Then, if these correlation functions are zero when n is odd, they coincide. Otherwise, they define two different equilibrium states which have been called the up-magnetized and down-magnetized pure phases.

Let us now restrict our attention to the one point correlation function $\rho(\sigma_x)$. In a ferromagnetic system with pair potentials the theorem of Lee and Yang holds [8 , 9] and we deduce that when the external magnetic field $h \neq 0$, the thermodynamic free energy is an analytic function of h. A discontinuity in the derivative of p(T,h) can only exist at h = 0, and therefore only in this case can $\rho^+(\sigma_x) \neq \rho^-(\sigma_x)$. The spontaneous magnetization is defined as the right derivative of p(T,h) at h = 0,

$$m(T) = \left.\frac{\partial p(T,h)}{\partial h}\right|_{h=0^+}$$

Since p(T,h) = p(T,-h) from the symmetry of the partition function under the transformation $\sigma_x \to -\sigma_x$, we deduce that the left derivative at $h = 0^-$ coincides with $-m(T)$, and we see that a first order phase transition exists when the spontaneous magnetization $m(T)$ is strictly positive.

From the facts already mentioned we know that $\rho^{\pm}(\sigma_x)$ are translation invariant, i.e. constants, and that $\rho^+(\sigma_x) \geqslant \rho(\sigma_x) \geqslant \rho^-(\sigma_x)$ where $\rho(\sigma_x)$ is the magnetization corresponding to any Gibbs state. This allows us to conclude that

$$\rho^+(\sigma_x) = -\rho^-(\sigma_x) = m(T)$$

when the magnetic field $h = 0$. By applying G.K.S. or F.K.G. inequalities we can now say that $m(T)$ is a decreasing function of T. This justifies the definition of the critical temperature T_c, as the temperature below which the spontaneous magnetization becomes positive. This will be a convenient definition for our discussion. Then $m(T)=0$ for any $T > T_c$ and $m(T) > 0$ for $T < T_c$.

III. Statement of the Results

Let us now state the results concerning the two-dimensional Ising ferromagnet which will be discussed in the following paragraphs. In the discussion we shall also mention information on the behaviour of more general lattice systems that can be obtained from the methods used in proving these results.

For the two dimensional Ising model with nearest neighbour ferromagnetic interactions we have

a) In the region $h \neq 0$ there is only one Gibbs state $\rho_{h,T}(\sigma_{x_1} \cdots \sigma_{x_n})$ for any T.

b) In the region $h = 0$ and $T \geqslant T_c$ the Gibbs state $\rho_{0,T}(\sigma_{x_1} \cdots \sigma_{x_n})$ is also unique.

c) If $h = 0$ and $T < T_c$ any translation invariant equilibrium state $\rho_{0,T}(\sigma_{x_1} \cdots \sigma_{x_n})$ is such that

$$\rho_{0,T}(\sigma_{x_1} \cdots \sigma_{x_n}) = \lambda \rho^{+}_{0,T}(\sigma_{x_1} \cdots \sigma_{x_n}) + (1-\lambda)\rho^{-}_{0,T}(\sigma_{x_1} \cdots \sigma_{x_n})$$

for some $0 \leqslant \lambda \leqslant 1$.

These results are complemented by the following continuity properties

d) In the region $h \neq 0$ the unique correlation functions $\rho_{h,T}(\sigma_{x_1} \cdots \sigma_{x_n})$ are analytic functions of the parameters h and T.

e) When $h \rightarrow 0$, one has

$$\lim_{h \rightarrow 0^+} \rho_{h,T}(\sigma_{x_1} \cdots \sigma_{x_n}) = \rho^{+}_{0,T}(\sigma_{x_1} \cdots \sigma_{x_n})$$

$$\lim_{h \rightarrow 0^-} \rho_{h,T}(\sigma_{x_1} \cdots \sigma_{x_n}) = \rho^{-}_{0,T}(\sigma_{x_1} \cdots \sigma_{x_n})$$

f) The correlation functions $\rho^{+}_{h,T}(\sigma_{x_1} \cdots \sigma_{x_n})$ and $\rho^{-}_{h,T}(\sigma_{x_1} \cdots \sigma_{x_n})$ are infinitely differentiable with respect to h and T, even at $h = 0$ for $T > T_c$. They are also infinitely differentiable with respect to T and h, when $h \rightarrow 0^+$ or $h \rightarrow 0^-$ for $T < T_c$.

IV. Uniqueness of the Equilibrium State

Results (a), (d) and (e) are due to Ruelle [10] and hold for any ferromagnetic system with two body translation invariant interactions. They are related to the Lee-Yang circle theorem. It is proved that the zeros of the partition function do not move too much under small perturbations of the hamiltonian even if one allows many body interactions. The domain of analyticity of the free energy is then a circle as close as one wants to the unit circle when the perturbation tends to zero. The free energy is then differentiable with respect to this perturbation if $h \neq 0$. From this follows the unicity of the invariant equilibrium state and its analyticity when $h \neq 0$. Then $\rho^+ = \rho^-$ and by the remark above the Gibbs state is also unique. Property (e) is deduced from the fact that $\rho_{h,T}(\sigma_{x_1} \dots \sigma_{x_n})$ is an increasing function of h and that the correlations associated with ρ^- minorize and those associated with ρ^+ majorize any other correlations at $h = 0$.

Lebowitz and Martin-Löf [11] have proved that when the derivative $\frac{\partial p(\beta, h)}{\partial h}$ of the free energy with respect to h exists, the two states $\rho^+_{h,T}(\sigma_{x_1} \dots \sigma_{x_n})$ and $\rho^-_{h,T}(\sigma_{x_1} \dots \sigma_{x_n})$ coincide. Then proposition (a) follows directly from this fact and the Lee-Yang circle theorem. Also proposition (b) is deduced from this fact for any $T > T_c$, if the critical temperature is defined as above. Their proof is based on F.K.G. inequalities and holds also for ferromagnetic systems with invariant pair interactions. One observes that the function $\sum_{i=1}^{n} n_{x_i} - \prod_{i=1}^{n} n_{x_i}$ is an increasing function of the n_x and therefore F.K.G. inequalities apply to it. Hence

$$0 \leq \rho^+(n_{x_1} \dots n_{x_n}) - \rho^-(n_{x_1} \dots n_{x_n}) \leq \sum_{i=1}^{n} \{\rho^+(n_{x_i}) - \rho^-(n_{x_i})\}$$

But if the derivative of ρ with respect to h exists $\rho^+(n_x) = \rho^-(n_x)$ and therefore the states ρ^+ and ρ^- coincide, and that shows the proposition.

V. Critical Temperature and Spontaneous Magnetization

In the case of the two-dimensional Ising model, Lebowitz [12] has shown that the critical temperature T_c, that we have defined, coincides with the Onsager value T_0 for the critical temperature, given by

$$\sinh \quad 2\beta_0 J = 1$$

Let us recall the Onsager-Yang expression of the spontaneous magnetization for the Ising model

$$m_0(T) = (1 - (\sinh \quad 2\beta J)^{-4})^{\frac{1}{8}} \quad \text{for } T < T_0$$

$$m_0(T) = 0 \qquad \text{for } T \geqslant T_0$$

This expression has been computed from

$$m_0(T)^2 = \lim_{|x-y| \to \infty} \rho^P(\sigma_x \sigma_y)$$

where $\rho^P(\sigma_x \sigma_y)$ is the two-spin correlation function with periodic boundary conditions. Since $\rho^+(\sigma_x \sigma_y)$ is larger than $\rho^P(\sigma_x \sigma_y)$ and tends to $m(T)^2$ when $|x-y| \to \infty$, we deduce that $m(T) \geqslant m_0(T)$ and therefore that $T_c \geqslant T_0$. It remains then to show also that $T_c \leqslant T_0$, or equivalently that $m(T)=0$ for $T > T_0$. And here comes another result on the exact calculations on the Ising model, namely the fact that $\rho^P(\sigma_x \sigma_y)$ decays exponentially to $m_0(T)^2$, i.e.

$$|\rho^P(\sigma_x \sigma_y) - m_0(T)^2| \leqslant A \exp(-K \ |x-y|)$$

where A and K are some positive constants. (For references about this fact see [9,10]).

Now, assuming $T \geqslant T_0$ in which case $\rho^P(\sigma_x) = 0$, one remarks that

$$\left.\frac{\partial^2 p^P_\Lambda(\beta, h)}{\partial h^2}\right|_{h=0} = \frac{1}{|\Lambda|} \sum_{x,y\in\Lambda} \rho^P_\Lambda(\sigma_x \sigma_y) \leqslant \frac{1}{|\Lambda|} \sum_{x,y\in\Lambda} \rho^P(\sigma_x \sigma_y)$$

where p_Λ^P means the free energy in the box Λ with periodic boundary conditions, is bounded because of the exponential decay of $\rho^P(\sigma_x \sigma_y)$. This bound is uniform in Λ. Furthermore for every h

$$\frac{\partial^2 p_\Lambda^P(\beta,h)}{\partial h^2} \leqslant \left.\frac{\partial^2 p_\Lambda^P(\beta,h)}{\partial h^2}\right|_{h=0}$$

because this second derivative is an increasing function of h for $h \leqslant 0$ and a decreasing function of h for $h < 0$. This corresponds to the fact that the magnetization, which is the first derivative of $p(\beta,h)$ is a concave function of h for $h > 0$ and convex for $h < 0$. This fact, intuitively expected, can be established exactly by the Griffiths, Hurst and Sherman (G.H.S.) inequalities [13]. As the second derivative of $p_\Lambda^P(\beta,h)$ is bounded for all h, it follows that the first derivative of the thermodynamic free energy $p(\beta,h)$ is continuous, and hence that $\rho^+_{h,T}(\sigma_x) = \rho^-_{h,T}(\sigma_x)$ for all $T \geqslant T_o$. It follows then, as a direct consequence of the definitions, that $T_o \geqslant T_c$.

In the work by Lebowitz [12] that we are describing it is shown also how G.K.S. inequalities can be used to bound all other derivatives of $p_\Lambda^P(\beta,h)$ with respect to h and respect to β by convergent sums of the two point correlation functions. This proves proposition (f) for $T > T_c$. Proposition (f) for $T < T_c$ has been proved by Gallavotti and Lebowitz [14] by the same kind of arguments applied to $\rho^+(\sigma_x \sigma_y) - m_o(T)^2$, which is the Ursell function that bounds the derivatives of $p_\Lambda^+(\beta,h)$. The fact that this Ursell function has an exponential decay will be obvious in the next paragraph.

It must be pointed out here that, although these bounds on the Ursell functions are uniform in Λ, so that one gets the infinite differentiability of the correlation functions, they rapidly increase when the order of the derivative becomes large. It is not possible to deduce from these arguments whether the corresponding Taylor series has a non zero radius of convergence, or in other terms whether $p(\beta,h)$ and $\rho_{T,h}(\sigma_{x_1} \dots \sigma_{x_n})$ can be analytically continued across the line $h = 0$. This is a very interesting question for $T < T_c$ about which the situation at present is unclear. As it is known that there is analiticity at high temperature (see section VI) it is natural to conjecture that it persists up to $T = T_c$.

We will justify now that the Onsager value of the spontaneous magnetization coincides with the right derivative at $h = 0$ of the free energy, i.e.

$m_o(T) = m\,(T)$

This equality has been proved by Benettin, Gallavotti, Jona-Lasinio and Stella [15], but it is itself a corollary of proposition (c). As the proposition says that the two spin correlation function does not depend on the boundary conditions, we have the equality $\rho^P(\sigma_x \sigma_y) = \rho^+(\sigma_x \sigma_y)$, and therefore $m_o(T) = m(T)$ by the definition of $m_o(T)$, the cluster property of ρ^+ and the fact that $m(T) = \rho^+(\sigma_x)$.

In the same way the exact computations of the correlation function for an even number of points at $h = 0$ existing in the literature (see for instance [15]) are then confirmed to be independent of the boundary conditions.

VI. Two phase coexistence at Low Temperature

In what concerns the region of coexistence of phases a first result has been obtained by Gallavotti and the author [17] . It was shown that at sufficiently low temperature $T \leq T_o < T_c$, every invariant equilibrium state is a convex linear combination of only two extremal states, the states ρ^+ and ρ^-. The proof is based on contour techniques similar to those employed in the Peierls argument [19] for proving the existence of phase transitions. (See also refs. [1,18]).

The interest of these techniques lies in the fact that they are, as is the Peierls argument, rather close to the physics of the problem. The proof just mentioned holds in fact for the Ising ferromagnet in any dimension $\nu \geqslant 2$, and it can be extended to other ferromagnetic systems.

Let us now make a digression and consider the Ising antiferromagnet where the hamiltonian is the same as in the Ising model but $J < 0$. In this system the Peierls argument has been extended by Dobrushin [20] to show the existence of at least two different periodic states $\rho^{(1)}$ and $\rho^{(2)}$. For a very readable account on the Peierls argument in the two cases we are considering the reader is advised to look at ref. [7] . Suppose the square lattice broken into two sublattices L_1 and L_2, $L_1 \cup L_2 = L$, with the property that all nearest neighbours of a site on sublattice L_1 belong to sublattice L_2 and viceversa. The states $\rho^{(1)}$ and $\rho^{(2)}$ correspond to the boundary conditions obtained by putting on the boundary of the square $\sigma_x = +1$ if x is on the L_1 sublattice and $\sigma_x = -1$ if x is on the L_2 sublattice or viceversa. The state $\rho^{(2)}$ is obtained then by translating $\rho^{(1)}$ by the vector connecting to nearest neighbour points and both states have the periodicity of the sublattices. They differ already in the one point correlation function provided that $|h| < 4\,|J|$ and the temperature T is sufficiently low. The method used to deal with the Ising ferromagnet applies also in this case, and proves that in the above mentioned region, any translation invariant equilibrium state is of the form

$$\rho(\sigma_{x_1}\cdots\sigma_{x_n}) = \lambda\rho^{(1)}(\sigma_{x_1}\cdots\sigma_{x_n})+(1-\lambda)\,\rho^{(2)}(\sigma_{x_1}\cdots\sigma_{x_n})$$

with $0\leq\lambda\leq 1$. The state ρ can then be interpreted as a crystalline phase in the lattice, and the formula above is the mathematical expression for the spontaneous symmetry breaking occuring in this system. The proof along these lines has been established by Di Liberto [21] .

We like to mention in this context the recent interesting work of Pigorov and Sinai [22] which by a distillation of the essence of the notion of contours and Peierls argument (see also [23]) arrive at a rather general description of the behaviour and number of periodic states of lattice systems at low temperature. Let us mention here that the behaviour of general lattice systems at higher temperature is known from previous results (see [4]). It follows that if the temperature is sufficiently high the Gibbs state is unique and is an analytic function of β, h and the interactions. Similar results hold for $|h|$ large (the low activity region).

VII. Two Phase Coexistence below T_c

Let us now return to the two-dimensional Ising ferromagnet. From what we have said it is only lacking a small interval near the critical temperature, in the region of coexistence of phases, to complete the phase diagram of the system. The fact that there are exactly two pure phases for this system in the whole region $T < T_c$ and $h = 0$ has been proved in a work by Messager and the author [24] . We shall give here only the idea of the proof and refer for the details to [24] or to a more didactic exposition which can be found in [25].

We first remark that the statement of the theorem is equivalent to the uniqueness of translationally invariant even correlation functions. This follows from an argument based on the fact that the equilibrium states form a Choquet simplex. Let us suppose that $\rho(\sigma_{x_1} \dots \sigma_{x_n})$ is an extremal invariant equilibrium state at $h = 0$. Then, also $(-1)^n \rho(\sigma_{x_1} \dots \sigma_{x_n})$ is an equilibrium state, which is obtained by exchanging the +1 and -1 spins in the boundary. Therefore $\frac{1}{2} \rho(\sigma_{x_1} \dots \sigma_{x_n}) + \frac{1}{2} (-1)^n \rho(\sigma_{x_1} \dots \sigma_{x_n})$ is also an equilibrium state. It coincides with

$\frac{1}{2} \rho^+ (\sigma_{x_1} \dots \sigma_{x_n}) + \frac{1}{2} \rho^-(\sigma_{x_1} \dots \sigma_{x_n})$ because they are equal by hypothesis when n is even and they are zero when n is odd. But the decomposition of this state has to be unique in a Choquet simplex. Therefore, if the translation invariant correlation functions are unique, only the two states $\rho^+(\sigma_{x_1} \dots \sigma_{x_n})$ and $\rho^-(\sigma_{x_1} \dots \sigma_{x_n})$ are extremal in the set of invariant equilibrium states.

The uniqueness of even correlation functions is known for $T > T_c$ because then the Gibbs state is unique. Therefore, it might be useful to apply the duality properties of the Ising model in order to deduce the uniqueness in the interval $T < T_c$, as it is known that duality is a symmetry relation of the partition function and certain correlation functions between high and low temperatures related by

$$\mathrm{th}\, \beta J = \exp - 2\, \beta^* J \quad .$$

This idea has in fact been used in [12] to prove that the Onsager-Yang value of the spontaneous magnetization is exact.

In order to express the duality relations one introduces spin variables associated with the bonds of the lattice. If a denotes a bond with end points x_1 and x_2 we put $\sigma_a = \sigma_{x_1} \sigma_{x_2}$. In this way any even product $\sigma_{x_1} \cdots \sigma_{x_{2n}}$ of spin variable can be written as a product of bond spin variables σ_a. Consider for instance the product $\sigma_x \sigma_y$. If Γ is an arbitrary path connecting the two sites x and y and $a_1, a_2, \ldots, a_n$ are the bonds associated to Γ, one has the path independent relation

$$\sigma_x \sigma_y = \sigma_{a_1} \sigma_{a_2} \cdots \sigma_{a_n}$$

This relation can be generalized to any product containing an even number 2n of points, by grouping them in n pairs. The duality relation for the partition and correlation functions at h=0 are

$$Z^F_\Lambda(\beta) = K_\Lambda \; Z^+_\Lambda\,(\beta^*)$$

$$\rho^F_{\Lambda,\beta}\,(\sigma_{a_1} \cdots \sigma_{a_n}) = \rho^+_{\Lambda,\beta}\,(\mu_{a_1} \cdots \mu_{a_n})$$

where K_Λ is a factor, unimportant in our discussion, and $\mu_a = \exp - 2\, \beta^* J\, \sigma_a$. Notice that these relations are valid when one takes free boundary conditions (F) at T and closed boundary conditions at T^* . The (+) and (-) boundary conditions give the same value to the even correlation functions and we call them closed boundary conditions. The free boundary conditions correspond to taking in the expression for the energy only the spin variables associated with points inside Λ, i.e. no interaction with spins outside Λ. If one examines the duality transformation when an arbitrary boundary condition is given one finds that the expressions for the dual correlation functions appear as a linear combination of the correlation functions of the dual system with different boundary conditions, the coefficients being +1 or -1. One cannot say that they correspond to a real state of the dual system. The reason why this is so will become clear when we consider the three dimensional Ising model where non invariant states exist at low temperature. By duality they will give non invariant correlations which cannot exist at high temperature. This is the main difficulty which appears in the problem we are considering. The idea to overcome this difficulty is to restrict our treatment to the free and closed boundary conditions, and to study the invariant equilibrium states through the free energy of perturbed systems, the free energy having the advantage of being independent of the boundary conditions.

Let us introduce a system S_1 in which the hamiltonian is given by

$$\beta H_\Lambda^{(1)} = \beta H_\Lambda + \lambda \sum_{x \in \Lambda_1} \sigma_{M+x}$$

where H_Λ is the Ising hamiltonian at $h = 0$, M is a given set of bonds of the lattice, $M+x$ its translate by a vector associated to a point of a of the lattice, Λ_1 the set of all a such that the bonds of the set $M+a$ are contained in Λ and λ is a real coefficient. We denote by $p_1(\beta, \lambda)$ the free energy of the system S_1.

We next introduce a system S_2 in which the hamiltonian is given by

$$\beta H_\Lambda^{(2)} = \beta H_\Lambda + \lambda \sum_{x \in \Lambda_1} \tau_{M+x}$$

and call $p_2(\beta, \lambda)$ the pressure of the system S_2. The derivative of $p_2(\beta, \lambda)$ with respect to λ at $\lambda = 0$ gives the dual expression of the correlation functions.

The system S_2 is not the dual of the system S_1, except in the case $\lambda = 0$ in which both systems reduce to the Ising model. However, the duality relation between the correlation functions suggest to prove that

$$\lim_{\lambda \to 0} \left| \frac{p_1(\beta, \lambda) - p_1(\beta, 0)}{\lambda} - \frac{p_2(\beta^*, \lambda) - p_2(\beta^*, 0)}{\lambda} \right| = 0$$

This would prove that the derivative at $\lambda = 0$ of $p_1(\beta, \lambda)$ exists if, and only if, the derivative of $p_2(\beta, \lambda)$ exists. As the derivative of $p_2(\beta^*, \lambda)$ exists when $\beta^* < \beta_c$, it would follow that $p_1(\beta, \lambda)$ has also a derivative at $\lambda = 0$ for all $\beta < \beta_c$. And this would imply the uniqueness of the translationally invariant correlation functions $\rho(\sigma_M)$ below the critical temperature.

At this point it is natural to think that $p_2(\beta^*, \lambda)$ is an approximation to the pressure of the dual of the system S_1 having the same derivative at the point $\lambda = 0$ and allowing us to justify the formula above.

We shall denote by S_3 the dual of the system S_1. Let us now compute the corresponding partition function. We start with the partition function of S_1

$$
\begin{aligned}
Z_\Lambda^{(1)}(\beta, \lambda) &= \sum_\sigma e^{-\beta H_\Lambda(\sigma)} e^{-\lambda \sum_{x \in \Lambda_1} \sigma_{M+x}} \\
&= Z_\Lambda^F(\beta)\, \rho_\Lambda^F\left(e^{-\lambda \sum_{a \in \Lambda_1} \sigma_{M+x}}\right) \\
&= Z_\Lambda(\beta)\, \rho_\Lambda\left(\prod_{x \in \Lambda_1} \{1 + \tau \sigma_{M+x}\}\right) \\
&= Z_\Lambda(\beta)\, \rho_\Lambda\left(\sum_{S \subset \Lambda_1} \tau^{|S|} \sigma_{M(S)}\right) \\
&= (\mathrm{ch}\,\lambda)^{|\Lambda_1|} K_{\Lambda_1} Z_\Lambda^+(\beta^*)\, \rho_{\Lambda\,\beta^*}^+\left(\sum_{S \subset \Lambda_1} \tau^{|S|} \mu_{M(S)}\right)
\end{aligned}
$$

where we have represented by $M(S)$ the product of the sets of bonds $M+x$, for all $x \in S$ in the sense of the symmetric difference of ensembles, and $\tau = \mathrm{th}\,\lambda$. We get then as the partition function of the system S_3

$$
\begin{aligned}
Z^{(3)}(\beta^*, \lambda) &= \exp -\beta H_\Lambda(\sigma^\Lambda) \sum_{S \subset \Lambda_1} \tau^{|S|} \mu_{M(S)} \\
&= \exp -\beta H_\Lambda(\sigma^\Lambda) \sum_{S \subset \Lambda_1} \tau^{|S|} \exp\{-\beta J \sum_{a \in M(S)} \sigma_a\}
\end{aligned}
$$

which is related by duality to $Z_\Lambda^{(1)}(\beta, \lambda)$.

Here appears an auxiliary system S_4 whose partition function is given by

$$
Z^{(4)}(\tau, \sigma^\Lambda) = \sum_{S \subset \Lambda_1} \tau^{|S|} \exp -\beta J \sum_{a \in M(S)} \sigma_a
$$

for which τ represents the activity, and $U(S) = \beta J \sum_{a \in M(S)} \sigma_a$

represents the interaction energy, which depends on the configuration σ^Λ appearing in the Ising hamiltonian. Let $\rho_{\tau,\sigma^\Lambda,\Lambda_1}(a)$ be the density at the point a of such a system, then

$$Z^{(4)}(\tau,\sigma^\Lambda) = \exp\left\{\sum_{a\in\Lambda_1}\int_0^\tau \frac{dt}{t}\rho_{t,\sigma^\Lambda,\Lambda_1}(a)\right\}$$

This tells us that the energy of the system S_3 should be

$$\beta H^{(3)}_\Lambda(\sigma^\Lambda) = \beta H_\Lambda + \sum_{a\in\Lambda_1}\int_0^\tau \frac{dt}{t}\rho_{t,\sigma^\Lambda}(a)$$

where $\rho_{t,\sigma^\Lambda}(a)$ is the infinite volume density of S_4. It also gives the form of the perturbative term of system S_3.

The proof continues by justifying the exactness of this perturbative term, and that it appoaches the perturbative term of system S_2 in such a way that the equality of the derivatives of $p_1(\beta,\lambda)$ and $p_2(\beta,\lambda)$ at $\lambda = 0$ follows.

Let us only mention here that this can be proved by using the results about the unicity and analyticity of the correlation functions at small activity, as we are interested only in the region $\tau = \mathrm{th}\lambda \to 0$, and τ is the activity of the system S_4. The necessary results were proved in ref. [26] and it can be justified that the hypothesis which are needed in ref. [26] hold for the system S_4. In this way one arrives at a complete proof of proposition (c).

Interesting properties of the pure phases ρ^+ and ρ^- have been pointed out by several authors after a first study by Minlos and Sinai [27] . Some of them are discussed in Gallavotti's lectures and in ref. [18] . Let us also mention the recent work by Coniglio, Nappi, Peruggi and Russo [28] in which the typical configurations appearing in the pure phases are discussed in terms of the percolation properties of these states.

VIII. Sharpness of the Interface Surface

Let us now say something about other possible states of the system. Non translational invariant Gibbs states can appear, even if the interaction is translation invariant, due to several physical reasons. They can correspond to correlation functions having some periodic structure, as in the antiferromagnetic Ising model. In this case they can be obtained as a decomposition into extremal Gibbs states of the invariant equilibrium states. They can also correspond to the presence of separation lines between the different phases of the system when several phases coexist. Then the correlation functions have different values in the different regions of the space.

The first class of states can be excluded in our case, because, as we have seen, $\rho^{+}(\sigma_{x_1} \cdots \sigma_{x_n})$ and $\rho^{-}(\sigma_{x_1} \cdots \sigma_{x_n})$ are extremal in the set of all Gibbs states at the same T and h.

The second class of states have been proved to exist for the Ising model in three dimensions at low temperature. An equivalent statement is obtained by saying that the surface of separation between the two phases is rigid in three dimensions. The physical reason for this phenomenon is that it costs too much energy to deform a two-dimensional surface. A proof of this fact has been given by Dobrushin [29] and it rests on contour techniques and an elaboration of Peierls argument to make rigorous this physical idea. Another simpler proof based on inequalities has been provided by van Beijeren [30] and gives at the same time some interesting information which will be commented on in the following paragraph. The interest of the first proof lies in the generality of the method, and for instance in the antiferromagnetic case would show the appearance of the crystaline (in the lattice sense) domains when there is a symmetry breaking.

Let us first say what are the boundary conditions considered in the works just mentioned. Let Λ be a cube centered at the origin and say that $x \geq 0$ to indicate that the point x in the lattice Λ has a non-negative vertical componenet, and that $x < 0$ when its vertical component is negative. Put σ_x on the boundary equal to +1 for all points on the boundary such that $x \geq 0$ and $\sigma_x = -1$ for all points on the boundary such that $x < 0$. This boundary condition will be denoted by (+ -) and the corresponding Gibbs state will be denoted by the subscript 3 to indicate that we are dealing with a three-dimensional Ising ferromagnet. Let $m_2(T)$ be the spontaneous magnetization of the two-dimensional Ising ferromagnet at the same temperature T. The van Beijeren result is the following:

$$\rho^{(+\,-)}_{T,3}(\sigma_x) \geq m_2(T) \qquad \text{if } x \geq 0$$

$$\rho^{(+\,-)}_{T,3}(\sigma_x) \leq -m_2(T) \qquad \text{if } x < 0$$

This shows that such states appear in the three-dimensional Ising model for any temperature below the critical temperature $T_c^{(2)}$ of the two-dimensional model. Similar results hold for higher dimensions.

Numerical calculations due to Weeks, Gilmer and Leamy [31] indicate that such states are not present above a certain temperature $T_R^{(3)}$ (located above 0.57 $T_c^{(3)}$, where $T_c^{(3)}$ is the critical temperature of the 3-dimensional model), although between $T_R^{(3)}$ and $T_c^{(3)}$ there is still a phase transition in the system. At the moment there is no definitive answer to this problem.

IX. Interface Profile in Two Dimensions

In the two dimensional Ising ferromagnet Gallavotti [32] and Abraham and Reed [33,34] have found some evidence on the absence of such non-invariant states. They consider again, in the two-dimensional case, the boundary conditions (+ -) that we have described before. Gallavotti's discussion is based on general methods while the results of Abraham and Reed follow from explicit calculations in the two-dimensional Ising model. A consequence of the first mentioned work is that the states $\rho^{(+ -)}(\sigma_{x_1} \ldots \sigma_{x_n})$ are translation invariant if the temperature is sufficiently low $T \leq T_0 < T_c$.

Some information about how rigid the interface line of separation should be is also obtained in these works. Let us state in connection with this question, the main theorem of Abraham and Reed. For this, let us introduce the notation σ_p to indicate the spin at a point with vertical coordinate equal to p and any finite horizontal coordinate, and let L be the side of the square Λ. The conclusion, valid for all values of T, is then the following:

Let $p = \alpha L^{\delta}$, $\delta \geqslant 0$; then

$$\lim_{L\to\infty} \rho_L^{(+ -)}(\sigma_p) = 0 \qquad \text{if } \delta < \tfrac{1}{2}$$

$$= m(T)\, \mathrm{sgn}\,\alpha \qquad \text{if } \delta > \tfrac{1}{2}$$

$$= m(T)\, \mathrm{sgn}\,\alpha\, \Phi(b|\alpha|) \text{ if } \delta = \tfrac{1}{2}$$

where $\mathrm{sgn}\,\alpha$ denotes the sign of α ,

$$b = [\sinh\ 2(\beta J - \beta^* J)\,]^{\frac{1}{2}}$$

and

$$\Phi(x) = \frac{2}{\sqrt{\pi}} \int_0^x e^{-u^2} du \quad .$$

We see therefore that the interface line makes large oscillations in such a way that the magnetization is zero in the middle of the square. These oscillations have an amplitude of the order of $\sqrt{L}$.

What happens for general boundary conditions and arbitrary correlation functions is not known, it is believed however that all Gibbs states of the two-dimensional Ising ferromagnet are translation invariant.

REFERENCES

1) G. GALLAVOTTI, Instabilities and Phase Transitions in the Ising Model. A Review. Revista Nuovo Cim. 2-2, 133 (1972).

2) R.L. DOBRUSHIN, The Description of a Random Field by Means of Conditional Probabilities. L. Funct. Anal. Appl. 2, 291 (1968).

3) O.E. LANFORD, D. RUELLE, Observables at Infinity and States with Short Range Correlations in Statistical Mechanics. Comm. Math. Phys. 13, 194 (1969).

4) D. RUELLE, Statistical Mechanics. Rigorous Results. Benjamin, New York, 1969.

5) O.E. LANFORD, in Statistical Mechanics and Mathematical Problems. A. LENARD, Editor. Springer-Verlag, Berlin, 1973.

6) C.M. FORTUIN, P.W. KASTELEYN, J. GINIBRE, Correlation Inequalities of Some Partially Ordered Sets. Comm. Math. Phys. 22, 89 (1971).

7) R.B. GRIFFITHS, Rigorous Results and Theorems. In Phase Transitions and Critical Phenomena, C.DOMB, M.S. GREEN, Editors, Vol.1 Academic Press, London, 1972.

8) T.D. LEE, C.N. YANG, Statistical Theory of Equations of State and Phase Transitions. Phys. Rev. 87, 410 (1952).

9) D. RUELLE, Extension of the Lee-Yang Circle Theorem. Phys. Rev. Letters, 26, 303,(1971).

10) D. RUELLE, On the Use of "Small External Fields" in the Problem of Symmetry Breakdown in Statistical Mechanics. Ann. of Phys. 69, 364 (1972).

11) J.L. LEBOWITZ, A. MARTIN-LOF, On the Uniqueness of the Equilibrium States for Ising Spin Systems. Comm. Math. Phys. 25, 276 (1972).

12) J.L. LEBOWITZ, Bounds on the Correlations and Analyticity Properties of Ferromagnetic Ising Spin Systems. Comm. Math. Phys. 28, 313 (1972).

13) R.B. GRIFFITHS, C.A. HURST, S. SHERMAN, J. Math. Phys. 11, 790, (1970).

14) G. GALLAVOTTI, J.L. LEBOWITZ, Some Remarks on Ising-Spin Systems. Physica, 70, 219 (1973).

15) G. BENETTIN, G.GALLAVOTTI, G. JONA-LASINIO, A.L. STELLA, On the Onsager Value of the Spontaneous Magnetization. Comm. Math. Phys. 30, 45 (1973).

16) B. MC COY, T. WU, The Two-Dimensional Ising Model. Harvard University Press, Cambridge, Mass. 1973.

17) G. GALLAVOTTI, S. MIRACLE-SOLE, Equilibrium States of the Ising Model in the Two Phase Region. Phys. Rev. 5B, 2555 (1972).

18) G. GALLAVOTTI, A. MARTIN-LOF, S. MIRACLE-SOLE, in Statistical Mechanics and Mathematical Problems. A. LENARD, Editor, Springer-Verlag, Berlin (1973).

19) R. PEIERLS, Proc. Cambridge Phil. Soc. 32, 477 (1936).

20) R.L. DOBRUSHIN, Funct. Anal. and Appl. 2,31 (1968).

21) F. DI LIBERTO, Comm. Math. Phys. 29

22) S.A. PIGOROV, Y.G. SINAI, To appear.

23) S.A. PIGOROV, Y.G. SINAI, Phase Transitions of the First Kind for Small Perturbations of the Ising Model. Funct. Anal. and Appl. 8, 21 (1974).

24) A. MESSAGER, S. MIRACLE-SOLE, Equilibrium States in the Two-Dimensional Ising Model in the Two Phase Region. Comm. Math. Phys. 40, 187 (1975).

25) A. MESSAGER, S. MIRACLE-SOLE, in Lectures on Mathematical Physics, G. GALLAVOTTI, Editor. University of Camerino, 1976.

26) G. GALLAVOTTI, S. MIRACLE-SOLE, Correlation Functions of a Lattice Gas. Comm. Math. Phys. 7, 274 (1968).

27) R.L. MINLOS, Y.G. SINAI, Trans. Moscow Math. Soc. 19, 121 (1968), Math. USSR Sbornik, 2, 335 (1967).

28) A. CONIGLIO, C. NAPPI, F. PERUGGI, L. RUSSO, Percolation Problems in the Two-Dimensional Ising Model. University of Naples, Preprint.

29) R.L. DOBRUSHIN. Gibbsian States Describing the Coexistence of Phases in the Three-Dimensional Ising Model. Theory Probab. and Appl. 17, 619 (1972).

30) H.VAN BEIJEREN, Interface Sharpness in the Ising System. Comm. Math. Phys. 40, 1 (1975).

31) J.D. WECKS, G.H. GILMER, H.J. LEAMY, Structural Transition in the Ising Model Interface. Phys. Rev. Letters, 31, 549 (1973).

32) G. GALLAVOTTI, Comm. Math. Phys. 27, 103 (1972).

33) D.B. ABRAHAM, P. REED, Phase Separation in the Two-Dimensional Ising Ferromagnet. Phys. Rev. Letters, 33, 377 (1974).

34) D.B. ABRAHAM, P. REED, Interface Profile of the Ising Ferromagnet in Two Dimensions. 3e Cycle de Physique en Suisse Romande, Cully, 1975.

STATISTICAL MECHANICS OF EQUILIBRIUM SYSTEMS: SOME RIGOROUS RESULTS

JOEL L. LEBOWITZ

Belfer Graduate School of Science
Yeshiva University
New York, N.Y.

Supported in part by NSF Grant and by AFOSR Grant

STATISTICAL MECHANICS OF EQUILIBRIUM SYSTEMS: SOME RIGOROUS RESULTS

JOEL L. LEBOWITZ

Belfer Graduate School of Science
NEW YORK, N.Y.

I. Introduction

Statistical Mechanics relates the properties of macroscopic objects to the properties of the microscopic atoms making up these objects. In many situations of physical interest the description of the microscopic elements may be taken to be given by classical mechanics. This means that a system of N point particles in ν-dimensions will be described microscopically by a point $\underline{x} = (\underline{q}, \underline{p})$ in a $2\nu N$ dimensional phase space where $\underline{q} = (q_1, \ldots, q_N)$, $\underline{p} = (p_1, \ldots, p_N)$, q_i, p_i are ν-dimensional vectors representing the position and momentum of the i-th particle.

The characteristic size of macroscopic objects is very large on a microscopic scale, containing typically $N \sim 10^{23}$ atoms, and their properties and mode of description are correspondingly very different from those of objects containing only a few particles. In particular macroscopic objects can be in "equilibrium" when nothing apparently changes macroscopically even though the atoms are moving around wildly. When in equilibrium macroscopic objects have a characterization in terms of a small number of parameters, such as temperature, density, ..., certainly much smaller than the number of parameters, $2\nu N$, necessary to characterize the microscopic state of the system. (Even when not in equilibrium the appropriate variables are quite different than those used for the microscopic dynamical descriptions but I shall not have much to say about that in these lectures). The way statistical mechanics bridges the gap between the microscopic and macroscopic description is to identify macroscopic states with probability measures on the phase space of the system and the values of macroscopic variables with expectation values of appropriate functions $f(\underline{x})$, $\underline{x} = (x_i)$, with respect to such measures. It is the measures which are characterized by a few parameters.

Since as I mentioned earlier the characteristic feature of macroscopic systems is that they are very large, their characteristics are best seen when one considers formally infinite systems. Large finite systems behave in a quantitatively similar manner if one asks the right questions. This is an essential basic ingredient in thermodynamics where the description is entirely macroscopic.

For a simple one component system containing n-moles (or $N = nN_0$ particles) in a volume V and having a total energy U, thermodynamics postulates the existence of an entropy function $S(U,N,V)$ (or for mixtures containing k species, $S(U,N_1, \ldots, N_k,V)$) from which all the equilibrium properties of the system can be derived. This function has the following properties:

i) $S(U,N,V) = V\, s(u,\rho), \quad \rho = \frac{N}{V}, \quad u = U/V$ (1.1)

ii) $s(u,\rho)$ is jointly concave in u and ρ.

Clearly "surface and shape effects" are ignored here and only "intensive variables", ρ and u, are considered relevant for the description of the bulk properties of a macroscopic system. The pressure p, temperature T are obtained from $s(u,\rho)$ by taking appropriate derivatives.

Thermodynamics gives no a priori way of computing the function S for any system. Equilibrium statistical mechanics gives, among other things, such a prescription through the use of the microcanonical ensemble which is the appropriate one for a system with fixed u and ρ. Instead of discussing this ensemble, however, it is more convenient (easier) to consider the thermodynamically entirely equivalent situation in which the macroscopic system has a fixed T, N and V. Its full thermodynamic description is then given by the Helmholtz free energy function

$$F(T,N,V) = U(T,N,V) - TS(U,N,V) \tag{1.2}$$

where U is obtained as a function of T,N,V by "inverting" the relation

$$\frac{1}{T} = \beta = \frac{\partial S(U,N,V)}{\partial U} = \frac{\partial s(u,\rho)}{\partial u} > 0 \,. \tag{1.3}$$

More precisely we again have

$$F(T,N,V) = V\, f(\beta, \rho), \qquad \beta \equiv T^{-1} \tag{1.4}$$

and

$$f(\beta, \rho) = \operatorname*{Min}_{s} \left[u(s, \rho) - Ts\right] \tag{1.5}$$

where $u(s, \rho)$, the energy per unit volume, is uniquely related to $s(u, \rho)$ by (1.3) at all $T > 0$. The pressure p, energy density u, and chemical potential μ, of the system are given by

$$p = -\frac{\partial F}{\partial V} = \rho^2 \frac{\partial}{\partial \rho} [f(\beta, \rho)/\rho] , \qquad (1.6)$$

$$u = \frac{\partial}{\partial \beta} [\beta f] , \quad \mu = \frac{\partial f}{\partial \rho} .$$

It follows from the definition (1.5) and the concavity of $s(u,\rho)$ that $f(\beta, \rho)$ is convex in ρ and concave in β, i.e.

$$f(\alpha \rho' + (1-\alpha)\rho'') \leq \alpha f(\rho') + (1-\alpha)f(\rho''), \quad 0 \leq \alpha \leq 1, \qquad (1.7)$$

where the dependence of f on β has not been indicated explicitly. The convexity in ρ implies that the pressure p is monotone non-decreasing as a function of ρ, while the concavity in β implies that the specific heat, $\partial u/\partial T$, is non-negative.

Another equivalent thermodynamic formulation is to consider the temperature T, and chemical potential μ, as the independent variables in which case the appropriate thermodynamic potential is the pressure

$$p(T,\mu) = V^{-1} [U - TS + \mu N] = \operatorname*{Max}_{\rho} [\mu\rho - a(\beta,\rho)] . \qquad (1.8)$$

The corresponding statistical mechanical ensemble is the grand canonical one which we shall discuss later.

II. The Canonical Ensemble: Thermodynamic Limit of the Free Energy

The statistical mechanical ensemble, appropriate for a system with a fixed number of particles, volume and temperature is the canonical ensemble. This is generally described as being obtained by keeping the system in contact with a thermal reservoir at a fixed temperature or as the ensemble appropriate for a subsystem of a much larger system described by a microcanonical ensemble. It is specified, for a system with a Hamiltonian $H(\underline{x})$, confined to a region Λ with volume $V(\Lambda)$ by the ensemble density

$$\mu^c(\underline{x}) = Ce^{-\beta H(\underline{x})}, \qquad C = \int_{(\Lambda x R^\nu)^N} e^{-\beta H(\underline{x})} d\underline{x} = N!\, h^{\nu N}\, Z(\beta, N, \Lambda) \tag{2.1}$$

Here C is a normalization constant and h has been introduced to conform to the usual definition of the partition function Z. The connection with thermodynamics is given by

$$\begin{aligned} F(T,N,V) \Longleftrightarrow A(\beta,N,\Lambda) &= -\beta^{-1} \ln Z(\beta,N,\Lambda) \\ &= V(\Lambda)\, a(\beta, \frac{N}{V(\Lambda)}; \Lambda). \end{aligned} \tag{2.2}$$

The $\Longleftrightarrow$ means that as $\Lambda \to \infty$ with $\frac{N}{V(\Lambda)} \to \rho$,

$$a(\beta, \frac{N}{V(\Lambda)}; \Lambda) \to a(\beta, \rho) \equiv f(\beta, \rho).$$

Note that the connection between F and A becomes precise only in the thermodynamic limit, $\Lambda \to \infty$. It is thus really not even meaningful to say that the pressure is given by the derivative of $A(\beta, N, \Lambda)$ with respect to the volume $V(\Lambda)$, c.f. (1.6), without specifying the shape of Λ and how it is to be changed. Once the limit, $\Lambda \to \infty$, is taken however the problem disappears and the pressure can be identified unambiguously with $\rho^2 \partial\, [a(\beta, \rho)/\rho]\, / \partial \rho$.

It is thus seen that to relate statistical mechanics to (bulk) thermodynamics it is essential to take the limit $\Lambda \to \infty$. This is not because we are interested in infinite systems as such; we are not. It is simply the correct way to consider physical situations in which surface and shape effects are unimportant. As Griffiths has pointed out this is what an experimentalist does every time he reports a specific heat measurement per gram of some material without specifying the shape of the sample used in his experiments.

An important central question is therefore whether for reasonable Hamiltonians this so called thermodynamic limit actually exists independent of the shape of Λ (as long as it is reasonable) and furthermore does $f(\beta,\rho)$ so obtained have the right convexity properties: concave in β and convex in ρ. (These convexity properties ensure thermodynamic stability, i.e. we want the pressure p to be monotone non-decreasing in ρ for otherwise we would have that, in a system under constant pressure, any fluctuation in the density will be amplified.) Another question is whether the $f(\beta,\rho)$ obtained from the canonical ensemble is correctly related to $s(u,\rho)$ obtained from the microcanonical ensemble, i.e. as in (1.5).

It turns out that these questions can be answered in the positive for "all" physically relevant Hamiltonians leaving out relativistic and gravitational effects. The former because we don't know how to deal with them and the latter because thermodynamics, where the free energy (or entropy) is an extensive quantity (which expresses the fact that if you bring two systems together with the same temperature and density then the energy per unit volume of a particle is unchanged) does not hold when internal gravitational effects are significant; bringing together two suns will have a drastic effect on their internal structure. As Onsager phrased it: thermodynamics applies to systems which are large compared to the size of an atom but small compared to the size of the moon. This still leaves sizes of order $10^{20} \sim 10^{30}$ or so which is where our lives take place.

Let us consider then a system of N particles in a box $\Lambda \subset \mathbb{R}^\nu$ with a Hamiltonian

$$H_N(\underline{x}) = \sum p_i^2/2m + U_N(\underline{q}), \quad U_N(\underline{q}) = \sum_{i<j} v(q_i - q_j), \quad q_i \in \Lambda, \quad p_i \in \mathbb{R}^\nu \tag{2.3}$$

Note first that (for a classical system) the momentum integration is trivial

$$Z(\beta,N,\Lambda) = \frac{1}{N!h^{3N}} \int_{(\Lambda \times R^\nu)^N} e^{-\beta H} \, d\underline{q} \, d\underline{p} = \frac{\lambda^{-\nu N}}{N!} \int_{\Lambda^N} e^{-\beta U_N(\underline{q})} d\underline{q} \tag{2.4}$$

where $\lambda = (2\pi m/\beta h^2)^{\frac{1}{2}}$.

For the proof of the existence of the thermodynamic limit we require generally:

(I) H - stability: this gives a lower bound on the interactions

$$U_N(\underline{q}) \geq -BN \quad \text{for} \quad \underline{q} \in R^{N\nu}, \quad B \text{ independent of } N.$$

(II) Tempering: this provides an upper bound on the interactions between two groups of particles, N and N', which are separated by a distance greater than some r_0. Using an obvious notation, we require that

$$W_{N,N'}(\Lambda,\Lambda') = U_{N+N'}(\Lambda\cup\Lambda') - U_N(\Lambda) - U_{N'}(\Lambda') \leq w\frac{N\,N'}{r^{\nu+\epsilon}} \tag{2.5}$$

$$\text{for Dist.}(\Lambda,\Lambda') > r_0, \quad \epsilon > 0, \quad w \geq 0.$$

Strong Tempering means that $w = 0$, e.g. the potential $v(r) \leq 0$ for $r > r_0$.

For pair potentials to satisfy tempering it is sufficient that

$$v(r) \leq \frac{w}{r^{\nu+\epsilon}} \quad \text{for} \quad r > r_0 \;. \tag{2.6}$$

For pair potentials to satisfy stability it is sufficient that $v(r) = \Phi_+(r) + \Phi^+(r)$ where $\Phi_+(r) \geq 0$ and $\Phi^+(r)$ is of positive type

$$\Phi^+(r) = \int \widetilde{\Phi}_+(k)\, e^{i\underline{k}\cdot\underline{r}} dk; \quad \widetilde{\Phi}_+(k) \geq 0, \quad \int \Phi_+(k)\, d\underline{k} \leq 2B \;.$$

<u>Proof</u>:

$$\sum v(q_i - q_j) = \tfrac{1}{2}\int d\underline{k}\, \Phi_+(k)\, \left[(\sum e^{i\underline{k}\cdot q_i})^2 - N \right] + \sum \Phi_+(q_i - q_j) \geq -BN.$$

Potentials of Lennard-Jones type $(v(r) = 4\epsilon\left[(\frac{\sigma}{r})^{12} - (\frac{\sigma}{r})^{6}\right])$ are stable since they can be written in the above form. Quite generally any pair potential satisfying the conditions

$$v(r) \geq \frac{C'}{r^{\nu+\epsilon}}, \quad r < r_0, \quad |v(r)| \leq \frac{w}{r^{\nu+\epsilon}}, \quad r > r_0, \quad C' > 0,\ \epsilon > 0. \tag{2.7}$$

will satisfy both (I) and (II).

The requirement of H-stability immediately gives us a lower bound on $a(\beta, \rho : \Lambda)$. We have by (I)

$$Z(\beta, N, \Lambda) \leq \frac{\lambda^{-\nu N}}{N!} \left[V(\Lambda)\right]^N \exp\left[\beta B N\right], \tag{2.8}$$

and thus

$$a(\beta,\rho;\Lambda) \geqslant a_I(\beta,\rho) + \rho B \tag{2.9}$$

where $a_I(\beta,\rho)$ is the ideal gas, $(U_N(\underline{q}) = 0)$ free energy density,

$$a_I(\beta,\rho) = \beta^{-1}[\rho\,\ell n\,\rho + \rho\,\nu\,\ell n\,\lambda + \text{const.}].$$

The basic inequality used in proving the existence of the thermodynamic limit is the following. Let

$$\Lambda_1 \cap \Lambda_2 = 0, \quad \Lambda_1 \cup \Lambda_2 \subset \Lambda$$

and let $N = N_1 + N_2$. Then

$$Z(\beta,N,\Lambda) = \frac{\lambda^{-\nu N}}{N!}\int \exp\left[-\beta U_N(q_1, \ldots, q_N)\right] dq_1 \ldots dq_N$$

$$\geqslant \frac{\lambda^{-\nu(N_1+N_2)}}{N!}\,\frac{N!}{N_1!N_2!}\int dq_1 \ldots dq_{N_1}\exp[-\beta U_{N_1}]\,.$$

$$\int q_{N_1+1} \ldots dq_{N_1+N_2}\exp[-\beta U_{N_2}]\exp -\beta W_{N_1,N_2}(\Lambda_1\Lambda_2)$$

$$\geqslant Z(\beta,N_1,\Lambda_1)\,Z(\beta,N_2,\Lambda_2)\exp\left[-\beta W^{\text{Max}}_{N_1,N_2}(\Lambda_1,\Lambda_2)\right]. \tag{2.10}$$

In particular if the potential is strongly tempered and $d(\Lambda_1,\Lambda_2) \geqslant r_0$ then $Z(\beta,N,\Lambda) \geqslant Z(\beta,N_1,\Lambda_1)\;Z(\beta,N_2,\Lambda_2)$ or

$$a(\beta,\rho;\Lambda) \leqslant \frac{V(\Lambda_1)}{V(\Lambda)}\,a(\beta,\rho_1;\Lambda_1) + \frac{V(\Lambda_2)}{V(\Lambda)}\,a(\beta,\rho_2;\Lambda)$$
$$\rho = \frac{V(\Lambda_1)}{V(\Lambda)}\,\rho_1 + \frac{V(\Lambda_2)}{V(\Lambda)}\,\rho_2\,. \tag{2.11}$$

We shall consider now such strongly tempered potentials and prove the existence of the limit for a sequence of cubes Γ_j of sides $2^j L$ and particle numbers $N_j = \rho\,2^{j\nu} L^\nu$. For simplicity we let the particles be confined to an interior cube Γ_j', whose sides are a distance $\frac{1}{2}r_0$ from the sides of Γ_j so that 2^ν boxes Γ_j make

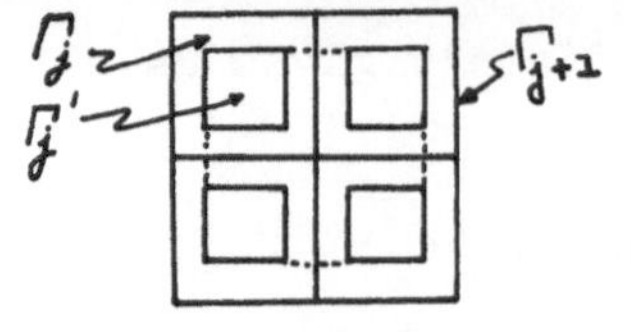

Fig. 1.

up one box Γ_{j+1} and the union of the 2^ν interior boxes Γ_j' are entirely inside Γ_{j+1}'; see illustration for $\nu = 2$.

We now have that

$$Z(\beta, N_j, \Gamma_j) \geqslant 2^\nu \; Z(\beta, N_{j-1}, \Gamma_{j-1}) \tag{2.12}$$

or

$$a(\beta, \rho; \Gamma_j) \leqslant a(\beta, \rho; \Gamma_{j-1}) \, . \tag{2.13}$$

Hence the $a(\beta, \rho; \Gamma_j)$ form a decreasing sequence which by (2.9) has a lower bound and so the limit exists

$$a(\beta, \rho; \Gamma_j) \longrightarrow a_\Gamma(\beta, \rho) \quad .$$

Let me conclude this part with a few remarks:

i) $a_\Gamma(\beta, \rho)$ is convex, by (2.11), in ρ and hence continuous for $\rho < \rho_{max}$ (the close packing density if there are hard cores.)

ii) Since $a(\beta, \rho; \Gamma_j)$ is concave in β, $a_\Gamma(\beta, \rho)$ is also concave in β.

iii) By packing general domain Λ_k with cubes Γ_j and using the continuity of $a_\Gamma(\beta, \rho)$ one establishes the existence of the limit $a(\beta, \rho) = a_\Gamma(\beta, \rho)$ for general domains Λ_j going to infinity in the sense of Van Hove; $V(\Lambda_j) \rightarrow \infty$, $V(\partial\Lambda_j; h)/V(\Lambda_j) \longrightarrow 0$, and $N_j/V_j \longrightarrow \rho$: $V(\partial\Lambda_j; h)$ is the volume of the region lying within a distance h of the boundary, $\partial\Lambda$, of Λ.

iv) From convexity $\frac{\partial a(\beta, \rho)}{\partial \rho}$ and hence the pressure exists except for a denumerable number of densities. For reasonable potentials such as Lennard-Jones the pressure can also be shown to be a continuous function of the density ρ, i.e. we do not have an "anti" first order phase transition in which the pressure has a jump discontinuity as the density is varied. This is not universal, however: it is possible to construct potentials (rather strange many body types) for which $a(\beta, \rho)$ exists but for which p does have a jump discontinuity.

v) The inequality obtained in (2.10) which bounds the interaction W by its maximum value is not the best possible one. It is indeed not good enough for obtaining the thermodynamic limit of neutral Coulomb systems in which the interaction falls off only as fast as 1/r. (The stability for these systems is a quantum phenomenon and requires that at least one of the species, positive or negative charges, be Fermions. Classically we can get stability if the "ions" also have hard cores.) What is required then is the observation that, using the first inequality in (2.10) we have

$$Z(\beta,N,\Lambda) \geq Z(\beta,N_1,\Lambda_1)Z(\beta,N_2,\Lambda_2) \langle \exp[-\beta W]\rangle_0$$
$$\geq Z(\beta,N_1,\Lambda_1)Z(\beta,N_2,\Lambda_2) \exp[-\beta\langle W\rangle_0] \quad (2.14)$$

The subscript zero here indicates that the average is to be taken with respect to an ensemble in which there is no interaction between Λ_1 and Λ_2 and the second inequality follows from the convexity of the exponential function.

vi) We have used a particular type of boundary condition, rigid hard walls, to define the canonical ensemble and the corresponding partition function and free energy. It is not hard to show that the limit $a(\beta,\rho)$ is quite insensitive to a variety of boundary conditions, e.g. we could have an external potential, $\sum \Psi_\Lambda(q_i)$, in the Hamiltonian with $\Psi_\Lambda(q)$ different from zero (but bounded) in the neighbourhood of the wall. For some types of boundary conditions, however, additional conditions on the pair potential may be necessary to ensure the same limiting $a(\beta,\rho)$. Thus for periodic boundary conditions it is necessary that the $v(r)$ fall off as $r^{-(\nu+1+\varepsilon)}$.

It is easier to study this problem in the case of lattice gases and spin systems. I will have more to say about this when discussing those systems, as will other speakers at this meeting.

Illustrative Examples

a) For a 1-dimensional system of hard rods of diameter d:

$$Z(\beta,N,L) = \frac{\lambda^{-N}}{N!}[L - Nd]^N$$

$$p(\beta,\rho) = \frac{kT\rho}{1-\rho d}, \quad \rho_{max} = d^{-1}.$$

b) Lattice Gas-noninteracting: $V(\Lambda)$ sites containing N particles, $N \leq V(\Lambda)$

$$Z(\beta,N,\Lambda) = V(\Lambda)!/[V(\Lambda) - N]!\,N!$$

$$a(\beta,\rho) = \beta^{-1}[\rho\ln\rho + (1-\rho)\ln(1-\rho)]$$

$$p(\beta,\rho) = -kT\ln(1-\rho).$$

III. Phase Transitions for Systems with Long Range Potentials

Phase transitions are generally considered to occur whenever $f(\beta, \rho)$ has some singular (non-analytic) behavior at some values of β and ρ. Very typical is the first order liquid-vapour phase transition with a critical point which occurs in all systems.

It was a great achievement of van der Waals to realize that the universal existence of a gas-liquid phase transition at low temperatures, terminating at some critical temperature T_c, can be understood qualitatively as arising in a simple way from some general features of the interaction between molecules. To this end van der Waals, in 1873, visualized the interaction potential between a pair of molecules a distance r apart, $v(r)$, as consisting of two separate additive parts, a short-range repulsive part $q(r)$ and a 'long range' attractive part $w(r)$,

$$v(r) = q(r) + w(r) . \tag{3.1}$$

The short-range part keeps the particles apart and is responsible for detailed correlations. The long-range part on the other hand sees only the gross macroscopic density profile of the fluid and is responsible for the condensation from the gas into the liquid below the critical temperature T_c.

van der Waals' considerations led him to his famous equation of state. For $T > T_c$ the van der Waals equation of state gives a good qualitative representation of the isotherms of a real fluid: for $T < T_c$, however, each isotherm includes a section where the compressibility is negative, in violation of the thermodynamic stability principle which, as we have seen, is also a direct consequence of statistical mechanics.

Maxwell showed that the co-existence region could be included in the theory by using the van der Waals equation of state for both liquid and vapour phases and using the thermodynamic equilibrium condition that the two phases must have equal pressures and chemical potentials.

A very interesting derivation of van der Waals' equation of state with Maxwell's rule was given by van Kampen in 1964. In this derivation the volume occupied by the system is divided into a large number of cells, each small compared with the range of the long-range attractive force, but large enough to contain many particles. Avoiding the pitfall of assuming a uniform distribution of particles over cells van Kampen obtained the distribution over cells by minimizing the free energy. His method leads to the Maxwell modified van der Waals equation of state, which implies a first-order phase transition.

Van Kampen's treatment, while containing the physics of the problem, was however not mathematically rigorous. In particular the conditions to be satisfied by the interactions were not specified and various limiting processes were only hinted at but not carried out explicitly.

An entirely different approach to the van der Waals' equation of state was taken by Kac, Uhlenbeck and Hemmer. Their work concerned a one-dimensional system for which the short-range repulsive potential $q(r)$ is infinite for $r < r_0$ and vanishes for $r > r_0$ (hard rods of diameter r_0) and an attractive interaction $w(r)$ which contained an inverse range parameter γ

$$w(r,\gamma) = -a\gamma e^{-\gamma r} \tag{3.2}$$

Using a formalism based on Wiener processes especially adapted to this problem which was first considered by Kac in 1959, Kac, Uhlenbeck and Hemmer were able to prove rigorously the validity of the van der Waals equation of state, together with the Maxwell rule, in the van der Waals limit $\gamma \longrightarrow 0$.

The limit process $\gamma \longrightarrow 0$, provides a clear distinction between the short range of $q(r)$ and the long (infinite as $\gamma \longrightarrow 0$) range of $w(r,\gamma)$. This limit was first used by Baker in 1961 in his study of spin systems.

Lebowitz and Penrose combined the ideas of van Kampen with the use of the 'van der Waals limit' $\gamma \longrightarrow 0$. They considered systems with interparticle potentials of the form (3.1) with $w(r)$, a Kac potential of the form

$$w(r,\gamma) = \gamma^{\nu}\varphi(\gamma r), \qquad \gamma > 0 \tag{3.3}$$

where ν is the dimensionality of the space considered. This reduces to (3.2) when $\nu = 1$ and $\varphi(x) = -ae^{-x}$. By imposing certain conditions on $q(r)$ (which have since been relaxed), and $\varphi(r)$, Lebowitz and Penrose showed that in the limit $\gamma \longrightarrow 0$, taken in such a way that the range γ^{-1} remains small compared to the size of the system,

$$\lim_{\gamma \to 0} p(\beta,\rho,\gamma) \equiv p(\beta,\rho,0+) = p^{0}(\beta,\rho) + \tfrac{1}{2}\alpha\rho^{2} + \text{Maxwell's rule}$$

$$\equiv \quad \mathrm{MC}\left\{p^{o}(\beta,\rho) + \tfrac{1}{2}\alpha\rho^{2}\right\} . \tag{3.4}$$

Here $p^{o}(\beta,\rho)$ is the pressure of the reference system, one for which $w(r,\gamma) = 0$ (which may itself be having a transition of its own) and

$$\alpha = \int w(r,\gamma)d\underline{r} = \int \varphi(x)dx . \tag{3.5}$$

The right side of (3.5) is a Riemann integral over all of ν-dimensional space, whose existence is one of the conditions $\varphi(x)$ has to satisfy. The extension of (3.4) to quantum systems was done by Lieb.

The results of Lebowitz and Penrose were generalized further by Gates and Penrose, who showed in particular that there are some Kac potentials $\varphi(x)$ for which (3.4) does not hold. These potentials are of an oscillatory type and apparently produce in the system, for some values of ρ and T, an oscillatory local density rather than a separation into only two phases. It is the latter situation which leads to (3.4).

It turns out using properties of convex functions that proving (3.4) is equivalent to proving that

$$\lim_{\gamma\to 0} a(\rho,\gamma) \equiv a(\rho,0) = \mathrm{CE}\left\{a^{o}(\rho) + \tfrac{1}{2}\alpha\rho^{2}\right\} \tag{3.5}$$

(where we do not show explicitly the dependence on β). Here $a^{o}(\rho)$ is the thermodynamic free energy of the reference system and CE - convex envelope - means essentially that the Gibbs double tangent construction is applied to the function $a^{o}(\rho) + \frac{1}{2}\alpha\rho^{2}$ (see Fig. 2) which is drawn for $\alpha < 0$.

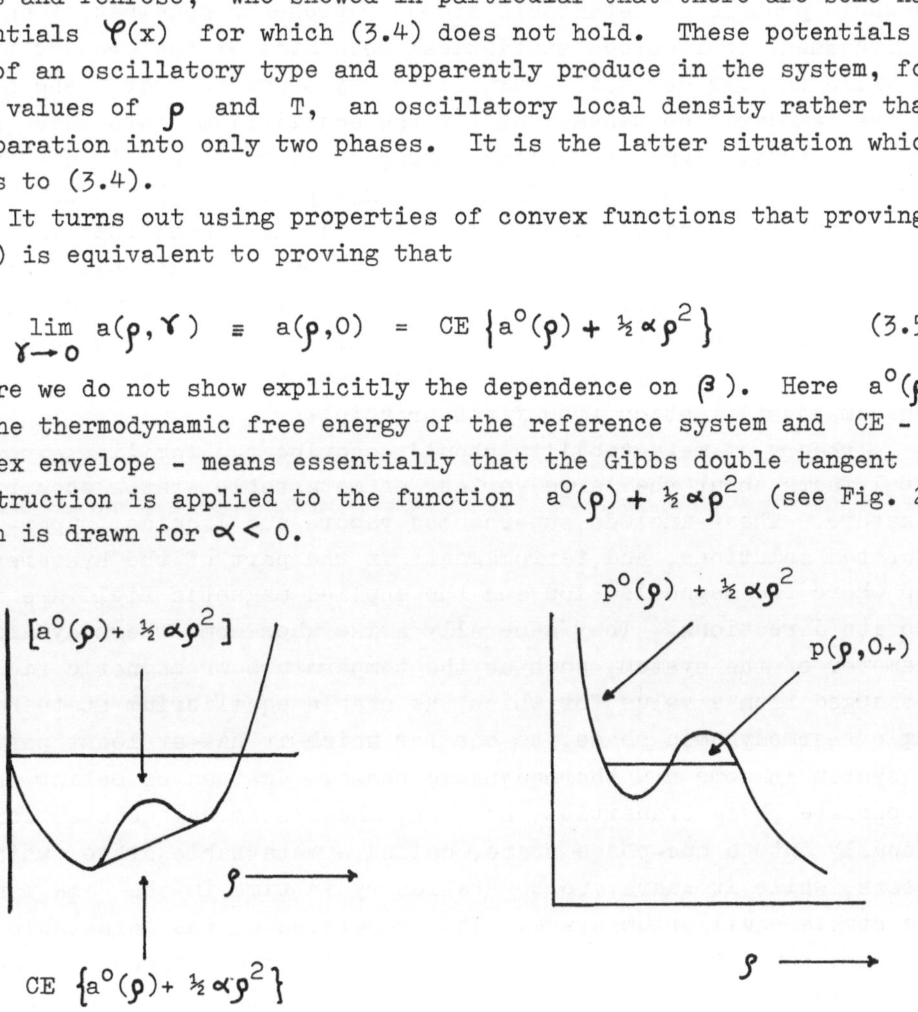

Fig. 2. The free energy and equation of state in the van der Waals limit $\gamma \to 0$.

More precisely the CE $\{\Psi(\rho)\}$, of any function $\Psi(\rho)$, is the largest convex function which is less or equal to $\Psi(\rho)$.

Metastability for Long Range Potentials

We have seen that a rigorous analysis of the equilibrium properties of a system with 'long range' attractive potentials yields below T_c a range of densities in which the free energy is given by the Gibbs double tangent construction, see Fig. 2. No meaning at all is given to the part of $a^o(o) + \frac{1}{2}\alpha\rho^2$ in Fig. 2 which lies above CE $\{a^o(o) + \frac{1}{2}\alpha\rho^2\}$, part of which, where $d^2 [a^o + \frac{1}{2}\alpha\rho^2]/d\rho^2 > 0$, was interpreted by Maxwell to represent metastable states. Indeed the whole problem of metastable states represents somewhat of an embarrassment to rigorous statistical mechanics at the present time. For while the van der Waals-Maxwell theory suggests that these states are the 'analytic continuations' of the equilibrium state there are many who argue, Langer and Fisher among them, that this is one of the qualitative features of the infinite range potential limit which does not persist for finite range potentials. It is argued that in first order phase transitions in real systems there is an essential singularity blocking analytic continuation. Even if this argument should turn out to be incorrect the question still remains of how to define (with or without analytic continuation) metastable states precisely, with some justification from first principles.

A theory of metastability should describe the familiar experimental facts about the large variety of metastable states occurring in nature. These include supercooled vapors and liquids, supersaturated solutions, and ferromagnets in the part of the hysteresis loop where the magnetization and the applied magnetic field are in opposite directions. They generally arise when some thermodynamic parameter of the system, such as the temperature or magnetic field, is changed from a value for which the stable equilibrium state has a single thermodynamic phase, to one for which it has at least part of the system in some new thermodynamic phase. Instead of making the appropriate phase transition, however, the system may go over continuously into a one-phase state, called a metastable state, which appears, while it lasts, to be stationary in time in the same manner as a stable equilibrium state. The properties of the metastable state are found to be reproducible; that is, they 'appear' to be completely determined by the values of the thermodynamic parameters, in just the same way as those of a stable equilibrium state. The distinguishing feature of a metastable state is that, eventually, either through some external disturbance or a spontaneous fluctuation which nucleates the missing phase, the system begins an irreversible process which leads it inexorably to the corresponding stable equilibrium state. Thermodynamically, the irreversibility of this transition corresponds to a decrease in free energy or an increase in entropy.

This indicates that we may characterize metastable thermodynamic states by the following properties:

Only one thermodynamic phase is present (a)

A system that starts in this state is likely to take a long time to get out (b)

Once the system has gotten out, it is unlikely to return (c)

One might add the statement that thermodynamics applies to the metastable state - for example, the usual theory would apply if a substance in such a state were taken around a Carnot cycle.

A complete theory of metastability must then describe both the static properties of these states as well as the dynamics of their persistence and decay. Some of the basic ideas underlying this dynamics are already contained in Maxwell's discussion of metastable states: Maxwell recognized the importance of nucleation; he saw that to set up the metastable state we must be sure that none of the new phase is present.

To develop a theory of metastable states we begin by making precise the notion, inherent in the previous discussion, of imposing a restriction on the system which keeps its density roughly uniform. In general, such a restriction may be represented by confining the configuration of the system to a suitable region R in configuration space. In order for this region to correspond to a metastable state, the restrictions defining it should correspond to the imposition of a roughly uniform density, in accordance with the criterion (a), and it should also have properties corresponding to the conditions (b) and (c) mentioned earlier: If the dynamical state is initially in R, it is unlikely to escape quickly, and once it has escaped, it is unlikely to return.

To compute the conditional probabilities implicit in (b) we shall, as is usually done in statistical mechanics, use the Gibbs ensemble made up by taking an equilibrium ensemble and selecting from it at some initial time, $t = 0$, all those systems whose configurations are in R. We call this ensemble a restricted equilibrium ensemble. The conditional probability $p(t)$ of the configuration being outside R at time t is then equal to the fraction of the members of this subensemble that are no longer in R at time t.

This treatment of metastable states hinges on finding a suitable region R in configuration space. The ideal choice would, perhaps, be the one minimizing the escape rate - that is, the probability per unit time for the configuration of the system to move out of R. We did not attempt the difficult task of optimizing the region R. Instead, we made our choice on physical grounds and showed that this choice leads to a very small escape rate (so that the minimum escape rate must be at least as small). The basic idea of our method is the following: We take the "cubical" region Ω of volume $V(\Omega) = |\Omega|$, in which our system containing N particles is confined and divide it up as before into M cubical cells ω_i; $i = 1, \ldots, M$, of equal volume $|\omega| = |\Omega|/M$.

Let n_i be the number of particles in ω_i, $\sum_{i=1}^{M} n_i = N$, $\rho_i = n_i/\omega$, $\frac{1}{M}\sum \rho_i = \rho$. We may consider now the space $\{\rho_k\}$. A point in this space corresponds to a specification of the average density of each cell. We define the region R in the configuration space by restricting the ρ_i to a certain region in the $\{\rho_k\}$-space. We choose in particular the constraints

$$\rho^- < \rho_i < \rho^+, \quad \text{where} \quad \rho^- < \rho < \rho^+ \tag{3.6}$$

such that in the whole interval $[\rho^-, \rho^+]$, $\partial^2[a^0 + \tfrac{1}{2}\alpha\rho^2]/\partial\rho^2 > 0$, no unstable part is included.

At time $t = 0$ we assume that our system is represented by a canonical ensemble restricted to R, i.e.

$$\mu(x, t=0) = \begin{cases} (h^{3N}\, N!\, Z_R)^{-1}\, e^{-\beta H(\underline{x})}, & \underline{x} \in R \\ 0, & \text{otherwise.} \end{cases} \tag{3.7}$$

$$Z_R = \frac{1}{N!\,h^{3N}} \int_R e^{-\beta H(\underline{x})}\, d\underline{x} .$$

We let $p(t) = 1 - \int_R \mu(x,t)dx$; $\frac{dp(t)}{dt}$ is the escape rate and it is readily shown that for all t, $\frac{dp(t)}{dt} \leqslant \left.\frac{dp(t)}{dt}\right|_{t=0} \equiv \lambda$.

We want to show that we can choose R in such a way that λ can be made arbitrary small in a certain limit even though the probability that the equilibrium system will be in R, Z_R/Z is vanishingly small. It turns out that this can be accomplished if we can show that the free energy for a given set of densities ρ_i, $A(\{\rho_i\}, \omega, M)$ has the property that

$$A(\{\rho'_k\}, \omega, M) - A(\{\rho\}, \omega, M) \geqslant |\omega|\,[C + o(1)] , \tag{3.8}$$

where $\{\rho'_k\}$ is a point on the boundary of $\{\rho_k\}$-space, as specified in (3.6), and C is some positive constant. Here o(1) means a quantity which goes to zero as $|\omega| \to \infty$ in the limit considered. When (3.8) holds, we show that,

$$\lambda \leqslant K \frac{|\Omega|}{\omega^{1/\nu}} e^{-\beta |\omega| [C + o(1)]} \tag{3.9}$$

where K is a constant which remains finite in the limit and ν is the dimensionality of the space considered.

It follows from (3.9) that the escape rate per unit volume $\lambda/|\Omega|$ will be very small when the different length scales in our problem satisfy the relations

$$r_0 << |\omega|^{1/\nu} << \gamma^{-1} << |\Omega|^{1/\nu} \tag{3.10}$$

where r_0 is the "range" of q(r) or the inter-particle spacing. Formally $\lambda/|\Omega|$ will go to zero when the last three quantities in (3.10) go to infinity keeping the right relations among themselves.

Remarks

i) The "constant" C involves $\partial^2[a^0 + \frac{1}{2}\alpha\rho^2]/\partial\rho^2$ and therefore goes to zero at the end of the metastable region.

ii) It should be emphasized that the key to our control over λ is based on the fact that $r_0 << |\Omega|^{1/\nu}$. This is crucial for introducing an intermediate size cell $|\omega|$ such that it contains very many particles and yet is small compared to the range of the part of the interaction which drives the phase transition. Since ω is large the densities $\{\rho_k\}$ are "macroscopic" variable with small "normal" fluctuations and so ρ_k will very likely be near its average value ρ, not near ρ^- or ρ^+. Yet because $|\omega|^{1/\nu} << \gamma^{-1}$ there is "no advantage" to the system (as far as lowering its free energy is concerned) in segregating into two phases inside each cell. What the system would like to do is have some regions, of sizes at least as large as γ^{-1}, at the liquid density and others at the vapour density. This however it cannot do because of our constraint. We do not know, at the present time, how to carry out such a program for real potentials, like the Lennard-Jones one.

IV. Grand Canonical Ensemble

The grand canonical ensemble is appropriate for systems open to both heat and particle exchange with reservoirs at temperature T and chemical potential μ ,

$$\mu^{g.c}(\underline{x}_N;\beta,\mu,\Lambda) = \frac{Z(\beta,N,\Lambda)}{\Xi(\beta,\mu,\Lambda)} e^{\beta\mu N} \mu^{c}(\underline{x}_N;\beta,N,\Lambda) \quad (4.1)$$

where

$$\Xi(\beta,z,\Lambda) = \sum e^{\beta\mu N} Z(\beta,N,\Lambda) = \sum z^N Q(\beta,N,\Lambda),$$

$$z = \lambda^{-\nu}\exp[\beta\mu], \quad (4.2)$$

$$Q(\beta,N,\Lambda) = \frac{1}{N!}\int_{\Lambda^N} \exp[-\beta U_N]\, dq_1 \ldots dq_N;$$

$$Q(\beta,N = 0,\Lambda) = 1.$$

We define the "pressure" $\pi(\beta,z;\Lambda)$ for fixed Λ by

$$\pi(\beta,z;\Lambda) = \beta^{-1}[\ln \Xi(\beta,z;\Lambda)]/V(\Lambda) \quad (4.3)$$

It is then possible to prove, under the same conditions of stability and temperedness described before, that in the limit, $\Lambda \to \infty$, with β and z fixed

$$\pi(\beta,z;\Lambda) \longrightarrow \pi(\beta,z) = p(T,\mu) \quad (4.4)$$

where $p(T,\mu)$ is given in terms of $a(\beta,\rho)$ by eq. (1.8). The average density $\langle N \rangle(\beta,z;\Lambda)/V(\Lambda)$ is given by

$$\rho(\beta,z;\Lambda) = z\frac{\partial}{\partial z}\pi(\beta,z;\Lambda) . \quad (4.5)$$

It is clear from (4.4) that the passage to the thermodynamic limit is "simpler" in the grand canonical ensemble formalism, where the basic variables, β and z, are kept fixed when $\Lambda \to \infty$ than in the canonical ensemble where N has to be adjusted too as the limit is taken. For this reason it is simpler to discuss phase transitions,; corresponding to nonanalyticities in $\pi(\beta,z)$, in this formalism which we shall now do. We begin by considering systems with hard cores, i.e. $v(r) = \infty$ for $r < d$. For such systems there is clearly a minimum number, $N_m(\Lambda)$, such that $Q(\beta, N, \Lambda) = 0$ for $N > N_m(\Lambda)$. Hence for such systems

$$\Xi(\beta, z; \Lambda) = \sum_{\alpha=1}^{N_m} (1 - z/z_\alpha) \tag{4.6}$$

$$\beta\pi(\beta, z; \Lambda) = V(\Lambda)^{-1} \sum \ell n(1 - z/z_\alpha) \tag{4.7}$$

$\pi(z;\Lambda)$ will thus be analytic in any region $D(\Lambda)$ of the complex z-plane in which there are no zeros of the grand partition function. It is then possible to show that given any domain D in the z-plane which is free of zeros of Ξ for all sufficiently large values of Λ and which contains part of the real axis, then $\lim_{\Lambda \to \infty} \pi(z;\Lambda) = \pi(z)$ exists and is analytic in D. For systems without hard cores a factorization similar to (4.6) is still possible, leaving the essential results unaffected. Use is made here of Vitali's theorem which may be stated for our purposes as follows: If a sequence of functions e.g. $\pi(z;\Lambda)$ is analytic and bounded in some region D, $|\pi(z;\Lambda)| < M$, and approaches a limit as $\Lambda \to \infty$ in part of D, e.g. on the segment of the real positive z-axis inside D, then $\pi(z;\Lambda)$ tends uniformly to a limit $\pi(z)$ inside D, and $\pi(z)$ is analytic in z. There is some arbitrariness in the imaginary part of (4.7) but this is unimportant.

It is thus clear that all thermodynamic functions will be completely regular in z and in some sense uninteresting, along any stretch of the real positive z-axis which does not contain a limit point of zeros of $\Xi(z;\Lambda)$ as $V \to \infty$. The nature of the limiting distribution of zeros of Ξ is therefore of obvious interest for this really determines $\pi(z)$. This has been investigated and solved by Lee and Yang, for lattice gases with attractive pair interactions (Ising ferromagnets) about which you have already heard from Miracle-Sole and about which I shall have more to say later. Before doing that however I shall first discuss some results about the absence of zeros for small values of z (and β) which are valid for both lattice and continuum systems with general potentials. These results establish the existence of a gas phase at low densities and high temperatures.

Starting with the grand partition function we may write

$$\beta\pi(\beta, z; \Lambda) = \sum_{\ell=1}^{\infty} b_\ell(\beta; \Lambda) z^\ell \tag{4.8}$$

$$\rho(\beta, z; \Lambda) = \sum \ell\, b_\ell(\beta; \Lambda) z^\ell \tag{4.9}$$

where

$$b_\ell(\beta : \Lambda) = \frac{-1}{\ell V(\Lambda)} \sum_\alpha [z_\alpha(\beta;\Lambda)]^{-1}$$

$$= \frac{1}{\ell! V(\Lambda)} \int_\Lambda \int dr_1 \ldots dr_\ell\, \mathcal{U}_\ell(r_1, \ldots r_\ell) \tag{4.10}$$

with $\mathcal{U}_\ell$ the well known Mayer cluster functions. It is clear from equations (4.6) and (4.7) that the series will converge for $|z| < R(\Lambda)$ where $R(\Lambda)$ is the distance from the origin of the z-plane to the nearest zero of $\Xi(\beta, z; \Lambda)$. It follows from our discussion in the last section that for $|z| < R(\infty) = \lim_{\Lambda\to\infty} R(\Lambda)$, $\pi(\beta, z)$ will also be analytic in z and that

$$\beta\pi(\beta, z) = \sum b_\ell(\beta) z^\ell, \text{ and } \rho(\beta, z) = \sum \ell\, b_\ell(\beta) z^\ell \qquad |z| < R(\infty) \tag{4.11}$$

with $b_\ell(\beta) = \lim_{\Lambda\to\infty} b_\ell(\beta; \Lambda)$. This implies particularly that there is no phase transition, as z is changed and β kept fixed, of any kind for $|z| < R(\infty)$ [$R(\Lambda)$ and $R(\infty)$ may depend, of course, on β] and hence that one is definitely in the gas phase for this range of z. It is, of course, not true that there has to be a phase transition when the fugacity is equal to $R(\infty)$. The radius of convergence of the series in (4.11), which we call R, is at least $R(\infty)$. The physical pressure $\pi(\beta, z)$ need not, however, coincide with that obtained from equation (4.11) for real positive $z > R(\infty)$.

The question now naturally arises of whether one can find $R(\infty)$ (or R) or at least prove that $R(\infty)$ is bounded below by some positive number. In principle R could be zero and there would be no range of values of z for which the Mayer expansion would be valid. (This is indeed believed to be the case for Coulombic forces and for nonequilibrium properties such as viscosity or heat conductivity.) The existence of a finite radius of convergence for the kind of systems we are considering was proven first by Groeneveld for non-negative interaction potentials $v(r) \geqslant 0$ and later by Ruelle and Penrose for general stable potentials. The result is

$$R \geqslant R(\infty) \geqslant R_0(\beta) \equiv [K \exp(2\beta B + 1)]^{-1} \tag{4.12}$$

where

$$K = \int |e^{-\beta v(r)} - 1|\, d\underline{r} \tag{4.13}$$

and B is the constant appearing in H-stability.

Penrose also showed that for systems with hard cores $R(\Lambda)$ has an upper bound,

$$R(\Lambda) \leqslant |N_m(\Lambda)/V(\Lambda)\ell b_\ell(\Lambda)|^{1/\ell} \tag{4.14}$$

(and a similar result for potentials without a hard core). A different sequence of upper bounds on R for $v(r) \geqslant 0$ was obtained by Groeneveld. Similar results hold for mixtures.

For systems with non-negative potentials Groeneveld showed also that the Mayer cluster functions $U_\ell(r_1, \ldots, r_\ell)$ alternate in sign $(-1)^\ell U_\ell \geqslant 0$ and that $|b_\ell(\Lambda)| \leqslant |b_\ell|$. It follows from this that for such potentials $R(\infty) = R$ and that $\pi(z)$ will have a singularity on the negative real z-axis for $z = -R$. For these potentials Groeneveld shows that $R \leqslant K^{-1}$ while equation (4.12) gives $R \geqslant (eK)^{-1}$.

As already mentioned, a breakdown in the convergence of the Mayer fugacity expansion need have nothing to do with any physical singularity which can occur only for real positive z. This may be illustrated by what happens for a lattice gas, when the only interaction between the particles is an infinite repulsion between two particles occupying the same lattice site, then, c.f. Section II,

$$\beta\pi(z:\Lambda) = \beta\pi(z) = \ell n(1 + z) \quad . \tag{4.15}$$

We see here that the fugacity series will diverge for $|z| = 1$ but $\pi(z)$ is analytic for all real positive z.

Elimination of z in Equations (4.8) and (4.9) leads to the virial expansions of the pressure

$$\beta p(\beta,\rho) = \rho\left[1 - \sum \tfrac{k}{k+1}\beta_k\rho^k\right] \tag{4.16}$$

where the β_k's are the irreducible Mayer cluster integrals. Using the results already obtained for the convergence of the fugacity expansion it is possible also to obtain a lower bound for the convergence of this series. Again the divergence of the virial expansion need not signify any physical singularity in the pressure, despite the fact that for the few commonly known explicit equations of state, e.g., the ideal lattice gas considered before where $\beta p = -\ell n(1 - \rho)$, the divergence of the virial expansion occurs at close packing density. Lebowitz and Penrose give indeed an explicit example of a system for which the virial expansion diverges due to a singularity in $p(\rho)$ on the imaginary ρ-axis.

<u>Correlation functions</u> The k-particle positional distribution function in a grand canonical ensemble is defined as

$$\rho_k(r_1, \ldots, r_k; \beta, z, \Lambda) = \sum_{n=0}^{\infty} \frac{z^{k+n}}{n!} \int_\Lambda \int dr_{k+1} \ldots dr_{k+n}$$

$$e^{-\beta U(r_1 \ldots r_{k+n})} / \Xi \tag{4.17}$$

The Ursell functions F_ℓ are defined in terms of ρ_k by the relations

$$F_1 = \rho_1, \quad F_2(r_1,r_2) = \rho_2(r_1,r_2) - \rho_1(r_1)\rho_1(r_2)$$
$$F_3(r_1,r_2,r_3) = \rho_3(r_1,r_2,r_3) - \rho_2(r_1,r_2)\rho_1(r_3) - \rho_2(r_1,r_3)\rho_1(r_2)$$
$$- \rho_2(r_2,r_3)\rho_1(r_1) + 2\rho_1(r_1)\rho_1(r_2)\rho_1(r_3), \text{ etc.} \tag{4.18}$$

It follows from the definition of the F's that

$$\int_\Lambda \int F_\ell dr_2 \ldots dr_\ell = z^{\ell-1} \frac{d^{\ell-1}}{dz^{\ell-1}} \rho_1(r_1 ; \beta, z, \Lambda) \tag{4.19}$$

where $\ell \geq 2$.

The Ursell functions have expansions in powers of the fugacity which are very similar to those for the pressure

$$F_k = \sum_{\ell=0}^{\infty} \frac{z^{k+\ell}}{\ell!} \int \underset{\Lambda}{\cdots} \int dr_{k+1} \ldots dr_{k+\ell}\, \mathcal{U}_{k+\ell} . \tag{4.20}$$

Ruelle and Penrose showed using the Kirkwood-Salsburg equations for the ρ_k, that the correlation functions (generic name for both distribution and Ursell functions) are analytic in z and approach well-defined limiting functions $F_k(r_1, \ldots, r_k; \beta, z)$ as $\Lambda \to \infty$ which are also analytic in z for $|z| < R_0(\beta)$. Similar results hold also for the reduced density matrices of quantum systems (Ginibre).

Ruelle showed further that for $|z| < R_0$ the Ursell functions have the cluster property

$$\int |F_k(r_1, \ldots r_k; \beta, z)| \, dr_2 \ldots dr_k < \infty \tag{4.21}$$

It follows that (4.19) holds also for the infinite volume limit functions.

In the case where the potential $v(r)$ has finite range, $v(r) = 0$ for r_0, (or decays exponentially) the convergence of the Mayer expansion (4.20) for $|z| < R_0$ actually implies that the F_k's cluster exponentially. That is, if we separate the arguments, $r_1, \ldots r_k$, into two groups with a minimum separation r then

$$|F_k| \leq C \; e^{-\mathcal{K}r} \quad . \tag{4.22}$$

Even stronger results can be obtained when the particles are separated into three or more groups.

As mentioned already these results are generally restricted to $|z| < R_o(\beta)$. In this case it is also easy to prove analyticity in real β. For $|z| > R_o$ there is no general way of proving analyticity of $\pi(\beta, z)$ or of the F_k's. Indeed even the existence of the thermodynamic limit of the correlation functions now requires a slightly stronger condition, called superstability, on the potential energy

$$U(q_1, \ldots q_N) \geq -BN + C\frac{N^2}{V(\Lambda)} \quad , \quad \text{for} \quad q_i \in \Lambda \quad . \tag{4.23}$$

Even with this condition however all one can hope to prove is the existence of the correlation functions for certain subsequences of domains $\Lambda_j \rightarrow \infty$, and certainly no independence of boundary conditions. This is so because for $z > R_o(\beta)$ we may be at a point β_o, z_o where the system is undergoing a first order phase transition and the correlation functions are then not expected to be unique. To clarify these matters it is best to consider spin systems, as was done in the lectures of Gallavotti and Miracle-Sole. I will therefore turn to these systems now.

V. Lattice Systems

I shall restrict myself now to the case of lattice systems. This includes first of all lattice gases which are entirely analogous to continuum systems (as far as their configurations go) once it is understood that the particle coordinate q takes on values in the lattice $\mathbb{Z}^{\nu}$ rather than in $\mathbb{R}^{\nu}$. All our previous considerations therefore apply directly to this system. In addition I shall also discuss spin systems on a lattice, the simplest of which, spin-½ Ising model, is formally isomorphic to the lattice gas. More general spin systems will also be considered.

As was pointed out by other lecturers, quite a lot of information is known about these systems particularly when the interaction between the spins is of a ferromagnetic nature. This information comes from various sources: Lee-Yang theorem, Onsager solution of two dimensional system, Peierls type argument, Griffiths inequalities and their generalization, transfer matrix and subharmonicity arguments, and most recently from some beautiful work by Froehlich, Simon and Spencer. These authors have introduced a new method for proving the existence of phase transitions in spin systems (the old and only other method being variations of the Peierls′ argument). Using this method they proved, for the first time, the existence of phase transitions in classical spin systems with a continuous symmetry (e.g. classical Heisenberg model) whenever the dimension $\nu \geq 3$. These results have been generalized (non-trivially) to quantum spins by Dyson, Lieb and Simon.

This large store of information about lattice systems can be used to begin answering what to me is one of the interesting questions in statistical mechanics: the relationship between the macroscopic and microscopic properties of equilibrium systems. By macroscopic properties I mean those which can be obtained from the free energy of the system, $a(\beta,\rho)$ or $\Pi(\beta,z)$, considered as a function of the "external" parameters β , ρ, μ in a fluid or the corresponding spin quantities where the natural variables are the magnetic field h and β or magnetization m and β . By microscopic properties I mean the correlation functions considered also as a function of the same variables. We expect intuitively, based on experience, that there are "regular" regions in the thermodynamic phase space, e.g. in the h-T plane, where the system exists in a single phase and where the dependence of the free energy on h and T is analytic. These analytic regions are separated by smooth transition surfaces of non-analytic behavior. On these transition surfaces the system may co-exist in two or more phases. We expect that the correlation functions have corresponding behavior, nice analyticity and clustering properties in the one phase regions, non-analyticity, long range correlations and non-uniqueness in the transition region.

This program is almost completed for the two dimensional, nearest neighbour, ferromagnetic, Ising spin-½ system (see lecture of Miracle-Sole) and there are many results along these lines for spin-½ ferro-magnetic Ising systems. Some work with E. Presutti extends some of these results to unbounded one-component ferromagnetic spin systems (used in field theory). The extension is not entirely trivial and required the development of the general statistical mechanics of such unbounded spins. I shall later give a brief description of this work but will begin with a general formulation which includes the old fashioned Ising model-lattice gas system.

General Formulation

Consider for definiteness a simple cubical ν-dimensional lattice $\mathbb{Z}^\nu$, at each site of which there is a vector 'spin' variable S_i, $i \in \mathbb{Z}^\nu$, $S_i \in \mathbb{R}^n$. Each S_i has associated with it an intrinsic, or free, probability distribution, (the same for all S_i) $\rho(S)dS$: $\rho(S)$ depends only on the magnitude of S, $|S|$, and satisfies the bound

$$\int e^{bS^2} \rho(S)dS < \infty \quad , \qquad \text{all } b < b_o \qquad (5.1)$$

where, unless otherwise specified, $b_o = \infty$. When $\rho(S) = K\,\delta(|S| - 1)$, K a normalization constant, we have, for n = 1,2,3, the Ising spin-½, the rigid rotator and the classical Heisenberg models respectively. Typical examples of unbounded distributions are the "polynomial" measures much used in field theory

$$\rho(S) = \exp\left[-V_j(S)\right] \quad \text{where } V_j \text{ is an even polynomial in } |S|$$

$$j = 2n,$$

$$V_j(S) = a_o S^j + a_2 S^{j-2} + \ldots + a_{j-2} S^2 + a_j, \quad a_o > 0.$$ For $j \geq 4$, (5.1) is satisfied with $b_o = \infty$, while for $j = 2$, the

Gaussian case, $b_o = a_o$.

General bounded distributions are obtained if $\rho(S)$ is required to have compact support,

$$\rho(S) \quad = \quad 0 \quad \text{if} \quad |S| > C, \quad \text{a positive constant .} \qquad (5.2)$$

This includes of course the fixed length case considered earlier. (I shall sometimes use $\sigma_i = \pm 1$ to designate the Ising spin variable, the isomorphism with the lattice gas system is then obtained by setting $\rho_i = \frac{1}{2}(\sigma_i + 1) = (0, 1)$ indicating whether the site i is empty or occupied.)

The Hamiltonian of the system in a finite domain $\Lambda \subset \mathbb{Z}^\nu$ with 'boundary conditions' (b.c.) b_Λ corresponding to specifying the values of the spin variables outside Λ has, for pair interactions, the form

$$H(S_\Lambda \; ; b_\Lambda) \quad = \quad -\tfrac{1}{2} \sum_{i \neq j \in \Lambda} J(i-j)\, S_i\, S_j - \sum_{i \in \Lambda} h_i\, S_i,$$

$$S_\Lambda = \left\{ S_i ; \quad i \in \Lambda \right\} \qquad (5.3)$$

Here S_Λ stands for the set of spins in Λ and $J(i-j)$ in the most general case, is a symmetric $n \times n$ matrix, $J(i-j)\, S_i\, S_j = \sum_{\gamma\delta}^{n} J_{\gamma\delta}\,(i-j) S_i^\gamma\, S_j^\delta$ and if we assume that the interaction has a finite range, $J(x) = 0$ for $|x| > R$: h_i is the "magnetic field" at site i, in the simplest case $h_i = h + \sum_{j \notin \Lambda} J(i-j)\bar{S}_j$, $h_i S_i = \sum_{\gamma=1}^{n} h_i^\gamma\, S_i^\gamma$, where $\{\bar{S}_j\}$ is a specified set of values of S_i for $i \notin \Lambda$. These constitute the set of boundary conditions $\{b_\Lambda\} = b$, i.e. they specify the fields h_i acting on spins near the boundary of Λ. (More generally only the probability distribution of the spins outside Λ need be specified.) Commonly used b.c. in statistical mechanics are the 'zero' b.c., corresponding to setting $\bar{S}_i = 0$ (this is essentially the Dirichlet b.c. in field theory) and the periodic b.c. We can take the latter into account for Λ a "rectangle", by modifying the definition of $J(\underline{x})$. For bounded spins and $n = 1$, $\pm$ boundary conditions correspond to $\bar{S}_j = \pm 1$ for $j \notin \Lambda$.

In order to emphasize the fact that boundary conditions correspond to a specification of S_i outside Λ, i.e. S_i for $i \in \Lambda^c$ the complement of Λ, I shall sometimes write $H(S_\Lambda \,|\, S_{\Lambda^c})$ for $H(S_\Lambda : b_\Lambda)$.

The Gibbs probability distribution of spins in Λ at a temperature $T = \beta^{-1}$ is, for a given external field h and b.c. b_Λ,

$$\mu(dS_\Lambda, \beta, h; \Lambda, b_\Lambda) = Z^{-1} \exp[-\beta H(S_\Lambda, h, b_\Lambda)] \prod_{i\in\Lambda} \rho(S_i) dS_i$$
$$= \mu(dS_\Lambda \mid S_{\Lambda^c}) \qquad (5.4)$$

where

$$Z(\beta, h; \Lambda, b_\Lambda) \equiv \exp[|\Lambda| \Psi(\beta, h; \Lambda, b_\Lambda)]$$
$$= \int \exp[-\beta H] \prod_{i\in\Lambda} [\rho(S_i) dS_i] \qquad (5.5)$$
$$= Z(\Lambda \mid S_{\Lambda^c}) \quad .$$

Here $|\Lambda|$ equals the number of sites in Λ, and Ψ is (except for a factor $-\beta$) the free energy per site.

Let $\mu(dS_A, \beta, h; \Lambda, b_\Lambda)$ be the projection of the measure (5.4) onto $A \subset \Lambda$, $S_A = \{S_i;\ i \in A\}$. We are interested in the behavior of the $\mu(dS_A, \beta, h; \Lambda, b_\Lambda)$, or (essentially) equivalently the correlation functions, as $\Lambda \to \mathbb{Z}^\nu$ through some sequence of domains Λ_j. We are also interested in the properties of the free energy per site, $\Psi(\beta, h; \Lambda, b_\Lambda)$ in this limit: here we want to require that the Λ_j be reasonable, e.g. the fraction of sites within a distance R of $\partial\Lambda_j$ $|\partial_R \Lambda_j| / |\Lambda_j|$ go to zero as $j \to \infty$. We shall denote these limits by $\mu(dS_A, \beta, h; b)$ and $\Psi(\beta, h:b)$. Some of the questions which naturally arise are the existence, uniqueness, analyticity (in β and h), and cluster properties of these limits. Uniqueness refers both to different ways of letting $\Lambda \to \mathbb{Z}^\nu$ for given b.c. and to the dependence of the limit on b.c.

Thermodynamic Limit

Among the above questions the easiest one to tackle is the existence and uniqueness of the limit of $\Psi(\beta, h; \Lambda, b_\Lambda)$. It is easy to show, using standard methods, that when J(x) falls off at least as fast as $|x|^{-\nu+\varepsilon}$ then for bounded spins and any reasonable sequence of domain shapes Λ_j, $\Psi(\beta, h; \Lambda, b_\Lambda) \to \Psi(\beta, h)$ independent of b.c. . The problem becomes however more difficult in the case where the spins are not bounded and will be discussed later.

It is clear that at least for bounded spins the projected measures $\mu(dS_A; \Lambda, b_\Lambda)$ will always have subsequences which approach limits as $\Lambda \to \infty$. These states which are the infinite volume limit of Gibbs states with boundary conditions b will satisfy the Dobrushin, Lanford and Ruelle (DLR) equations (see lectures by Miracle-Sole),

$$\mu(dS_{\Lambda}:b) = \int \mu(dS_{\Lambda} \mid S_{\Lambda^c})\, \mu(dS_{\Lambda^c}:b) \quad . \qquad (5.6)$$

The converse statement; any state satisfying (5.6) may be obtained as the limit of finite volume Gibbs states given by Eq. (5.4) with suitably general boundary conditions (permitting a specified probability distribution for the S_i, $i \in \Lambda^c$), is also true under mild regularity assumptions on the solutions of Eq. (5.6).

The questions regarding the existence and uniqueness of the infinite volume limits of Gibbs states are therefore equivalent to questions regarding the solutions of the DLR equations, e.g. iff (5.6) has a unique (state) solution (for some values of β and h) then $\mu(dS_A, \beta, h; \Lambda, b_{\Lambda})$ will have a limit independent of (reasonable) b.c. .

It has been shown by Israel that for bounded spins there is a unique equilibrium state at sufficiently high temperatures and large magnetic fields. This state, which being unique is of necessity translation invariant, has correlation functions analytic in β and h, which (for finite range potentials) decay exponentially. Similar results were proven a long time ago for Ising spins and for continuum systems where "large magnetic field" corresponds to "small fugacity". Similar results should hold also for unbounded spins: they have been proven explicitly for some cases of interest in field theory and implicitly for the general case (5.1).

Since these results do not hold at phase transitions we cannot expect to extend them beyond such a limited region in the β-h plane without further restrictions on the interactions J(x) and the $\rho(S)$. A particularly interesting class of systems, already mentioned many times here because of their importance both in statistical mechanics and in field theory, are ferromagnetic spin systems. For these systems nice results hold for all $h \neq 0$. Most of these results are contained in the lecture by Miracle-Sole on the Ising spin-½ system and their extension to bounded spins which satisfy the Lee-Yang theorem is immediate. I shall therefore be very brief.

Ferromagnetic Ising Spins

The one component spin system, $S_i \in R$, will be called ferromagnetic if $J(i-j) \geq 0$. Consider now the case where $h_i = h$ for all $i \in \Lambda$, i.e. we have zero $S_i = 0$ for $i \in \Lambda^c$, $b_{\Lambda} = b_0$ (or periodic b.c.). It was then shown by Newman that if the one site partition function, e.g. where Λ consists just of the site $\{o\}$

$$Z(\beta,h;\{o\}) = \int e^{-\beta hS} \rho(S)dS \neq 0, \quad \text{for} \quad \mathrm{Re}\, h \neq 0, \tag{5.7}$$

then

$$Z(\beta,h:\Lambda,b_0) \neq 0 \quad \text{for} \quad \mathrm{Re}\, h \neq 0. \tag{5.8}$$

This is a generalization of the Lee-Yang theorem for the spin-½ case where $Z(\beta,h;\{o\}) = 2\cosh[\beta h]$ which certainly satisfies (5.7). It is also possible to show (Simon-Griffiths) that $\rho(S) = \exp[-V_4(S)]$ (discussed by Gallavotti) also satisfies (5.8). The Gaussian $\exp[-V_2(S)]$ obviously does too. Many compact measures, e.g. the spin-j Ising system,

$$\rho(S) = (2j+1)^{-1}\left[\delta(S-2j) + \delta(S-2j-2) + \dots + \delta(S+2j)\right] \tag{5.9}$$

2j a positive integer

and the flat distribution,

$$\rho(S) = \begin{cases} \tfrac{1}{2}, & |S| \leqslant 1 \\ 0, & |S| > 1, \end{cases} \tag{5.10}$$

are also known (Griffiths) to satisfy (5.7) and hence (5.8).

It is also known that the rigid rotator, $n = 2$, and the classical (and quantum) Heisenberg spin system $n = 3$, with $J(x) \leqslant 0$, satisfy (5.8). (The Newman theorem does not apply.) The proof for the quantum case was given by Asano and the classical result follows by a limit procedure (Dunlop-Newman). (The proof that the one component systems we mentioned satisfy (5.8) was also proven before Newman's theorem).

It follows from (5.8) that $\Psi(\beta,h:\Lambda,b_0)$ will be analytic in h for $\mathrm{Re}\, h \neq 0$ and hence, by the same analysis as in Section IV, so will the infinite volume $\Psi(\beta,h)$ which is independent of boundary conditions. For the spin-½ Ising system it is easy to show (Lebowitz-Penrose) that $\Psi(\beta,h)$ is real analytic in β for $\mathrm{Re}\, h \neq 0$ and the same presumably holds for all systems satisfying (5.8).

Quite independent of (5.8) and not even requiring that $\rho(S)$ be even, it is possible to show, for the one component Ising ferromagnet that the equilibrium state of the infinite system is unique whenever $\frac{\partial\Psi(\beta,h)}{\partial h}$ exists, i.e. at all values of β and h at which $\Psi(\beta,h)$ is differentiable with respect to h. The proof of this statement, which has to be qualified for the unbounded spin case by requiring some regularity condition on the equilibrium measures, (see later), relies on the FKG inequalities.

Unbounded Spins

It is convenient here to set $\rho(dS) = \exp[-\Phi(S)]\,\lambda(dS)$ and include $\Phi(S)$ in the energy $H(S_\Lambda)$ as a self interaction. We make then the following assumptions

a) $\int \lambda(dS) \exp[-\alpha S^2] < 0$, for $\alpha > 0$ (5.11)

b) Superstability. There exists $A > 0$, $C \in R$ such that

$$H(S_\Lambda) \geqslant \sum_{x \in \Lambda} [A\, S_x^2 - C] \tag{5.12}$$

where S_Λ is a configuration in Λ.

c) Regularity If Λ_1, Λ_2 are disjoint then their interaction energy $W(S_{\Lambda_1} | S_{\Lambda_2}) = U(S_{\Lambda_1} \cup S_{\Lambda_2}) - U(S_{\Lambda_1}) - U(S_{\Lambda_2})$ has the bound

$$|W(S_{\Lambda_1} | S_{\Lambda_2})| \leqslant \tfrac{1}{2} K \sum_{x \in \Lambda_1} \sum_{y \in \Lambda_2} |S_x|\, |S_y|\, |x-y|^{-\nu-\epsilon} \tag{5.13}$$

where $|x| = \max_{1 \leqslant i \leqslant \nu} |x^i|$, $|S| = \left[\sum_{i=1}^{d} (S^i)^2\right]^{\frac{1}{2}}$.

Theorem 1 Let (5.11) - (5.13) hold and let $\underline{S} \in \mathcal{H}_a$: ($S$ a spin configuration on $\mathbb{Z}^\nu$)

$\mathcal{H}_a = \{\underline{S} \mid S_y^2 \leqslant a \ln |y| \text{ for } |y| > 1\}$. Let $\{\Lambda_n\}$ be a sequence of increasing domains tending to $\mathbb{Z}^\nu$ in the sense of Van Hove, then $\lim_{n \to \infty} \Psi(\Lambda_n | S_{\Lambda^c}) = \Psi$ exists and is independent of the sequence $\{\Lambda_n\}$ and of the b.c. S_{Λ^c}.

Remark: The dependence on β and other parameters is not indicated explicitly in Ψ. The thermodynamic limit of the "periodic" b.c. free energy can also be shown to exist and be equal to Ψ.

A probability measure ν on the configuration space $\{S\}$ is said to be regular if it satisfies the following condition: There exists $\gamma > 0$, $\delta \geqslant 0$, such that for every Δ bounded in $\mathbb{Z}^\nu$ and $N^2 > 0$ the following holds:

$$\nu[B(N^2|\Delta)] \leqslant \exp[-|\Delta|\,(\gamma N^2 - \delta)]$$

where

$$B(N^2|\Delta) = \left\{\underline{S} \mid \sum_{x \in \Lambda} S_x^2 \geqslant N^2 |\Delta| \right\}.$$

For Λ bounded in $\mathbb{Z}^\nu$ we denote by $\Psi(\Lambda|\nu)$ the free energy in Λ for boundary conditions specified by the measure ν as

$$\Psi(\Lambda|\nu) = \int \nu(d\underline{S})\, \Psi(\Lambda|S_{\Lambda^c}) \ .$$

Theorem 2 Let (5.11) - (5.13) hold and let Λ_n be as in theorem 1, then for ν regular $\lim_{n\to\infty} \Psi(\Lambda_n|\nu) = \Psi$.

Theorem 3 Let the conditions of Theorem 1 be satisfied and let $\mu_{\Lambda_n}(dS_{\Lambda_n} | S_{\Lambda_n^c})$ be finite volume equilibrium states: it is always possible to choose subsequences $\{n_i\}$ (which may depend on the b.c.) such that $\mu_{\Lambda_{n_i}}(dS_{\Lambda_{n_i}} | S_{\Lambda^c_{n_i}}) \to \mu$ a regular equilibrium measure on $\{S\}$.

The set of spin $\underline{S}$ constituting the boundary conditions is assumed here to belong to $\mathcal{H}_a$. It is these measures μ which are unique and hence we need not use any subsequences for ferromagnetic one component systems whenever $\Psi(\beta,h)$ is differentiable in h.

REFERENCES

GENERAL

(1) J.L. LEBOWITZ, Statistical Mechanics - A Review of Selected Rigorous Results, An. Rev. of Chem. Phys., 389 (1968). This article contains extensive references through 1967 for all the subjects covered in these lectures.

(2) The Equilibrium Theory of Classical Fluids, H.L. FRISCH and J.L. LEBOWITZ eds, W.A. Benjamin, N.Y. 1964. Contains reprints and original articles.

(3) D. RUELLE, Statistical Mechanics, W.A. Benjamin, New York, 1969.

(4) Critical Phenomena and Phase Transitions, Vols. 1-6, C. DOMB and M.S. GREEN, eds., Academic Press, 1971. Contains many articles relevant to lectures. See in particular article by R. GRIFFITHS.

(5) Proceedings of the 1971 Battelle Rencontres on Statistical Mechanics, A. LENARD, Ed., Springer Verlag, Lecture Notes in Physics, 1972.

SECTION II

(1) M.E. FISHER, Arch. Ratl. Mech. Anal. 17, 377 (1964). This paper contains a complete detailed description of the thermodynamic limit of the free energy in both canonical and grand canonical ensemble.

(2) J.L. LEBOWITZ and E.H. LIEB, Phys. Rev. Lett.
E.H. LIEB and J.L. LEBOWITZ, Adv. in Math. 1973.
These articles prove the existence of the thermodynamic limit for Coulomb Systems.

(3) M.E. FISHER and J.L. LEBOWITZ, Comm. Math. Phys. discuss periodic boundary conditions for classical systems.

SECTION III

(1) J.L. LEBOWITZ and O. PENROSE, J. Math. Phys. 7, 98 (1966);
O. PENROSE and J.L. LEBOWITZ, J. Stat. Phys. 3, 211 (1971).
The first of these papers deals with equilibrium properties and the second with metastable states.

(2) P.C. HEMMER and J.L. LEBOWITZ, Systems with Weak Long-Range Potentials, in Vol. 5 of Reference I.4. This paper contains a review of both rigorous and non-rigorous results.

(3) O. PENROSE and J.L. LEBOWITZ. Towards a Rigorous Molecular Theory of Metastability; review article, to appear.

(4) G.A. BAKER, Phys. Rev. 122, 1477 (1961). First use of van der Waals' limit, $\gamma \to 0$.

(5) M. KAC, G.E. UHLENBECK and P.C. HEMMER, J. Math. Phys. 4, 216 (1963); 4, 229 (1963); 5, 60 (1964). Exact derivation of van der Waals' equation for a special system.

(6) N.G. van Kampen, Phys. Rev. 135A, 362 (1964); Physica 48, 313 (1970). Basic idea of dividing system into cells containing many particles yet small compared to γ^{-1}. Discusses surface tension.

(7) E. LIEB, J. Math. Phys. 7, 1016 (1966). Extension of results to quantum systems.

(8) D.J. GATES and O. PENROSE, Comm. Math. Phys. 15, 255; 16, 231; 17, 194 (1970). Generalization of results to oscillatory long range potentials plus other things.

(9) K. MILLARD and L. LUND, J. Stat. Phys. 8, 225 (1973). Completes PENROSE and LEBOWITZ argument bounding escape rate λ to all of metastable region.

SECTION IV. Same as for Sections I and II.

SECTION V

(1) J.L. LEBOWITZ, Uniqueness, Analyticity and Decay Properties of Correlations in Equilibrium Systems, Proceedings of the Kyoto Conference on Math. Phys. 1975. K. Araki, Ed. Springer Verlag, 1975. Contains summary of known results for spin-$\frac{1}{2}$ Ising systems.

(2) C.N. YANG and T.D. LEE and companion paper by T.D. LEE and C.N. YANG, Phys. Rev. 87, 410 (1952). They introduce analysis of zeros of $\Xi(z, \Lambda)$ and prove "circle" theorem for ferromagnetic spin systems.

(3) C. NEWMAN, Comm. Pure and Appl. Math., 1974, generalizes LEE-YANG circle theorem.

(4) S.A. PIROGOV and Ya.G. SINAI, Phase Transitions of the First Kind for Small Perturbations of the Ising Model, Func. Anal. Appl. 21, 8 (1974). They prove phase transitions at some value of magnetic field h when "small" but otherwise quite arbitrary many spin interactions are added to ferromagnetic pair interactions.

(5) J. FROHLICH, B. SIMON and T. SPENCER, Phase Transitions and Continuous Symmetry Breaking, Phys. Rev. Lett. (1976) and preprint. Prove phase transitions for n-component spin systems with nearest neighbor ferromagnetic interactions in $\nu \geq 3$ dimensions.

(6) F. DYSON, E.H. LIEB and B. SIMON, preprint prove phase transition for quantum spin systems in $\nu \geqslant 3$.

(7) J.L. LEBOWITZ and E. PRESUTTI, Statistical Mechanics of Systems of Unbounded Spins, to appear in Comm. Math. Phys.

PROBABILISTIC ASPECTS OF CRITICAL FLUCTUATIONS

G. GALLAVOTTI

Istituto di Matematica
Università degli Studi di Roma
Roma, Italia

Probabilistic Aspects of Critical Fluctuations

G. GALLAVOTTI

Istituto di Matematica,
Università Degli Studi Di Roma,
Roma, Italia.

I. Generalities

To be definite we shall consider the problem of critical fluctuations with the purpose of illustrating their relevance in the analysis of other difficult problems like field theory.

The Ising model on the d-dimensional lattice is described by the Hamiltonian

$$H(\underline{\sigma}) = -\sum_{\substack{|i-j|=1 \\ i,j\in\Lambda}} J\,\sigma_i\sigma_j - h\sum_i \sigma_i - B_\Lambda(\underline{\sigma})$$

where $\underline{\sigma} = (\sigma_i)_{i\in\Lambda}$, $\sigma_i = \pm 1$ is a "finite spin configuration" in the cube $\Lambda \subset Z^d$ centred at the origin and $B_\Lambda(\underline{\sigma})$ is a function which describes the "boundary conditions", i.e. defines a model for the walls of the box. For instance

$$B_\Lambda(\underline{\sigma}) = J\sum_i \sigma_i$$

has the physical interpretation that the box Λ is surrounded by an array of spins +1.

The Hamiltonian allows us to define the "equilibrium state in Λ" of the system as the family:

$$\langle \sigma_x \rangle_\Lambda = \frac{\sum_{\underline{\sigma}} \sigma_x\, e^{-\beta H_\Lambda(\underline{\sigma})}}{\sum_{\underline{\sigma}} e^{-\beta H_\Lambda(\underline{\sigma})}}$$

$$\langle \sigma_x \sigma_y \rangle_\Lambda = \frac{\sum_{\underline{\sigma}} \sigma_x \sigma_y \, e^{-\beta H_\Lambda(\underline{\sigma})}}{\sum_{\underline{\sigma}} e^{-\beta H_\Lambda(\underline{\sigma})}}$$

$$\langle \sigma_x \sigma_y \sigma_z \rangle_\Lambda = \frac{\sum_{\underline{\sigma}} \sigma_x \sigma_y \sigma_z \, e^{-\beta H_\Lambda(\underline{\sigma})}}{\sum_{\underline{\sigma}} e^{-\beta H_\Lambda(\underline{\sigma})}}$$

where the sums run over the $2^{|\Lambda|}$ spin configurations in Λ ($|\Lambda|$ = number of points in Λ).

If we assign a reasonable sequence $B = (B_\Lambda)_{\Lambda \in \text{ set of cubes}}$ of boundary terms we can try to take the limit

$$\lim_{\Lambda \to \infty} \langle \sigma_{x_1} \dots \sigma_{x_n} \rangle_\Lambda = \langle \sigma_{x_1} \dots \sigma_{x_n} \rangle .$$

If, for given values of (β, h), this limit depends upon the sequence (B) we say that the model has a phase transition at (β, h).

Of course we should specify better which are the allowed choices of the sequence B : a possible definition is that $B = (B_\Lambda)_{\Lambda \in \text{ set of cubes}}$ should be such that

i) $\max_{\underline{\sigma}} \dfrac{|B_\Lambda(\underline{\sigma})|}{|\Lambda|} \xrightarrow[\Lambda \to \infty]{} 0$

ii) $\exists \delta_\Lambda$ such that $\dfrac{\delta_\Lambda}{|\Lambda|^{1/d}} \xrightarrow[\Lambda \to \infty]{} 0$ and

$$B_\Lambda(\underline{\sigma}) \equiv B_\Lambda(\underline{\sigma}')$$

if $\sigma_\xi = \sigma'_\xi$ for all ξ further than δ_Λ from the boundary of Λ.

The existence of a phase transition at given (β, h) means that the system is so sensitive to boundary perturbations that their influence does not disappear even in the limit $\Lambda \longrightarrow \infty$.

Possibly passing to a subsequence we can and shall assume that the sequences B, which we consider, are such that all the limits of the correlation function

$$\lim_{\Lambda \longrightarrow \infty} \langle \sigma_{\xi_1} \dots \sigma_{\xi_n} \rangle_\Lambda = \langle \sigma_{\xi_1} \dots \sigma_{\xi_n} \rangle$$

exist simultaneously: the set of all the functions $\langle \sigma_\xi \rangle$, $\langle \sigma_\zeta \sigma_\eta \rangle$, obtained as limits of the omonimous finite volume functions for a given sequence B and given β,h is called an "equilibrium state" for the model at given (β, h).

A possible interpretation of an "equilibrium state" is that of a probability measure on the space of the "infinite configurations". The measurable events are those which can be approximated by "local events" described by the "cylinders":

$$E_{\xi_1 \dots \xi_n}^{\bar{\sigma}_1 \dots \bar{\sigma}_n} = \{\underline{\sigma} \mid \underline{\sigma} = (\sigma_{\xi})_{\xi \in Z^d}, \ \sigma_{\xi_i} = \bar{\sigma}_i, \ i = 1, \dots, n\}$$

where $\underline{\sigma} = (\sigma_{\xi})_{\xi \in Z^d}$, $\sigma_{\xi} = \pm 1$, is what we call an infinite configuration and $\bar{\sigma}_1, \dots, \bar{\sigma}_n$ are assigned values ± 1: say, let $\gamma_+ = (i_1, \dots i_p)$ and suppose that $\bar{\sigma}_{i_1}, \dots, \bar{\sigma}_{i_p} = +1$ while the others are -1. The probability of a cylinder is obviously

$$\mu(E_{\xi_1 \dots \xi_n}^{\bar{\sigma}_1 \dots \bar{\sigma}_n}) = \left\langle \prod_{j \in \gamma_+} \frac{(1 + \sigma_{\xi_j})}{2} \prod_{i \notin \gamma_+} \frac{(1 - \sigma_{\xi_i})}{2} \right\rangle .$$

As mentioned above the cylinders are not the only measurable sets: also those sets which can be obtained by finitely many unions and complementation operations are measurable and receive the natural value for their probability. Actually all the sets in the smallest σ-algebra generated by the cylinders can be measured: i.e. the measure μ can be extended to a countably additive probability measure on the sets which are in the smallest family of sets closed under the operations of complementation and countable union containing the cylinders. This fact, though not immediately obvious, is easy to prove but we shall not need such a result.

Therefore the equilibrium states can be regarded as probability measures (on the space of infinite configurations) obtained through the above described process of thermodynamic limit.

It should be stressed that conceptually it is not necessary to really take the limit $\Lambda \longrightarrow \infty$: all systems are actually finite and only those properties of infinite systems which can be interpreted in terms of properties of finite systems are physically meaningful.

Nevertheless it is sometimes simpler to pose the problems for infinite systems: for instance, when we wish, as we shall do shortly, to talk of macroscopic regions inside a macroscopic system it is convenient (but by no means absolutely necessary) to have a really infinite system and consider large regions in it. In this way the relative importance of the ordersof magnitude is more easily taken into account.

The situation resembles that which arises when we decide to make free use of the irrational numbers even though we are aware that in no application shall we ever need an irrational number: everything could be done (and is done) using only rationals.

Let μ be a probability measure which describes the equilibrium state of an infinite Ising model at temperature (β, h). We shall be interested in investigating the distribution of the magnetization in large, macroscopically large, regions.

There are several ways of formulating the problem according to which are the real questions interesting to us.

We can proceed as follows: divide Z^d into squares of side L. We parametrize these squares by the indices $(\underline{n}, L)$:

$$(\underline{n},L) = \left\{ \underline{\xi} \,\middle|\, \underline{\xi} \in Z^d,\ n_i L \leqslant \xi_i \leqslant (n_i+1)L, \quad i = 1,2,\ldots d \right\}$$

where $\underline{n} = (n_1,\ldots, n_d)$ is a point in Z^d itself.

We then define the deviation from the total average magnetization of the box $(\underline{n}, L)$ divided by a normalization factor to be chosen later:

$$V_{\underline{n}} = \frac{\sum_{\underline{\xi} \in (\underline{n},L)} (\sigma_{\underline{\xi}} - \langle \sigma_{\underline{\xi}} \rangle)}{L^{d\rho/2}}$$

where $0 < \rho$ has to be chosen. The variable $V_{\underline{n}}$ will be called a "block spin".

The new variables $(V_{\underline{n}})_{\underline{n} \in Z^d}$ will have a probability distribution μ' which is obtained from the probability distribution of the original variables $(\sigma_{\underline{n}})_{\underline{n} \in Z^d}$: this defines a transformation $K_{\rho,L}$:

$$\mu' = K_{\rho,L}(\mu)$$

It is clear that there is no need for the original variables to take only the values ± 1. The transformation $K_{\rho, L}$ can be defined likewise, as we shall assume, if the variables $\sigma_{\underline{n}}$ can take arbitrary real values.

Then it makes sense to iterate $K_{\rho, L}$ and:

$$K_{\rho,L}(K_{\rho,L'}) = K_{\rho,L'\cdot L}.$$

Given (β ,h) and a corresponding equilibrium state μ of the Ising model, we now ask the question: can we fix ρ so that

$$\lim_{L \to \infty} K_{\rho,}(\mu)$$

exists and is non trivial? The existence and non triviality of the above limit can be defined to mean that there is a probability measure μ_∞ such that

$$\langle V_{\xi_1} \cdots V_{\xi_s} \rangle_{\mu_\infty} = \int d\mu_\infty V_{\xi_1} \cdots V_{\xi_s} = \lim_{L \to \infty} \langle V_{\xi_1} \cdots V_{\xi_n} \rangle_{K_{\rho,L}(\mu)}$$

(with the obvious meaning for the symbols) and, furthermore, μ_∞ is not the δ-function on the 0-configuration (i.e. $\langle V_{\xi_1} \cdots V_{\xi_n} \rangle_{\mu_\infty} \neq 0$ for some n and $\xi_1, \ldots, \xi_n \in Z^d$) .

It is easy to convince oneself that if a value of ρ is "good" (i.e. $K_{\rho,L}(\mu)$ converges to a non trivial distribution) then any other value is not good.

The following basic result on the theory of the Ising model holds and is useful to clarify the meaning of the next theorem on the transformation $K_{\rho,L}$.

Theorem A:[1] If $h \neq 0$ there is no phase transition for the model (β, h) and, furthermore, there is an upper bound for the decay of the correlation functions of the form:

$$\left| \langle \sigma_{x_1} \cdots \sigma_{x_p} \rangle^T \right| \leq \frac{C}{n_1! \cdots n_\ell!} e^{-\varkappa \Delta(x_1 \cdots x_p)}$$

where $C, \varkappa > 0$ and $n_1, n_2, \ldots, n_\ell$ are the multiplicities of the different sites among $x_1, \ldots x_p$ and $\Delta(x_1 \ldots x_n)$ = length of the shortest graph connecting $x_1 \ldots x_n$ as a set in R^d.

The truncated correlation functions are:

$$\langle \sigma_{x_1} \rangle^T = \langle \sigma_{x_1} \rangle$$

$$\langle \sigma_{x_1} \sigma_{x_2} \rangle^T = \langle \sigma_{x_1} \sigma_{x_2} \rangle - \langle \sigma_{x_1} \rangle \langle \sigma_{x_2} \rangle$$

$$\begin{aligned} \langle \sigma_{x_1} \sigma_{x_2} \sigma_{x_3} \rangle^T = {} & \langle \sigma_{x_1} \sigma_{x_2} \sigma_{x_3} \rangle - \langle \sigma_{x_1} \sigma_{x_2} \rangle \langle \sigma_{x_3} \rangle - \\ & - \langle \sigma_{x_1} \rangle \langle \sigma_{x_2} \sigma_{x_3} \rangle - \langle \sigma_{x_2} \rangle \langle \sigma_{x_1} \sigma_{x_3} \rangle \\ & + 2 \langle \sigma_{x_1} \rangle \langle \sigma_{x_2} \rangle \langle \sigma_{x_3} \rangle \end{aligned}$$

$$\vdots$$

The general truncated function is defined as follows: let f be a function on Z^d and put

$$Z_\Lambda(\beta, h+f) = \sum_{\underline{\sigma}} e^{-\beta H_\Lambda(\underline{\sigma}) - \beta \sum_{\xi \in \Lambda} f(\xi)\sigma_\xi}$$

i.e. Z_Λ is the partition function when the external field is modified from h to h+f. Then provided $x_1, \ldots, x_n \subset \Lambda$:

$$\langle \sigma_{x_1} \cdots \sigma_{x_n} \rangle_\Lambda^T = \frac{1}{\beta^n} \frac{\partial^n \log Z_\Lambda(\beta, h+f)}{\partial f(x_1) \ldots \partial f(x_n)} \Bigg|_{f=0} .$$

Clearly $\langle \sigma_{x_1} \cdots \sigma_{x_n} \rangle_\Lambda^T$ is a well defined combination of products of correlation functions with coefficients which are Λ independent: hence it makes sense to define by the same combinations the truncated functions for the measure μ.

The following theorem completes the above:

Theorem B: [2] If $h = 0$ and β is small enough (say $< \bar{\beta}$) there is no phase transition for the model $(\beta, 0)$ and the same estimate as above holds for the truncated functions.

Furthermore there exists β_c ($\beta_c > \bar{\beta}$) such that there is a phase transition at $(\beta, 0)$ when $\beta > \beta_c$. If the dimension $d = 2$, it can be shown that the truncated functions decay exponentially but not necessarily with a bound like that of theorem A as far as the combinatorial factors are concerned.

Coming back to the block-spin distributions we can quote the following result.

Theorem C: If (β, h) is such that $h \neq 0$ or $h = 0$ but $\beta < \bar{\beta}$ ($\beta < \beta_c$ if $d = 2$), the limit $\lim_{L \to \infty} K_{\rho, L}(\mu) = \mu_\infty$ exists and if

$$\chi^2 = \sum_{\xi \in Z^d} (\langle \sigma_0 \sigma_\xi \rangle - \langle \sigma_0 \rangle \langle \sigma_\xi \rangle)$$

it turns out

$$\mu_\infty\Big(\prod_{\xi \in Z^d} dV_\xi\Big) = \prod_{\xi \in Z^d} \frac{e^{-\frac{V_\xi^2}{2\chi^2}}}{\sqrt{2\pi\chi^2}} \, dV_\xi .$$

This theorem tells us that outside the region of phase transition the magnetization fluctuations on a large scale are independently distributed and the distribution of each of them is essentially the same as that relative to a set of independent spins (see next exercise):

Exercise:[3] Prove that if $\mu(\prod_{\xi\in Z^d} d\sigma_\xi) = \prod_{\xi\in Z^d} f(\sigma_\xi)d\sigma$ when $f(\sigma_\xi)$ is a probability distribution with finite second moment

$$\alpha^2 = \int \sigma^2 f(\sigma)d\sigma$$

and finite third moment and zero mean $(\int f(\sigma)|\sigma|^3 d\sigma < \infty$, $\int f(\sigma)\sigma d\sigma = 0)$ then

$$\lim_{L\to\infty} K_{\frac{1}{2},L}(\mu) = \mu_\infty = \prod_{\xi\in Z^d} \frac{e^{-\frac{V_\xi^2}{2\alpha^2}}}{\sqrt{2\pi\alpha^2}} dV \quad .$$

(This is the well known central limit theorem and the hint for the proof is to analyze the characteristic functions, or the Fourier transforms of the joint block spin distributions.)

The physical reason for the above theorem is that away from the phase transition region of the parameter (β, h) the correlations of the equilibrium states decay exponentially fast and this essentially means that far spins are independently distributed around their average value and therefore large collections of them behave as a collection of independent random variables with values ± 1 and their distribution verifies the central limit theorem (see exercise above).

I shall try, in the next lectures, to sketch a proof of the above theorem: however, I shall present it in such a way that it will be adaptable to the case when $\beta = \beta_c$, $h = 0$ where the correlations between far spins decay more slowly than exponentially. To prepare the discussion about the block spin distribution at the critical point let me remind you about the conjectured properties of the equilibrium measures near the critical point $(\beta_c, 0)$. (See for instance [2]).

Let us agree that the correlation functions of the equilibrium state (β, h) will bear the subscript (β,h) and c will simply replace $(\beta_c,0)$.

It is assumed, on the basis of the phenomenology of the critical point, that:

i) $\langle \sigma_0 \sigma_{\xi_1} \ldots \sigma_{\xi_{2n-1}} \rangle_c \propto \chi_{2n}(0\, \xi_1 \ldots \xi_{2n-1})$, $n = 1,2,3, \ldots$

where χ_{2n} is a homogeneous function of degree ω_{2n} of the coordinates of $\xi_1, \ldots \xi_{2n-1}$ regarded as points in R^d:

$$\chi_{2n}(0\, \frac{\xi_1}{\sigma} \ldots \frac{\xi_{2n-1}}{\sigma}) = \sigma^{\omega_{2n}} \chi_{2n}(x_1 \ldots x_{2n}) \quad .$$

The symbol $\propto$ means that the sums of the sites over $\xi_1, \ldots \xi_{2n-1}$ in Z^d such that $|\xi_i| \leqslant R$ agree to leading order in R as $R \longrightarrow \infty$.

If $n = 1$

$$\chi_2(0\,\xi) = \frac{c}{|\xi|^{\omega_2}}$$

and ω_2 is usually written as

$$\omega_2 = d + \eta - 2 \ .$$

ii) $\langle \sigma_\xi \rangle_{(\beta_c, h)} = m(h) = m_o\, h^{1/\delta} (1 + O(h))$

and here this formula is meant to signify that

$$\frac{d^n}{dh^n} m(h) = m_o(\tfrac{1}{\delta})(\tfrac{1}{\delta} - 1) \ldots (\tfrac{1}{\delta} - n+1) h^{\frac{1}{\delta} - n} (1 + O(h)) \ .$$

iii) There are homogeneous functions of degree 1 on $R^d x\, R^d x \ldots R^d$: $(x_1 \ldots x_n) \longrightarrow \Delta(x_1 \ldots x_n)$ such that if $\Delta_o(x_1 \ldots x_n)$ is the length of the smallest graph connecting $x_1 \ldots x_n$

$$0 < a \leqslant \frac{\Delta(x_1 \ldots x_n)}{\Delta_o(x_1 \ldots x_n)} \leqslant b < \infty$$

for some a, b and

$$\left\langle \sigma_0\, \sigma_{\xi_1} \ldots \sigma_{\xi_{2n-1}} \right\rangle_{(\beta_c, h)} \propto e^{-\frac{\Delta(0\, \xi_1 \ldots \xi_{(2n-1)})}{L(h)}} \cdot \chi_{2m}(0\, \xi_1 \ldots \xi_{2n-1}) \ .$$

iv) Similarly if $\beta < \beta_c$:

$$\left\langle \sigma_0\, \sigma_{\xi_1} \ldots \sigma_{\xi_{2n-1}} \right\rangle_{(\beta, 0)} \propto e^{-\frac{\Delta(0 \xi_1 \ldots \xi_{2n-1})}{\lambda(\beta)}} \chi_{2m}(0\, \xi_1 \ldots \xi_{2n-1}) \ .$$

The symbol $\propto$ in iii), iv) means that the sums of both sides over $\xi_1 \ldots \xi_{2n-1}$ agree to leading order in $1/h$ or $1/\beta - \beta_c$.

v) There exist $\bar{\nu}$ and ν such that

$$L(h) = l_o\, h^{-\bar{\nu}}$$
$$\lambda(\beta) = \lambda_o(\beta_c - \beta)^{-\bar{\nu}} \ .$$

Because of assumptions iii), ii) and i) and the definition of the truncated functions it follows that:

$$\frac{d^{2n-1}m(h)}{dh^{2n-1}} = \sum_{\xi_1 \cdots \xi_{2n-1}} \langle \sigma_0 \sigma_{\xi_1} \cdots \sigma_{\xi_{2n-1}} \rangle^T_{(\beta_c, h)} \quad .$$

Hence there must be a simple relation between ω_{2n}, δ, $\bar{\gamma}$:

$$\bar{\gamma}((2n+1)d - \omega_{2n+2}) = 2n + 1 - \frac{1}{\delta} \quad .$$

The derivation of this formula is a useful exercise.

We can eliminate $\bar{\gamma}$ in terms of the more familiar exponent η by using $\omega_2 = d + \eta - 2$ by definition: it turns out that

$$\bar{\gamma} = \frac{\delta - 1}{\delta} \; \frac{1}{2 - \eta}$$

and, also, we can rewrite the formula for ω_{2n}:

$$\omega_{2n+2} = (2n+1)(d - (2-\eta)\frac{\delta}{\delta - 1}) + \frac{2-\eta}{\delta - 1} \quad .$$

Assumption i) for $n = 1$ can be verified only in one interesting instance: in the two dimensional Ising model and when 0 and ξ_1 are on the same row. The theory of Kadanoff on the two dimensional Ising model would allow us to prove i) when all the points lie on the same row, [4].

Assumptions ii) have never been verified except perhaps the case of $n = 0,1$ in $d = 2$ [5]. Assumption i) is a strict interpretation of the absence of characteristic length at $\beta = \beta_c$, $h = 0$, while assumptions iii), iv), which seem the most audacious, give a very strict interpretation of the following loose but frequent statements which are the basic ingredients for many theories of the scaling laws: "There is only one correlation length near $\beta = \beta_c$, $L = 0$" and "within the correlation length the spins are at criticality". Of course ii), iv) are not the only possible interpretations of the above phrases which are usually made (and used) only in connection with the single spin and pair correlation functions.

The way in which the correlation length is introduced in iv) is inspired by recent rigorous results on the cluster property in the Ising model in the $h \neq 0$ or in the high temperature region [1,2].

II. Macroscopic Fluctuations Theory

We shall first outline the simple calculations [2] necessary to derive theorem C and then we shall apply them to discuss the critical fluctuations.

Let $\underline{n}_1, \dots \underline{n}_s \in Z^d$ and let $\underline{n}_i \neq \underline{n}_j$, $i \neq j$, and

$$E(\omega_1, \dots, \omega_s) = \int e^{i \sum_{k=1}^{n} \omega_k V_{\underline{n}_k}} d\mu_{L,\rho}$$

where $\mu_{L,\rho} = K_{L,\rho}(\mu)$ and μ is the Ising model equilibrium state at values (β, h) of its parameters. This quantity is the Fourier transform of the joint distribution of s block spins.

By the definition of $Z_\Lambda(\beta, h + f)$ (see page) we see immediately that

$$E_\Lambda(\omega_1 \dots \omega_s) = e^{-i \sum_{k=1}^{n} \omega_k \frac{\sum_{\xi \in (\underline{n}_k, L)} \langle \sigma_\xi \rangle}{L^{\rho d/2}}} \frac{Z_\Lambda(\beta, h + \sum_{j=1}^{n} \frac{i\omega_j}{L^{\rho d/2}} \chi_j)}{Z_\Lambda(\beta, h)}$$

where E_Λ is the characteristic function of the block spins computed in the approximate equilibrium state obtained by regarding the system confined in a finite cube Λ (so $E(\omega_1 \dots \omega_s) = \lim_{\Lambda \to \infty} E_\Lambda(\omega_1 \dots \omega_s)$); furthermore the functions χ_j are

$$\chi_j(\xi) = 1 \quad \text{if } \xi \in (\underline{n}_j, L), \qquad j = 1, 2, \dots, n$$

$$\chi_j(\xi) = 0 \quad \text{otherwise.}$$

Therefore

$$E_\Lambda(\omega_1 \dots \omega_s) = e^{-i \sum_{k=1}^{n} \omega_k \frac{\sum_{\xi \in (\underline{n}_k, L)} \langle \sigma_\xi \rangle}{L^{\rho d/2}}} \exp\left[\int_0^1 dt \frac{d}{dt} \log Z_\Lambda(\beta, h + t \sum_{j=1}^{n} \frac{i\omega_j}{L^{\rho d/2}} \chi_j)\right]$$

and, using the definition of the truncated functions to write the Taylor expansion of the derivative inside the above integral, we find, after a straightforward computation:

$$E_\Lambda(\omega_1 \dots \omega_s) = \exp \sum_{\substack{k_1 \dots k_s \\ \sum_i k_i \geq 2}} {}^{0} J^{(\Lambda)\, k_1 \dots k_s}_{\underline{n}_1 \dots \underline{n}_s} \frac{(i\omega_1)^{k_1} \dots (i\omega_s)^{k_s}}{k_1! \dots k_s!} \frac{1}{L^{\rho d/2 \sum_j k_j}}$$

where

$$J^{(\Lambda)\,k_1 \,\dots\, \underline{k}_s}_{\quad \underline{n}_1 \,\dots\, \underline{n}_s} = \sum_{\substack{x_1 \subset (\underline{n}_1, L),\ |x_1| = k_1 \\ \vdots \\ x_s \subset (\underline{n}_s, L),\ |x_s| = k_s}} \langle \sigma_{x_1} \dots \sigma_{x_s} \rangle^T_\Lambda$$

where x_1 is a subset of $(\underline{n}_1, L)$ with possibly repeated sites and σ_{x_1} shortens $\sigma_{\xi_1} \dots \sigma_{\xi_\ell}$ if $x_1 = (\xi_1, \dots, \xi_\ell)$; the subscript Λ recalls that the correlation functions are relative to the finite volume equilibrium state.

Passing to the limit $\Lambda \longrightarrow \infty$, formally, we find

$$E(\omega_1 \dots \omega_s) = \exp \sum_{\substack{k_1 \dots k_s \geq 0 \\ \sum_i k_i \geq 2}} J^{\,k_1 \dots k_s}_{\,\underline{n}_1 \dots \underline{n}_s} \frac{(i\omega_1)^{k_1}}{k_1!} \dots$$

$$\dots \frac{(i\omega_s)^{k_s}}{k_s!} \frac{1}{L^{\rho d/2 \sum_j k_j}}$$

where $J^{\dots}_{\dots}$ is defined as $J^{(\Lambda)\dots}_{\dots}$ above by eliminating the subscript Λ on the right hand side.

It is now easy to see that if $\sum_{x_1 \dots x_\ell} |\langle \sigma_0 \sigma_{x_1} \dots \sigma_{x_\ell} \rangle^T| < \infty$

then, by choosing $\rho = 1$ the above series converges, term by term, to zero with the exception of the s terms

$$k_1 = 0,\ k_2 = 0,\ \dots,\ k_j = 2,\ k_{j+1} = 0,\ \dots\ k_s = 0,\quad j = 1,2,\dots s$$

and since

$$J^{\,2\,0\,\dots\,0}_{\,n_1, n_2 \dots n_s} = \sum_{x_1 x_2 \in (n_1, L)} (\langle \sigma_{x_1} \sigma_{x_2} \rangle - \langle \sigma_{x_1} \rangle \langle \sigma_{x_2} \rangle).$$

We see that

$$\frac{1}{L^d} J^{\,2\,0\,\dots\,0}_{\,n_1 \dots n_s} \xrightarrow[L\to\infty]{} \sum_{x \in Z} (\langle \sigma_0 \sigma_x \rangle - \langle \sigma_0 \rangle \langle \sigma_x \rangle) = \chi^2$$

and therefore

$$\lim_{L\to\infty} E(\omega_1 \dots \omega_s) = \exp - \sum_{j=1}^{s} \chi^2 \omega_j^2 .$$

The problems of showing that convergence, term by term, implies actual convergence (as well as the problem of taking first the limit $\Lambda \to \infty$) are not too hard in the case in hand and are based on correlation inequalities or analyticity arguments (see [2]). That the convergence of the Fourier transforms implies the convergence of the correlation functions is standard in probability [3] . The above argument also shows that at the critical point the above calculations fail: in fact χ^2, which is physically proportional to the susceptibility, is infinite at $(\beta_c, 0)$!

However, if we accept the assumptions about the critical behaviour of the last section, we realize that the above term by term argument can easily be made even at $\beta = \beta_c$ and the result of the calculation, which we leave an exercise to the reader, is if ρ is suitably chosen (i.e. $\rho = 1 + \frac{2-\eta}{d}$):

$$E(\omega_1 \dots \omega_s) = \exp \sum_{0 \neq \sum_i k_i = \text{even}} \frac{(i\,\omega_1)^{k_1}}{k_1!} \cdots \frac{(i\,\omega_s)^{k_s}}{k_s!} L^{(\frac{\sum_i k_i}{2} - 1)\zeta}$$

$$\int_{[1]^{k_1} \times [2]^{k_2} \times \dots \times [s]^{k_s}} d\underline{x}_1^{k_1} \dots d\underline{x}_s^{k_s} \; \chi_{\sum_i k_i}(\underline{x}_1 \dots \underline{x}_s)$$

where $[i]$ = unit square of R^d centred around the point $\underline{n}_i \in Z^d$ and

$$\zeta = \frac{\delta+1}{\delta-1}(2-\eta) - d \quad .$$

This formula is remarkable in view of the Buckingham-Gunton [5] scaling law which holds under assumptions weaker than i) - v) above:

$$\zeta \leq 0.$$

This shows that with the chosen ρ the limit term by term exists and is gaussian (i.e. only the terms with $\sum_i k_i = 2$ do not tend to zero) if

$$\zeta < 0$$

while all the terms are non zero if

$$\zeta = 0$$

and the limit distribution of the macroscopic magnetizations is not gaussian.

Notice that even in the gaussian case the terms like $k_1 = 1$, $k_2 = 1$, $k_3 = \dots = k_s = 0$ are now present (while away from the critical point they were zero in the limit $L \to \infty$): this means that the block spins are not independently distributed.

Of course the above argument works term by term only: before trying to make a rigorous proof of the convergence, however, it seems more important to try to justify rigorously the assumptions although we do not try to do anything in this direction here.

Let me finish this argument by remarking that

$\zeta = 0$

if $d = 2$, as it stems from the exact solutions. [6]

If $d = 3$ it is not completely clear how to interpret the numerical experiments. [7]

If $d = 4$ $\zeta = 0$ in the mean field theory and it is not known if, in the true theory, the value of $\zeta = 0$: it is widely believed that at $d = 4$ the critical exponents cannot be defined naively. One should in this case reformulate the whole discussion, allowing for an L-dependence of the various exponents (to take into account the so-called "logarithmic corrections").

III. Applications to Field Theory [8]

Euclidean quantum field theory tries to construct a probability measure on the space of functions on R^d which is of the following formal form:

$$\mu(d\phi) = C\left\{\exp - \int (z_1 |\nabla\phi(x)|^2 + z_3 \ \phi(x)^2 + z_2 \ \phi(x)^4)dx\right\}\text{"}d\phi\text{"}$$

where $z_1, z_2 > 0$, z_3 is real.

This measure is subject to the requirement that

$$\int_{R^d} \langle\phi(0)\ \phi(x)\rangle\ dx = m^{-2} > 0$$

$$\int_{R^d} \langle\phi(0)\ \phi(x)\rangle\ |x|^2\ dx = Z^2 > 0$$

$$-\int_{R^d} \langle\phi(0)\ \phi(x)\ \phi(y)\ \phi(z)\rangle^T\ dx\ dy\ dz = \lambda^2 > 0$$

where the values m, Z, λ are, respectively, the values of the "renormalized"[1] mass, wave function and coupling constant.

The above theory should be constructed in a non-perturbative way by introducing infrared and ultraviolet cut-offs.

The infrared cut-off is introduced by confining the fields in a box Λ and the ultraviolet one by discretizing the space. So if L is the number of lattice spacings into which the side of Λ is divided, let a be the size of the lattice step.

The volume of Λ is then

$$V(a, L) = L^d\ a^d$$

and we shall also introduce the quantity

$$D(a, L) = a^2L^{-d} \sum_{|n_1| \ \cdots |n_d| \leq \frac{L}{2}} |\underline{n}|^2 \sim a^2\ L^2$$

which measures the size of Λ.

We shall denote by C_L the box in Z^d with side L.

Then

$$\int_\Lambda (z_1 |\nabla\phi(x)|^2 + z_2\ \phi(x)^2 + z_3\ \phi(x)^4)dx =$$

---

[1] These names are not quite appropriate: for instance the m^{-2} is not the square of the physical mass. (It is, however, the sum of the "mass diagrams".)

$$\cong a^d \sum_{\underline{n} \in C_L} \left(z_1 \sum_{\underline{e}} \frac{(\phi_{\underline{n}+\underline{e}} - \phi_{\underline{n}})^2}{a^2} + z_3 \phi_{\underline{n}}^2 + z_2 \phi_{\underline{n}}^4 \right)$$

where $\underline{e}$ is a unit step along the lattice directions and $\phi_{\underline{n}}$ is the value of the field $\phi(\underline{n}\, a)$.

The above measure μ, to be constructed, is then by definition "approximated" by the "regularized model measure":

$$\mu^{(L,a)} \Big(\prod_{\underline{n} \in C_L} d\, \phi_{\underline{n}} \Big) = \text{const.} \exp\Big(\sum_{\substack{\underline{n},\underline{m} \in C_L \\ |\underline{n}-\underline{m}|=1}} \alpha_1 \phi_{\underline{n}} \phi_{\underline{m}} - $$

$$ - \sum_{\underline{n} \in C_L} (\alpha_2 \phi_{\underline{n}}^4 - \alpha_3 \phi_{\underline{n}}^2) \Big) \prod_{\underline{n} \in C_L} d\, \phi_{\underline{n}} \qquad (*)$$

where $\alpha_1, \alpha_2, \alpha_3$ are simply related to z_1, z_2, z_3; the fact that $z_1, z_2 > 0$ implies $\alpha_1, \alpha_2 > 0$.

The values of α_1, α_2, α_3 are to be determined by imposing that the functions y_1, y_2, y_3 defined below, have the appropriate values m^2, Z^2, λ^2, respectively:

$$y_1 = a^d \sum_{\underline{n} \in C_L} \langle \phi_{\underline{n}} \phi_{\underline{o}} \rangle \qquad (*)$$

$$y_2 = a^{d+2} \sum_{\underline{n},\underline{m} \in C_L} \langle \phi_{\underline{n}} \phi_{\underline{o}} \rangle |\underline{n}|^2 \qquad (*)$$

$$y_3 = - a^{3d} \sum_{\underline{m},\underline{p},\underline{q} \in C_L} \langle \phi_{\underline{o}} \phi_{\underline{m}} \phi_{\underline{p}} \phi_{\underline{q}} \rangle^T \qquad (*)$$

The theory of correlation inequalities implies that $m^2, Z^2, \lambda^2 \geqslant 0$ [8,9,10].

The question now is to find which are the triples (m^2, Z^2, λ^2) which can be fixed so that for all L, a there exist values α_1, α_2, α_3 which correspond to them via the $(*)$ formulae (allowed values").

If m^2, λ^2, Z^2 are allowed, then the correlation inequalities can be used [8,10] to guarantee that, passing possibly to a subsequence of $L \to \infty$, $a \to 0$ (so that $La \to \infty$), the measures $\mu^{(L,a)}$ converge (in some sense which se do not discuss here) to a limit measure with all correlation functions bounded in terms of the 2-fields correlation. (This is a statement about the "renormalizability" of the above ϕ^4

theory: it proves that by conveniently fixing the "bare" constants $\alpha_1, \alpha_2, \alpha_3$ as a function of the cut-offs L, a, we can make all correlation functions uniformly bounded in the value of the cut-off.) The correlation inequalities of interest here are the Lebowitz-Neumann inequalities which bound the n-field correlations in terms of the pair correlation: for more details see [8] and its references. The discussion of the allowed values of Z^2, λ^2, m^2 can be made as follows.

The boundary of the set $R_+ \times R_+ \times R$ of the allowed choices of $(\alpha_1, \alpha_2, \alpha_3)$ will contain the following four pieces:

$\mathfrak{M} = \{\underline{\alpha} \mid \alpha_1 = 0\}$ (for a precise definition see below)

$\mathfrak{M} = \{\underline{\alpha} \mid \alpha_2 = 0\}$ " " " " " "

$\mathfrak{M} = \{\underline{\alpha} \mid \alpha_1 = +\infty\}$ (for a precise definition see below)

$\mathfrak{M} = \{\underline{\alpha} \mid \alpha_2 = +\infty\}$ " " " " " "

Let $F: (\alpha_1, \alpha_2, \alpha_3) \longrightarrow (y_1, y_2, y_3)$ be the map defined by the $(*)$-formulae above. It is an easy task to find the set $F(\mathfrak{M}_1) \cup F(\mathfrak{M}_2) \cup F(\mathfrak{M}_3) \cup F(\mathfrak{M}_4)$: it will turn out that it is the boundary of a certain region in the space (y_1, y_2, y_3).

The conjecture is that this region is the region of the allowed values for (y_1, y_2, y_3). This is not obvious because the map F is not necessarily 1 - 1, at least this is not known, nor has it necessarily a differentiable inverse; also one may wonder what happens to the other pieces of the boundary of $R_+ \times R_+ \times R$ (e.g. $\alpha_3 = +\infty$).

We shall assume that the above region is the region of the allowed values and refer the reader to [8] to a deeper discussion of the hidden assumptions just mentioned.

To avoid talking about 3-dimensional sets we shall fix the value $y_1 = m^{-2}$ and we look at the intersection of $F(\mathfrak{M}_1) \cup F(\mathfrak{M}_2) \cup F(\mathfrak{M}_3) \cup F(\mathfrak{M}_4)$ with the plane $y_1 = m^{-2}$

i) Study of $F(\mathfrak{M}_1)$: If $\alpha_1 = 0$ the variables $\phi_{\underline{n}}$ are independent in $\mu^{(L,a)}$ and, therefore,

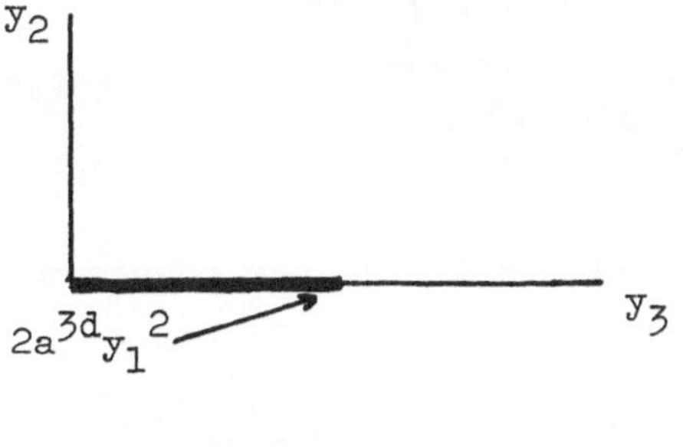

$$y_2 \equiv 0$$

while

$$y_1 = a^d \langle \phi_{\underline{o}}^{\,2} \rangle = m^{-2}$$

$$y_3 = -a^{3d}(\langle \phi_{\underline{o}}^{\,4} \rangle - 3\langle \phi_{\underline{o}}^{\,2} \rangle^2)$$

and the value of y_3 varies between the values of the variable

$$y = -\left[\frac{\int e^{-\alpha_2 \phi^4 + \alpha_3 \phi^2} \phi^4 \, d\phi}{\int e^{-\alpha_2 \phi^4 + \alpha_3 \phi^2} \, d\phi} - 3 y_1^{\,2} \right] a^{3d}$$

where α_2 and α_3 vary in such a way as to keep

$$y_1 = \frac{\int e^{+\alpha_3 \phi^2 - \alpha_2 \phi^4} \phi^2 \, d\phi}{\int e^{\alpha_3 \phi^2 - \alpha_2 \phi^4} \, d\phi} = m^{-2}.$$

The minimum value is $y_3 = 0$ (when $\alpha_2 = 0,\ \alpha_3 < 0$) and the maximum is $2y_1^{\,2}\, a^{3d}$ realized when $\alpha_3, \alpha_2 \to \infty$ in such a way that the distribution $e^{\alpha_3 \phi^2 - \alpha_2 \phi^4}$ approaches

$$\frac{\delta(\phi + \sqrt{y_1}) + \delta(\phi - \sqrt{y_1})}{2} .$$

ii) $\mu^{(a,L)}$ is a gaussian measure on $\mathfrak{M}_2$: hence the four point truncated correlations vanish and $y_3 \equiv 0$.

If $G_{\underline{n},\,\underline{m}} = -\alpha_1 \delta_{|\underline{n}-\underline{m}|,\,1} + \alpha_3 \delta_{\underline{n},\underline{m}}$, $\underline{n}, \underline{m} \in C_L$

has $g_{\underline{n},\underline{m}}$ as inverse matrix

$$y_1 = a^d \sum_{\underline{n} \in C_L} g_{\underline{o},\underline{n}}$$

$$y_2 = a^{d+2} \sum_{\underline{n}} g_{\underline{o},\underline{n}} \, |\underline{n}|^2 .$$

[Notice that this makes sense only if $-\alpha_3 > d\alpha_1$: in the other cases what follows should be properly interpreted by first fixing $\alpha_2 > 0$ and then letting it tend to 0.]

The extreme cases at fixed y_1 are $y_2 = 0$ (independent fields, $\alpha_1 = 0$) and $g_{\underline{n},\underline{m}} = \text{const.}$ (coherent fields, $\alpha_1 = +\infty$):

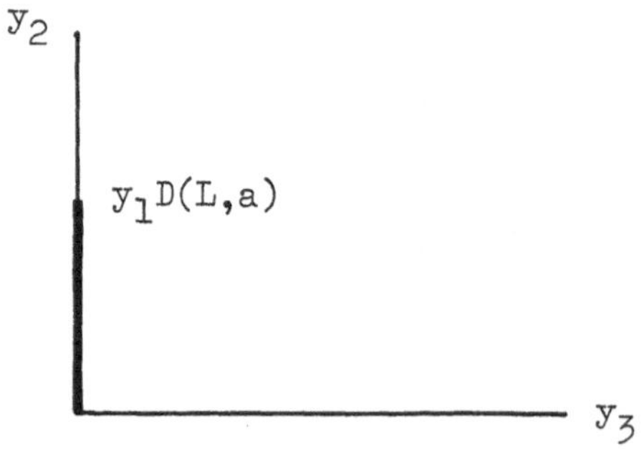

$$y_1 = a^d L^d \text{ const.}$$

$$y_2 = a^{d+2} \text{ const.} \sum_{\underline{n}} |n|^2$$

$$= y_1 \, D(L, a) .$$

iii) In this case $\alpha_1 = +\infty$ means that all the fields are coherent hence, as before,

$$y_2 \;=\; y_1\, D(a,\, L)$$

while y_3 varies from zero (gaussian case $\alpha_2 = 0$) to the limiting case when the coherent value of the field is distributed as

$$\frac{\delta(\phi - \sqrt{\langle\phi^2\rangle}) + \delta(\phi + \sqrt{\langle\phi^2\rangle})}{2}$$

then

$$y_3 \;=\; 2y_1{}^2\, a^d\, L^d \quad .$$

iv) This piece of the boundary is obtained by letting $\alpha_2 \rightarrow +\infty$: however, to get a well defined limit, we need to tie α_2 and α_3 so that the ratio

$$\frac{\alpha_3}{2\alpha_2} \;=\; \gamma^2$$

is kept constant (and > 0). In this case $e^{-\alpha_2\phi_n^4 + \alpha_3\phi_n^2}$ is replaced by

$$\frac{\delta(\phi_n + \gamma) + \delta(\phi_n - \gamma)}{2}$$

and

$$y_1 \;=\; a^d\, \gamma^2 \sum_{\underline{m} \in C_L} \langle \sigma_{\underline{o}}\, \sigma_{\underline{m}} \rangle_{(\alpha_1\gamma^2,o)}$$

$$y_2 \;=\; a^{d+2}\, \gamma^2 \sum_{\underline{m} \in C_L} (\sigma_{\underline{o}}\, \sigma_{\underline{m}} \rangle_{(\alpha_1\gamma^2,o)}$$

$$y_3 \;=\; -a^{3d}\, \gamma^4 \sum_{\underline{n},\underline{m},\underline{p} \in C_L} \langle \sigma_{\underline{o}}\, \sigma_{\underline{n}}\, \sigma_{\underline{m}}\, \sigma_{\underline{p}} \rangle_{(\alpha_1\gamma^2,o)}$$

So the final picture of the allowed region that we get is

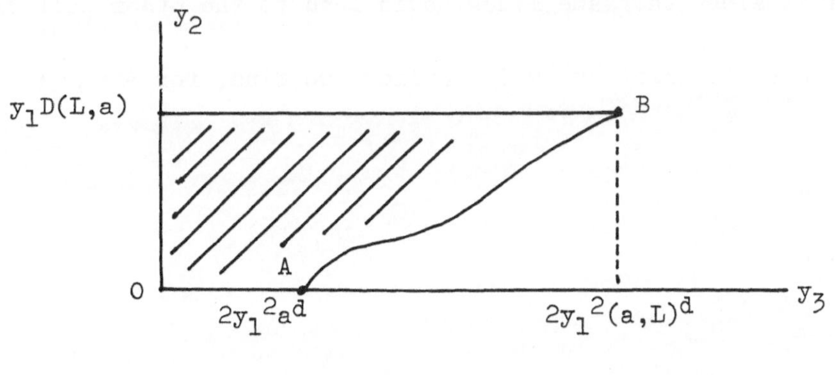

where the equation of the AB line is given parametrically in terms of $\beta = \alpha_1 \gamma^2 \in (0, +\infty)$:

$$\frac{y_2}{y_1} = \frac{a^2 \sum_{\underline{m} \in C_L} \langle \sigma_0 \sigma_{\underline{m}} \rangle_{(\beta,0)} |\underline{m}|^2}{\sum_{\underline{m}} \langle \sigma_0 \sigma_{\underline{m}} \rangle_{(\beta,0)}}$$

$$\frac{y_3}{y_1^2} = - \frac{a^d \sum_{\underline{m}\,\underline{p}\,\underline{q}} \langle \sigma_0 \sigma_{\underline{m}} \sigma_{\underline{p}} \sigma_{\underline{q}} \rangle^T_{(\beta,0)}}{\left(\sum_{\underline{m}} \langle \sigma_0 \sigma_{\underline{m}} \rangle_{(\beta,0)} \right)^2}$$

We can also easily see that when $a \to 0$, $L \to \infty$, (and $aL \to \infty$), the part of the curve which comes from the interval $|\beta - \beta_c| > \varepsilon > 0$, for any $\varepsilon > 0$ fixed (before taking the limit $L \to \infty$, $a \to 0$) collapses either into the lowest point A ($\beta < \beta_c$) or into the point B ($\beta > \beta_c$).

Exercise: hint - use the exponential clustering in the non-critical Ising model.

Therefore, removing the cut-offs, we realize that the whole curve is determined by the critical behaviour!

It is even more remarkable that the assumptions made in part II about the critical behaviour permit us to write down explicitly the equation of "half" of the curve $\overline{AB}$ in the limit $a \to 0$, $L \to \infty$. We can proceed as follows. We consider only the case $\beta < \beta_c$ because we did not describe the critical behaviour for $\beta > \beta_c$. The description of it along the same lines would lead to the other half of $\overline{AB}$

Applying the assumptions made earlier, we find, for $\beta < \beta_c$:

$$\sum_{\underline{m}\in C_L} \langle \sigma_{\underline{o}}\ \sigma_{\underline{m}} \rangle_{(\beta,o)} \propto \sum_{\underline{m}\in C_L} c_2 \frac{e^{-\mathcal{E}(\beta)|\underline{m}|}}{|\underline{m}|^{d-2+\eta}}$$

$$\sum_{\underline{m}\in C_L} \langle \sigma_{\underline{o}}\ \sigma_{\underline{m}} \rangle_{(\beta,o)} |\underline{m}|^2 \propto \sum_{\underline{m}\in C_L} c_2 \frac{e^{-\mathcal{E}(\beta)|\underline{m}|}}{|\underline{m}|^{d-4+\eta}}$$

$$\sum_{\underline{m},\underline{p},\underline{q}\in C_L} \langle \sigma_{\underline{o}}\ \sigma_{\underline{m}}\ \sigma_{\underline{\ell}}\ \sigma_{\underline{p}} \rangle_{(\beta,o)} \propto \sum_{\underline{m},\underline{p},\underline{q}\in C_L} e^{-\mathcal{E}(\beta)\Delta(\underline{o}\ \underline{m}\ \underline{n}\ \underline{q})} \chi_4(o\ \underline{m},\ \underline{p},\ \underline{q}) \ .$$

Furthermore

$$\mathcal{E}(\beta) = (\beta_c - \beta)^{-\nu}\, \varkappa_o$$

hence

$$y_2 \underset{a\to 0}{\propto} y_1 \left[\int_{|x|\le La} e^{-\varkappa_o \xi |x|} \frac{dx}{|x|^{d+\eta-4}} \Big/ \int_{|x|\le La} \frac{e^{-\varkappa_o \xi |x|}\, dx}{x^{d+\eta-2}} \right]$$

where ξ is a new name for $a^{-1}|\beta - \beta_c| = \xi$ and

$$y_3 \underset{a\to 0}{\propto} y_1^2 \quad a^{\omega_4 - 2(d+\eta-2)} \frac{\left(\int e^{-\varkappa_o \xi \Delta(0\ x_1 x_2 x_3)} dx_1\, dx_2\, dx_3\right)}{\left(c \int e^{-\varkappa_o \xi |x|} \frac{dx}{x^{d+\eta-2}}\right)^2}$$

which, as ξ varies between 0 and ∞, parametrizes the part of the curve $\overline{AB}$ associated with $\beta - \beta_c < 0$.

Notice that

$$\omega_4 - 2(d + \eta - 2) = d - (2 - \eta)\frac{\delta + 1}{\delta - 1} = -\zeta \ .$$

Hence if $\zeta < 0$ (i.e. if the Buckingham-Gunton scaling law holds with inequality sign (which we remember corresponded to a gaussian distribution of the block spins at the critical point), we find that the curve $\overline{AB}$ collapses into the line $y_3 = 0$ (zero renormalized coupling constant, i.e. trivial theory). If $\zeta = 0$ the line $\overline{AB}$ stays non trivial when $a \to 0$, $La \to \infty$ and this indicates that non trivial theories exist in such a case.

To really reach the above conclusions one should perform a similar analysis of the other piece of the curve $\overline{AB}$ corresponding to $\beta > \beta_c$ (as already mentioned). This is not yet done and we shall not enter into this problem.

It is quite clear that the above discussion can be used to approximate in terms of the critical properties of the Ising model the field theories whose renormalized parameters lie close to the line $\overline{AB}$ whose equation was just derived.

The discussion just made shows in a very concrete way the deep connection between two very interesting and unsolved problems: the critical point in the Ising model and relativistic quantum field theory of the type $z_3\ \phi^4$.

IV. Remarks

1) If $\zeta = 0$ the critical point $(\theta = 0)$ recedes at ∞ in the cut-off removal. So the allowed region of (y_1, y_2, y_3) depends only on the critical region where $\beta < \beta_c$ (high temperature side). This is probably due to the fact that by imposing $\int \langle \phi(0)\ \phi(x) \rangle dx < \infty$ we have excluded spontaneous symmetry breakdown, i.e. $\langle \phi(0)\ \phi(x) \rangle \to \mu^2 > 0$. A similar theory to the one developed above should be possible allowing a symmetry breakdown: this is an interesting open question.

In such a theory the "allowed region" will probably be determined by the critical region where $\beta > \beta_c$ (low temperature side).

2) The above theory assumes perfect scaling: it would be interesting to see what would happen if scaling is not perfect (i.e. logarithmic corrections or more length scales are allowed).

3) $\lambda \phi^4$ theory is known (rigorously) to exist in $d = 1,2,3$. It is an open problem to derive from this fact information about the critical behaviour of the $d = 2, 3$ dimensional Ising models ("converse problem").

4) Let us sketch the renormalizability question. Let f be a smooth function with compact support on R^d. Define

$$\phi(f) = a^d \sum_{\underline{n} \in C_L} f(\underline{n}a)\ \phi_{\underline{n}} \ .$$

Then since $\alpha_1, \alpha_2 > 0$ the Neumann inequalities hold [10]

i.e. $(\forall L, a)$:

$$\langle \phi\ (f)^{2n} \rangle \leq (2n-1)!!\ \ \langle \phi\ (f)^2 \rangle^n \ .$$

but

$$\langle \phi(f)^2 \rangle = a^{2d} \sum_{\underline{n},\underline{m} \in C_L} f(\underline{m}a) f(\underline{n}a) \ . \ \langle \phi_{\underline{n}}\ \phi_{\underline{m}} \rangle \leq$$

$$\leq (\max |f|) a^d \sum_{\underline{n} \in C_L} |f(\underline{n}a)| \left(a^d \sum_{\underline{m}} \langle \phi_{\underline{n}}\ \phi_{\underline{m}} \rangle \right) \underset{\substack{a \to 0 \\ La \to \infty}}{\propto}$$

$$\propto (\max |f|) \left(\int_{R^d} d\underline{x}\ |f(\underline{x})| \right) y_1 < \infty$$

hence all the Schwinger functions make sense and are uniformly bounded as $a \to 0$, $La \to \infty$ (at least if tested with smooth functions). So renormalization is achieved by fixing $0 < y_1 < \infty$.

5) Exercise: derive the equation of the curve AB when $d = 1$. (Hint: if $\xi_1 < \xi_2 < \xi_3 < \xi_4 \dots < \xi_{2n}$ the exact solution of the 1-dimensional Ising model is

a) $\langle \sigma_{\xi_1} \dots \sigma_{\xi_{2n}} \rangle = \langle \sigma_{\xi_1} \sigma_{\xi_2} \rangle \langle \sigma_{\xi_3} \sigma_{\xi_4} \rangle \dots \langle \sigma_{\xi_{2n-1}} \sigma_{\xi_{2n}} \rangle$

b) $\langle \sigma_{\xi_1} \sigma_{\xi_2} \rangle = (\mathrm{th}\, \beta)^{|\xi_1 - \xi_2|}$

The result is

$$\frac{y_2}{y_3^2} = y_1^3 \frac{1}{18} .$$

REFERENCES

1) M. DUNEAU, D. IOGOLNITZER, B. SOUILLARD, Comm. Math. Phys. 35, 307 (1974).

2) G. GALLAVOTTI, A. MARTIN-LÖF: Il Nuovo Cimento 25B, 425 (1975).

3) W. FELLER: An Introduction to Probability Theory and its Applications, Wiley, New York, 1967. B. GNEDENKO: The Theory of Probability, Moscow, 1973.

4) L.P. KADANOFF: Physics, 2, 263 (1966).

5) See R. GRIFFITHS in Phase Transitions and Critical Phenomena, Vol. 1, edit. by C. DOMB and M.S. GREEN. Academic Press, New York 1972. For noncritical properties of the Ising model see, for instance, G. GALLAVOTTI, "Instability and Phase Transitions in the Ising Model", La Rivista del Nuovo Cimento 2, 133 (1972).

6) D. ABRAHAM, Physics Letters 43A, 163 (1973).

7) C. DOMB in Phase Transitions and Critical Phenomena, Vol. 2, edit. by C. DOMB and M.S. GREEN, Academic Press, New York, 1972.

8) The content of this lecture is taken from the paper by R. SCHRADER, preprint.

9) J.L. LEBOWITZ, Comm. Math. Phys. 35, 87 (1974).

10) C. NEWMANN, to be published.

THE APPLICATION OF RENORMALIZATION GROUP TECHNIQUES TO QUARKS AND STRINGS

LEO P. KADANOFF

Department of Physics
Brown University
Providence, Rhode Island 02912, USA

The Application of Renormalization Group Techniques to Quarks and Strings

LEO P. KADANOFF

Department of Physics
Brown University
Providence, Rhode Island 02912, USA

Introduction

These lectures describe how modern renormalization techniques may be applied to models of elementary particle phenomena in which quarks and strings are placed upon a lattice.

We concentrate on two points: (1) How the theory may be made consistent with the apparently contradictory concepts of asymptotic freedom and quark trapping and (2) How one might begin to approach actual renormalization calculations for these problems.

Chapter 1 describes the formulation of lattice problems in statistical physics. Some statistical variables are described, including the standard gaussian "boson" random variables and anti-commuting "fermion" random variables. The discussion of variable types is continued in Chapter 2, which is essentially a description of the relationship between symmetries and variable types. Local and global symmetries and the roles of statistical variables as bases for representations of the symmetry are discussed. Finally, the Wilson model [1 , 2] of quarks and strings is explicitly written down.

In Chapter 3, renormalization group techniques are introduced as a methods of solving one-dimensional problems. Particular attention is given to the fermion variables, which can serve as a representation of quarks, and to the case in which the statistical variables define a homogeneous space for the symmetry group. Following Migdal[3,4], we describe how the solution of the one-dimensional problem for this case can be converted into a solution of a two-dimensional problem with a local (gauge) symmetry. In this chapter, we also describe how the renormalization method can be used to calculate Green's functions.

Chapter 4 describes the qualitative properties of K. Wilson's lattice model. The quarks are represented by fermion random variables and the strings by homogeneous variables. This formulation can include asymptotic freedom and the unobservability of free quarks. However, Lorentz invariance is not an automatic consequence of the theory. Instead, it can only arise because the theory is near a critical point. The renormalization group technique is then suggested as a natural way of attacking and viewing these near-critical problems. Furthermore, the simultaneous existence of asymptotic freedom and trapped quarks is explained as a consequence of the non-existence of a phase transition in the strong variables.

Chapter 5 introduces approximation techniques for attacking lattice problems. The potential moving method of Kadanoff, Houghton, and Yalabik is introduced and shown to be variational in its nature. [5],[6] To cover cases in which the potential is strong, the "Fourier Transform" of this potential-moving is also developed. The Migdal approximation is introduced and shown to be derivable from both the direct potential-moving and also its Fourier transform.[7]

I. Statistical Variables and Statistical Sums

This chapter is intended to introduce lattice problems in statistical mechanics and describe the different types of statistical variables which might be employed. We shall be particularly concerned with using the variables to describe the possible symmetries of the system. For this reason, particular stress will be laid upon variables which are transitive representations of the symmetry groups in question. Furthermore, we shall introduce (following Schwinger and Wilson) "fermion" variables Ψ and $\bar{\Psi}$, which are anti-commuting random variables.

1.1 Formulation of Lattice Problems

Begin with a d-dimensional lattice with lattice points $x = (x_1, x_2, \ldots x_d)$. At each lattice site define a statistical variable $\sigma(x)$. Then, one might define a problem in statistical mechanics by writing an effective Hamiltonian which depends upon pairs of nearest neighbor variables

$$A[K,\sigma] = \sum_{\langle x\, x' \rangle} K(\sigma(x), \sigma(x')). \tag{1.1}$$

Here $K(\sigma,\sigma')$ is a function which defines the coupling between nearest neighbors and $\langle xx' \rangle$ means a sum over all nearest neighbors on the lattice.

The action or Hamiltonian (1.1) is used as a weight function which defines statistical averages. More precisely, we define a statistical sum, Tr, as a sum over each of the individual variables $\sigma(x)$ as

$$\mathrm{Tr} = \prod_x \mathrm{tr}_{\sigma(x)} \tag{1.2}$$

where tr_σ means a sum over an individual σ-variable. Then the probability function for the entire system is given by

$$\rho(\sigma) = \frac{e^{A[K,\sigma]}}{Z[K]} \tag{1.3}$$

where the partition function $Z[K]$ is given by:

$$Z\left[K\right] = \mathrm{Tr}\ e^{A\left[K,\sigma\right]} \tag{1.4}$$

The probability function (1.3) can be used to define the average of any function of σ , $Q(\sigma)$, via

$$\langle Q \rangle = \mathrm{Tr}\ \rho(\sigma)\ Q(\sigma). \tag{1.5}$$

In particular, we shall be most interested in a Green's function

$$G(x,x') = \langle \sigma(x)\ \sigma^{*}(x') \rangle. \tag{1.6}$$

It has been conjectured that problems in high energy physics might have a representation akin to Eqs. (1.1) - (1.6). The particular statistical problems which might form a representation of high energy phenomena are those which involve critical or near-critical behavior. The reason we look for critical behavior is relatively direct. We gain some advantage by constructing problems on a lattice since the lattice spacing may serve as a suitable cut-off and prevent divergences. However, in general, a function like $G(x,x')$ will not show a true rotational invariance in a lattice problem. Instead, it will depend on the direction of $(x-x')$ relative to the lattice axes. But, in the particular case of critical phenomena, the solution may exhibit a higher symmetry than the Hamiltonian. For example, in an Ising model in which

$$\mathrm{tr}_{\sigma}\ f(\sigma) = \sum_{\sigma=-1}^{+1} f(\sigma)$$

it is well known that the Hamiltonian (1.1) contains critical points for all dimensionalities higher than 1 and that near these critical points, for large $|x-x'|$, $G(x,x')$ is rotationally invariant and of the form

$$G(x,x') \sim e^{-|x-x'|m}$$

where the inverse coherence length, m, goes to zero as the problem approaches criticality.

Therefore, it lies within the realm of possibility that if we define a Hamiltonian like (1.1) on a very fine lattice (with lattice constant much smaller than one fermi) and if we insist that this

Hamiltonian be near-critical, we might have an effectively Lorentz-invariant description of high energy systems. This calculation is to be carried out in a fake four-dimensional Euclidian world. The fourth component of x will then be analytically continued to pure imaginary values to get into a Lorentz-invariant description.

1.2 Boson and Fermion Variables

In the previous section, we introduced one type of statistical variable, the Ising variable $\sigma(x)$ with the algebraic property

$$[\sigma(x)]^2 = 1 \tag{1.7}$$

A full definition of the trace over this kind of variable may be given by saying that

$$\begin{aligned} \mathrm{tr}_\sigma\, 1 &= 1 \\ \mathrm{tr}_\sigma\, \sigma &= 0 \end{aligned} \tag{1.8}$$

In general a statistical variable $\sigma(x)$ is given by

1. algebraic rules for adding and multiplying the variables,and
2. a definition of the fundamental statistical sum tr_σ

Sometimes $\sigma(x)$ will just be a set of real or complex numbers, sometimes it will be a vector of variables, $\sigma_i(x)$, with an internal index i, sometimes it will be a matrix or tensor with, in effect, several internal indices. To represent all these cases we write $\sigma_i(x)$ as $\sigma(1)$, where the 1 implies both the space index x_1 and the internal index i_1. The string variables are a special case. They are defined as being between the spatial points x_1 and x_2 and as being matrices with internal indices i_1 and i_2. We then write these as $U(1,2) = U_{i_1,i_2}(x_1,x_2)$.

The very simplest variable in statistical physics are the "boson" variables $\phi(x)$, which take on all possible real values between $-\infty$ and $+\infty$. They are added and multiplied as real numbers. The basic trace operation is defined by

$$\mathrm{tr}_\phi = \int_{-\infty}^{\infty} \frac{d\phi}{\sqrt{2\pi}} \tag{1.9}$$

The actual boson operators of quantum field theory are complex fields formed from two of these basic "boson" variables as $\phi_1 + i\phi_2$. In this case

$$\mathrm{tr}_{\varphi} = \int \frac{d(\mathrm{Re}\,\varphi)\,d(\mathrm{Im}\,\varphi)}{2\pi} \tag{1.9a}$$

Fermions have a more complex statistical representation.[8],[14] The basic objects are statistical variables $\Psi(1)$ and $\overline{\Psi}(1)$ which have typical fermion anti-commutation properties:

$$\{\Psi(1), \Psi(2)\} = \{\overline{\Psi}(1), \overline{\Psi}(2)\} = \{\Psi(1), \overline{\Psi}(2)\} = 0. \tag{1.10}$$

Therefore, for a particular position and value of the internal indices, there are only four possible quantities which can be formed:

$$1,\ \Psi(1),\ \overline{\Psi}(1),\ \overline{\Psi}(1)\,\Psi(1) = -\Psi(1)\overline{\Psi}(1)$$

The basic trace operation is defined by giving the trace of these objects according to

$$\mathrm{tr}_{\psi}\, 1 = \mathrm{tr}_{\psi}\, \Psi = \mathrm{tr}_{\psi}\, \overline{\Psi} = 0 \tag{1.11a}$$

but

$$\mathrm{tr}_{\psi} \overline{\Psi}\Psi = -\,\mathrm{tr}_{\psi} \Psi\overline{\Psi} = 1 \tag{1.11b}$$

Thus, the only terms which contribute to Tr $e^{A[\Psi]}$ are then products over all spatial indices and internal indices of $\overline{\Psi}(1)\Psi(1)$. We shall see that this rather peculiar representation generates the standard fermi Green's functions.

Notice that the summation operations described by Eqs. (1.11) are not representable as sums with positive weights. For this reason, some of the standard theorems of statistical mechanics will fail for the Ψ's.

1.3 Free Bosons and Quarks

In this section, we analyze the properties of elementary models of non-interacting bosons or quarks. The basic variables for the bosons will be $\phi(1)$, where 1 is a label which includes the position variable x_1 and all the internal indices. The basic action is the Gaussian form

$$A[\phi] = -\frac{1}{2}\phi^2(\bar{1}) + \frac{1}{2}\phi(\bar{1})\,\Sigma(\bar{1},\bar{2})\,\phi(\bar{2}) \tag{1.12}$$

(Sums are assumed over repeated barred indices.) For the quarks, we use variables $\bar{\Psi}(1)$ and $\Psi(1)$ and write the basic action as

$$A[\Psi] = -\bar{\Psi}(\bar{1})\Psi(\bar{1}) + \bar{\Psi}(\bar{1})\Sigma(\bar{1},\bar{2})\Psi(\bar{2}) \tag{1.13}$$

For both cases, we can find the partition function and the free energy in a trivial fashion. First do the bosons. In general we are interested in calculating correlation functions of the form

$$G(1\ 2\ 3\ldots) = (\mathrm{Tr}_{\varphi} e^{A}\varphi(1)\varphi(2)\varphi(3)\ldots)/Z \tag{1.14}$$

Using (1.12) we can write

$$G(1\ 2\ 3\ldots) = -\frac{1}{Z}\,\mathrm{Tr}_{\varphi}\, e^{\frac{1}{2}\varphi(\bar{1})\Sigma(\bar{1}\ \bar{1}')\varphi(\bar{1}')}\varphi(2)\varphi(3)\ldots$$
$$\times\,\delta\frac{\partial}{\partial\varphi(1)}\; e^{-\frac{1}{2}\varphi^{2}(\bar{1})}\;. \tag{1.15}$$

Since Tr_{φ} is an integrating procedure we can integrate by parts to find

$$G(1\ 2\ 3\ldots) = \frac{1}{2}\Sigma(1\ \bar{1}')G(\bar{1}'2\ 3\ldots) + \frac{1}{2}\Sigma(\bar{1}\ 1)G(\bar{1}'2\ 3\ldots)$$
$$+\ \delta(1\ 2)G(3\ldots) + \delta(1\ 3)G(2\ldots) + \ldots \tag{1.16}$$

So, for example, with Σ symmetrical

$$G(1\ 2) = \Sigma(1\ \bar{1})G(1\ 2) + \delta(1\ 2) \tag{1.17}$$

which has the symbolic solution

$$G = \frac{1}{1-\Sigma} \tag{1.18}$$

Finally, we get the partition function

$$Z = \mathrm{Tr}\ e^{A} \tag{1.19}$$

by noting that if we change Σ infinitesimally, $\ln Z$ changes by

$$\delta\ln Z = G(\bar{1},\bar{2})\,\delta\Sigma(\bar{2},\bar{1}) \tag{1.20}$$

In the symbolic notation we have

$$\delta \ln Z = \text{trace} \frac{1}{1 - \Sigma} \delta \Sigma \qquad (1.21)$$

where trace means a diagonal sum over the coordinates 1 and 2. This equation can be integrated to read

$$\delta \ln Z = - \delta \, \text{trace} \ln (1 - \Sigma) \qquad (1.22)$$

so that, since Z = 1 at $\Sigma = 0$,

$$\ln Z = \text{trace} \ln G \qquad (1.23)$$

The quarks are no more complex. The action for a single quark field $\Psi(1)$ and $\bar{\Psi}(1)$ is

$$A_1 = -\bar{\Psi}(1)\Psi(1) + \bar{\Psi}(1)\eta(1) + \bar{\eta}(1)\Psi(1) \qquad (1.24)$$

with

$$\begin{aligned} \eta(1) &= \Sigma(1,\bar{2})\Psi(\bar{2}) \\ \bar{\eta}(1) &= \bar{\Psi}(\bar{2})\Sigma(\bar{2},1) \end{aligned} \qquad (1.25)$$

Here $\eta(1), \bar{\eta}(1), \Psi(1)$, and $\bar{\Psi}(1)$ all anticommute [1]. Because of this anticommutation, we can write

$$\begin{aligned} e^{A_1} = 1 &- \bar{\Psi}(1)\Psi(1) + \bar{\Psi}(1)\eta(1) + \bar{\eta}(1)\Psi(1) \\ &- \bar{\Psi}(1)\Psi(1)\bar{\eta}(1)\eta(1) \end{aligned} \qquad (1.26)$$

The four basic traces given by Eq. (1.11) may be summarized as

$$\text{tr}_{\Psi}: \begin{pmatrix} 1 \\ \Psi \\ \bar{\Psi} \\ \bar{\Psi}\Psi \end{pmatrix} = \begin{pmatrix} 0 \\ 0 \\ 0 \\ 1 \end{pmatrix} \qquad (1.27)$$

We can evaluate G(1,2) by looking at the internal trace over the 1 component. Therefore,

---

[1] Thus $\eta(1)$ and $\bar{\eta}(1)$ serve as "sources" acting at point 1 due to all of the other fields $\bar{\Psi}, \Psi$ not at 1 (we take $\Sigma(1,1)=0$).

$$G(1,2) = \frac{1}{Z} \mathrm{Tr}\, e^{A} \Psi(1)\overline{\Psi}(2)$$

$$= \frac{1}{Z} \mathrm{Tr}^{\neq 1}\, e^{A^{\neq 1}}\, \mathrm{Tr}_1\, e^{A_1} \Psi(1)\overline{\Psi}(2). \tag{1.28}$$

We note the results:

$$\mathrm{Tr}_1\, e^{A_1} = -1+\eta(1)\overline{\eta}(1) \tag{1.29a}$$

$$\mathrm{Tr}_1\, e^{A_1} \Psi(1)\overline{\Psi}(2) = -\eta(1)\overline{\Psi}(2) \qquad 1 \neq 2 \tag{1.29b}$$

$$\mathrm{Tr}_1\, e^{A_1} \Psi(1)\overline{\Psi}(2) = -1 \qquad 1 = 2 \tag{1.29c}$$

Using (1.29a) we can rewrite (1.29b) and (1.29c) as

$$\mathrm{Tr}_1\, e^{A_1} \Psi(1)\overline{\Psi}(2) = \mathrm{Tr}_1\, e^{A_1}\eta(1)\overline{\Psi}(2) \qquad 1 \neq 2 \tag{1.30a}$$

$$\mathrm{Tr}_1\, e^{A_1} \Psi(1)\overline{\Psi}(2) = \mathrm{Tr}_1\, e^{A_1}\, (1+\eta(1)\overline{\eta}(1)) \qquad 1 = 2 \tag{1.30b}$$

The Green's function can then be written, using (1.25)

$$G(1,2) = (1-\delta(1,2))\, \Sigma(1,\overline{1})G(\overline{1},2)$$

$$+ \delta(1,2) \left[1+ \Sigma(1,\overline{1})G(\overline{1},\overline{2})\, \Sigma(\overline{2},2)\right]. \tag{1.31}$$

We can also derive the "adjoint" equation

$$G(1,2) = (1- \delta(1,2))G(1,\overline{1})\, \Sigma(\overline{1},2)$$

$$+ \delta(1,2) \left[1+\Sigma(1,\overline{1})G(\overline{1},\overline{2})\, \Sigma(\overline{2},2)\right]. \tag{1.32}$$

From this equation for $1 \neq 2$

$$G(1,2) = G(1,\overline{1})\, \Sigma(\overline{1},2) \tag{1.33}$$

So that (remembering $\Sigma(1,1) = 0$)

$$- \Sigma(1,\overline{1})G(\overline{1},1) + \Sigma(1,\overline{1})G(\overline{1},\overline{2})\, \Sigma(\overline{2},1)$$

$$= - \Sigma(1,\overline{1})G(\overline{1},1) + \Sigma(1,\overline{1})G(\overline{1},1) = 0.$$

We then obtain

$$G(1,2) = \Sigma(1,\bar{1})G(\bar{1},2) + \delta(1,2)$$

which is identical to the boson result.

The higher order Green's functions are slightly different however. For example, for the fermion case

$$G_2(1\ 2,\ 1'2') = \langle \Psi(1)\Psi(2)\bar{\Psi}(2')\bar{\Psi}(1')\rangle$$

$$= G(1,1')G(2,2')-G(1,2')G(2,1'). \tag{1.34}$$

The minus sign is characteristic of the fermion nature of Ψ and $\bar{\Psi}$. Similarly, an extra minus sign appears in the derivation of the partition function for the fermions. This arises at the point

$$\delta \ell nZ = \frac{1}{Z}\ \mathrm{Tr}\ e^{A}\ \delta\, \Sigma(\bar{1}\ \bar{2})\bar{\Psi}(\bar{1})\Psi(\bar{2}) \tag{1.35}$$

since we must anticommute $\bar{\Psi}$ and Ψ to obtain

$$\delta \ell nZ = -\delta\, \Sigma(\bar{1}\ \bar{2})G(\bar{2}\ \bar{1}) \quad . \tag{1.36}$$

The rest of the analysis is the same as for bosons with the result

$$\ell nZ = -\mathrm{Trace}\ \ell nG. \tag{1.37}$$

1.4 Specific Examples

To see the possible relationship between these Green's functions and the propagators of high energy physics, we shall look at two specific examples. Start from a "scalar" example in which the fields have no internal indices. Then, a nearest neighbor interaction with coupling constant K can be represented by choosing

$$\Sigma(x_1,x_2) = \sum_{\alpha} K\ [\delta(x_1-x_2-\hat{e}_\alpha a_0)+ \delta(x_1-x_2+\hat{e}_\alpha a_0)]\ . \tag{1.38}$$

Here $\hat{e}_\alpha$ are the set of d lattice vectors

$$\hat{e}_1 = (1,0,0,\ldots)$$
$$\hat{e}_2 = (0,1,0,\ldots)$$
$$\hat{e}_3 = (0,0,1,\ldots) \quad . \tag{1.39}$$

Then Eq. (1.17) is immediately solved by Fourier transformation, which gives

$$G(p) = \frac{1}{1 - \Sigma(p)} \tag{1.40}$$

with

$$\Sigma(p) = K \sum_{\hat{e}} e^{ip\cdot\hat{e}a_o}$$

$$= 2K \sum_{\alpha=1}^{d} \cos p_\alpha a_o \ . \tag{1.41}$$

In turn then the coordinate space Green's function takes the form

$$G(x,x') = \int \frac{d\,p\,a_o{}^d}{(2\pi)^d} \; \frac{e^{ip\cdot(x - x')}}{1-2K\sum_\alpha \cos p_\alpha a_o} \quad . \tag{1.42}$$

We wish to focus on the limit in which $a_o \to 0$ while $x-x'$ remains fixed. In order that G not vanish in this limit, we require that 1-2dK be very small. In particular, we write

$$1-2dK = \frac{a_o{}^2}{2d}\, m^2 \ll 1 \ . \tag{1.43}$$

Then, the denominator in Eq. (1.42) may be expanded in a power series in a_o to give

$$G(x,x') = \int \frac{dp}{(2\pi)^d} \; \frac{d^{ip\cdot(x-x')}}{m^2+p^2} \; \frac{2d}{a_o{}^2} \; a_o{}^d \ . \tag{1.44}$$

Except for the multiplicative factor, $2da_o{}^d$, Eq. (1.44) is almost the same as the boson propagator in field theory. The major difference is the Euclidian nature of the metric in Eq. (1.44) compared with the Minkowski metric in high energy physics. However, the difference may be eliminated by an appropriate analytic continuation.

The second example involves $\Psi(1)$ and $\overline{\Psi}(1)$ with spinor internal indices. These indices appear in matrix multiplication of γ-matrices, γ_α , with α=1,2,...,d. These matrices obey the Euclidian version of the standard anticommutation relations, i.e.,

$$\{\gamma_\alpha, \gamma_\beta\} = 2\,\delta_{\alpha,\beta} \ . \tag{1.45}$$

We then follow Wilson [1] and choose an interaction which, instead of (1.38), has the structure

$$\Sigma(x_1,x_2) = \sum_\alpha K\left[(1-\hat{e}_\alpha\cdot\gamma)\delta(x_1-x_2-\hat{e}_\alpha a_0)\right.$$

$$\left. + (1+\hat{e}_\alpha\cdot\gamma)\delta(x_1-x_2+\hat{e}_\alpha a_0)\right] \quad (1.46)$$

and find that Eq. (1.40) holds true once more but, instead of (1.41) $\Sigma(p)$ obeys

$$\Sigma(p) = 2K \sum_\alpha \left[\cos p_\alpha a_0 - i\,\gamma_\alpha \sin p_\alpha a_0\right] . \quad (1.47)$$

Once again focus on the possibility of having a finite $G(x,x')$ in the limit $a_0 \to 0$. To achieve this result, pick K to be very close to $(2d)^{-1}$ and define a mass, m, by

$$1-2dK = \frac{m\,a_0}{d} . \quad (1.48)$$

Thence, by the same line of reasoning as before, we find the (almost) standard result

$$G(x,x') = \int \frac{dp}{(2\pi)^d} \frac{e^{ip\cdot(x-x')}}{m - i\gamma\cdot p} d\,a_0^{d-1}$$

$$= \int \frac{dp}{(2\pi)^d} \frac{e^{ip\cdot(x-x')}}{m^2+p^2} (m+i\gamma\cdot p) d\,a_0^{d-1} \quad (1.49)$$

II. Symmetries

In setting up a quark-string model we shall eventually choose an action $A(\sigma)$ which depends upon two kinds of variables; quark creation and annihilation variables $\bar{\Psi}(1)$ and $\Psi(1)$ and string variables $U(1,2)$. These statistical quantities will contain internal indices reflecting the basic symmetries of the problem. The quark variables contain spinor indices, color indices, and flavor indices, while $U(1,2)$ is a matrix in its color indices. To describe the symmetries in a general way, we consider $\bar{\Psi}(1)$, $\Psi(1)$ and $U(1,2)$ to be special cases of the general statistical variable $\sigma(1) = \sigma_{i_1}(x_1)$. The internal index i_1 is always introduced in order to reflect some kind of symmetry.

2.1 Global Symmetry

To see how this kind of symmetry arises, imagine an action $A\{\sigma\}$ which is invariant under some group of operations G^{α}. Let G^{α}_{ij} be a unitary matrix representation of this symmetry group. Then, $A\{\sigma\}$ will be invariant under the symmetries of the group if

$$A\{\sigma\} = A\{\sigma'\} \tag{2.1}$$

$\sigma'(x)$ being defined by

$$\sigma'_i(x) = \sum_j G^{\alpha}_{ij} \quad \sigma_j(x) \ . \tag{2.2}$$

Here we define a global symmetry by demanding that the matrix G^{α}_{ij} be independent of x. In general, we choose group representations which are unitary, i.e. which obey

$$\sum_{\alpha} G^{\alpha}_{ij} \, (G^{\alpha}_{kj})^{*} = \delta_{ik}$$

Since, in general, we have

$$A\{\sigma\} = \sum_{xx'} K(\sigma(x),\sigma(x'))$$

(nearest neighbors)

Eq. (2.1) is equivalent to:

$$K(\sigma_1', \sigma_2') = K(\sigma_1, \sigma_2).$$

Of course, (2.1) is not sufficient to guarantee the invariance of the final problem. We must have the basic statistical sum, Tr, also be invariant under the change of variables (2.2). This additional condition will be met if

$$tr_{\sigma(x)} = \prod_i tr_{\sigma_i}(x) \tag{2.3}$$

obeys

$$tr_{\sigma(x)} f(\sigma(x)) = tr_{\sigma(x)} f(G^{\alpha} \sigma(x)) , \tag{2.4}$$

for all group elements G^{α} and all possible functions f.

Equations (2.1) and (2.2) are the condition that the action be invariant under group operations while Eq. (2.4) is the requirement that the basic sums be so invariant. The action will certainly be invariant if $A\{\sigma\}$ is a function of scalars like

$$\sum_i \sigma_i(x)(\sigma_i(x'))^* .$$

Thus, for example, the conventional two-dimensional rotation symmetry of the x-y model is obtained by taking $\sigma_i(x)$ to be a two component vector and by choosing the coupling function K to depend upon the combination

$$\sigma_1(x)\ \sigma_1(x') + \sigma_2(x)\sigma_2(x') = \vec{\sigma}(x) \cdot \vec{\sigma}(x') .$$

This action will then be invariant under two-dimensional rotations with the rotation matrix

$$G^{\alpha} = \begin{pmatrix} \cos\alpha & \sin\alpha \\ -\sin\alpha & \cos\alpha \end{pmatrix} \tag{2.5}$$

The full statistical problem will be invariant if $tr_{\sigma(x)}$ takes the form

$$tr_{\sigma(x)} = \int d\sigma_1\ d\sigma_2\ \ w(\sigma_1^{\ 2} + \sigma_2^{\ 2}) \tag{2.6}$$

where w is any weight function.

2.2 Transitive Representations

In this way, we make the σ's a basis for a representation of the symmetry group. A particularly simple situation arises when the σ's each form a homogeneous space for that symmetry group [9]. In that case, the trace over $\sigma(x)$ is just a sum over group elements. Then we have the basic sum being defined by

$$\mathrm{tr}_{\sigma}\, f(\sigma) = \sum_{\alpha} f(G^{\alpha}\sigma) \tag{2.7}$$

In this case we say that σ is a transitive representation of the symmetry.

For our purposes, we can take Eq. (2.7) as the definition of a transitive representation. In the x-y model example, since when making the trace tr we are integrating over length, as well as over angles, the σ's are not a transitive representation, except when $w = \delta(\sigma_1{}^2+\sigma_2{}^2-1)$.

Take another example. The usual boson "Hamiltonian" containing terms of 2nd, 4th, 6th,... order in the $\phi(x)$ is invariant under the group of transforms

$$\phi(x) \rightarrow + \phi(x)$$

and

$$\phi(x) \rightarrow - \phi(x) \quad .$$

This symmetry does not define a $\phi(x)$ which is a transitive representation of the symmetry. However, if $\sigma(x) = \pm 1$, then $\sigma(x)$ is indeed a transitive representation of the symmetry. The two elements of the group are $G^0 = 1$, $G^1 = -1$.

Let us give several more examples of transitive variables. The cyclic group of order N can be represented by

$$G^{\alpha} = e^{2\pi i \alpha / N} \qquad \alpha = 0,1,2,\ldots,N-1 \ .$$

Then $\sigma(x)$ should take on these N values. The action can depend on $\sigma(x)\,\sigma^*(x')$ while the trace over σ is simply a sum over these N possible values of σ. The case N=2 is the Ising case discussed above.

The group SU_n has $\sigma_i(x)$ being an n component complex vector of unit length

$$\sum_i \sigma_i(x)\sigma_i^*(x) = 1 \tag{2.8}$$

each vector being capable of arising from the basic vector $\sigma_i^{\,0}=\delta_{i,1}$ by the matrix multiplication

$$\sigma_i(x) = \sum_j u_{ij}(x)\ \delta_{j,1} \ . \tag{2.9}$$

Here u is an n x n unitary matrix

$$u^\dagger u = 1 \tag{2.10}$$

with determinant equal to 1,

$$\det u = 1 \ . \tag{2.11}$$

The action will be invariant if it depends upon $\sum_i \sigma_i(x)\sigma_i^*(x')$. We have $\sigma_i' = \sum_{ij} u_{ij}\,\sigma_j$ while the invariant sum is defined symbolically by

$$\mathrm{tr}_\sigma\, f(\sigma) = \int d\,u\ f\,(u\,\sigma) \ . \tag{2.12}$$

Here $\int du$ can be defined as follows.[19] We write u as a sum of real and imaginary parts

$$u_{jk} = u^1_{jk} + iu^2_{jk} \ . \tag{2.13}$$

Then $\int du$ is given by

$$\int du = \int \prod_{ij} \Big[du^1_{ij}\ du^2_{ij}\ \delta(\delta_{ij} - \sum_k u_{ij}\ u^*_{kj})\Big]\ \delta(\det u - 1). \tag{2.14}$$

For SU_2 this integral is simple since u can be expressed in terms of the standard Pauli matrices as

$$u = b_0 - i\vec{b}\cdot\tau = \begin{pmatrix} b_0 - ib_3 & -ib_1 - b_2 \\ b_2 - ib_1 & b_0 + ib_3 \end{pmatrix} \ . \tag{2.15}$$

Then, by computing $uu^\dagger$ and det u, one finds that b_0 and $\vec{b}$ are real, and

$$\int du = \int d^3\,\vec{b}\ db_0\ \ \delta(b_0^{\,2} + b^2 - 1) \ . \tag{2.16}$$

The similar integrals for SU_3 are much harder to set up. One has an integral over 18 variables, with 10 δ-functions. Instead of writing the weight function we list some moments derived from

$$\langle x \rangle = \frac{\int du\, x\,(u)}{\int du} \quad . \tag{2.17}$$

We then find

$$\langle 1 \rangle = 1$$

$$\langle u_{ij} \rangle = \langle u^{\dagger}_{ij} \rangle = 0$$

$$\langle u_{ij}\, u_{k\ell} \rangle = \langle u^{\dagger}_{ij}\, u^{\dagger}_{k\ell} \rangle = 0$$

$$\langle u_{ij} (u^{\dagger})_{k\ell} \rangle = \frac{1}{3}\, \delta_{i\ell}\, \delta_{jk}$$

$$\langle u_{ij}\, u_{jk}\, u^{\dagger}_{mn} \rangle = 0$$

$$\langle u_{ij}\, u_{k\ell}\, u_{mn} \rangle = \frac{1}{6}\, \epsilon_{ikm}\, \epsilon_{j\ell n} \tag{2.18}$$

For instance $\langle u_{ij} \rangle = 0$ can be proved as follows: $\int du$ is invariant under the transformation $u \to u' = ue^{\pm 2\pi i/3}$, because in Eq. (2.14) the δ- functions are trivially invariant under this transformation, and it is easily seen that the determinant of the transformation is equal to one. Then $\langle u_{ij} \rangle$ is an average over the three cubic roots of unity, and hence, it must vanish.

Later on, there will be considerable use made of various transitive representations.

2.3 String Variables and Local Symmetries

For the global symmetry, the entire action $A\{\sigma\}$ is transformed with the aid of the same matrix G^{α}_{ij}. In particular, the new variables are

$$\Psi_i'(x) = \sum_j G^{\alpha}_{ij}\ \Psi_j(x)$$
$$\overline{\Psi}_i'(x) = \sum_j (G^{\alpha})^{+}_{ij}\ \overline{\Psi}_j(x) \quad . \tag{2.19a}$$

If there is a string variable $U_{ij}(x,x')$ defined for each ordered pair of nearest neighbors, xx', it transforms according to

$$U'(x,x') = G^{\alpha} U(x,x')(G^{\alpha})^{+} . \qquad (2.19b)$$

The color symmetry assumed in high energy physics is a far richer symmetry than this. Under transformations, $A\{\Psi, \bar{\Psi}, U\}$ is required to be invariant even when the transformation is different at every point in space. Thus, if $G^{\alpha(x)}$ is a color transformation matrix which depends upon the space-time point x, A is required to be unchanged under the transformation:

$$\Psi(x) \longrightarrow \Psi'(x) = G^{\alpha(x)} \Psi(x)$$

$$\Psi(x) \longrightarrow \bar{\Psi}'(x) = \bar{\Psi}(x) \left[G^{\alpha(x)}\right]^{\dagger}$$

$$U(x,x') \rightarrow U'(x,x') = G^{\alpha(x)} U(x,x') \left[G^{\alpha(x')}\right]^{\dagger} \qquad (2.20)$$

(We omit here the ij indices.)

This new kind of invariance is easily constructed. Just let A depend upon terms like [1]

$$\sum_{ij} \bar{\Psi}_i(x) U_{ij}(x,x') \Psi_j(x') \quad . \qquad (2.21a)$$

This kind of term is certainly invariant under the local color symmetry. It represents the motion of a quark from x′ to x, with the aid of the bit of string $U_{ij}(x,x')$. Other invariant terms which involve no motion are

$$\sum_{i} \bar{\Psi}_i(x) \Psi_i(x) \qquad (2.21b)$$

and

$$\sum_{ij} U_{ij}(x,x') U_{ji}(x',x) \quad . \qquad (2.21c)$$

One additional term is needed: A term which provides string-string interactions. Imagine that we construct a closed path in the lattice from x_1 to x_2 to .. x_n to x_1 and construct the combination

$$\sum_{\ell_1 \ell_2 \ldots \ell_n} U(1,2)\, U(2,3)\, U(3,4) \ldots U(n,1) \quad . \qquad (2.21d)$$

which is summed over internal indices ℓ_i. This term is also unchanged by any local color transformation. Our action will then contain terms like these invariant structures.

2.4 The Wilson Model

Now, we can put together the pieces and write the Wilson model of quarks and strings. Each $\Psi(1)$ and $\overline{\Psi}(1)$ is a four-component spinor with an additional flavor index f and a color index i = 1,2,3. The color symmetry is a SU_3 symmetry and is an exact local symmetry. The flavor index takes on the values f= up, down, strange, and (perhaps) charmed. The first three values represent the broken SU_3 symmetry of Gell-Mann and Ne'eman. Thus we write $\overline{\Psi}_{i,f}(x)$ and $\Psi_{if}(x)$. The spinor indices are never written explicitly but represented by the γ-matrices of Eq. (1.39).

Each ordered pair of nearest-neighbor sites on the lattice defines a string bit variable $U_{ij}(x,x')$. Here i and j are SU_3 variables. They run from 1 to 3. We make $U_{ij}(x,x')$ itself a unitary matrix with unit determinant. To limit the number of variables, we choose

$$U(x',x) = \left[U(x,x')\right]^{\dagger} \tag{2.22}$$

For each x,x' and j, $U_{ij}(x,x')$ is a vector (with index i) which is a basis function for a transitive representation of the SU_3 symmetry, so that, again, the trace will be a sum over group elements.

Now we will set up an action which includes all the symmetries mentioned so far. This action is a sum of two terms

$$A = A_{\psi} + A_U . \tag{2.23}$$

Here A_U depends only on the string variables while A_{ψ} depends on both quark and string variables. Visualize a d-dimensional simple "cubic" lattice. On this lattice the nearest neighbors are connected by vectors $\pm \hat{e}_{\mu} a_o$ where

$$\hat{e}_1 = (1,0,0,\dots)$$

$$\hat{e}_2 = (0,1,0,\dots)$$

$$\hat{e}_3 = (0,0,1,\dots) \tag{2.24}$$

We then write the term in the action involving only U's as a sum over basic squares (called placquettes) with corners x, $x+\hat{e}_\mu a_0$, $x+(\hat{e}_\mu+\hat{e}_\nu)a_0$, $x+\hat{e}_\nu a_0$, i.e.,

$$A_U = \frac{1}{2}\sum_{\mu\nu}\sum_x J_{\mu\nu} \text{ trace }\Big(U(x,x+\hat{e}_\mu a_0)$$

$$U(x+\hat{e}_\mu a_0, x+(\hat{e}_\mu+\hat{e}_\nu)a_0)$$

$$U(x+(\hat{e}_\mu+\hat{e}_\nu)a_0, x+\hat{e}_\nu a_0)$$

$$U(x+\hat{e}_\nu a_0, x)\Big) + \text{complex conjugate}. \tag{2.25}$$

Here $J_{\mu\nu}$ is the coupling constant for the square of type $\mu\nu$. In any physical result $J_{\mu\nu}$ will have to be independent of μ and ν. In Eq. (2.25), there is an implied matrix multiplication and there is a trace over the color indices of U.

The action (2.25) is the simplest color symmetric structure [10-13] that can be constructed from the U's. Notice that terms like (2.21c) cannot be usefully included because Eq. (2.22) implies that these terms are simply unity.

Next consider A_ψ. This part of the action contains terms like (2.21a) and also terms like (2.21b). By adjusting our normalization of $\psi_{if}(x)$ we can always make the coefficient of different terms like (2.21b) all be the same. The remaining terms are constructed from (2.21a). Their γ-matrix structure is taken to be the same as in Eq. (1.40). Thus, we write

$$A_\psi = -\sum_{xif}\overline{\Psi}_{if}(x)\,\Psi_{if}(x) + \sum_{xfij}\sum_\mu \overline{\Psi}_{if}(x)\,K_{\mu f}$$

$$\Big[(1-\hat{e}_\mu\cdot\gamma)U_{ij}(x,x+\hat{e}_\mu a_0)\,\Psi_{jf}(x+\hat{e}_\mu a_0)$$

$$+(1+\hat{e}_\mu\cdot\gamma)U_{ij}(x,x-\hat{e}_\mu a_0)\,\Psi_{jf}(x-\hat{e}_\mu a_0)\Big]. \tag{2.26}$$

The μ-dependence of $K_{\mu f}$ will have to disappear in physical applications. Its flavor dependence represents the breaking of the flavor symmetry.

Physically, the second term in Eq. (2.26) represents the motion of quarks and antiquarks to x from $x \pm \hat{e}_\mu a_0$ through the coupling with the string variables.

This model's color symmetry is the statement that the action remains unchanged under the replacements

$$U(x,x') \rightarrow U'(x,x') = u(x)\, U(x,x')\, u^{-1}(x')$$

$$\Psi_f(x) \rightarrow \Psi_f{}'(x) = u(x)\, \Psi_f(x)$$

$$\overline{\Psi}_f(x) \rightarrow \overline{\Psi}_f{}'(x) = \overline{\Psi}_f(x) \left[u(x)\right]^{-1} \quad . \qquad (2.27)$$

Here u(x) is a set of SU_3 matrices which are different for each point in space.

The variables in the action can be combined into a variety of color singlet creation and annihilation combinations, for example,

$$\sum_i \Psi_{if}(x)\ \overline{\Psi}_{if'}(x) \qquad (2.28)$$

and

$$\sum_{ijK} \epsilon_{ijK}\, \Psi_{if_1}(x)\, \Psi_{jf_2}(x)\, \Psi_{Kf_3}(x) \ . \qquad (2.29)$$

The combination (2.28) is connected with the SU_6 meson multiplet of dimension 35 and an SU_6 singlet. The combination (2.29) is connected with a 56 dimensional representation of SU_6 which includes the nucleons.

2.5 Connection with Field Theory

In the limit $a_0 \rightarrow 0$, the theory just outlined should reduce to the field theory description of a set of fermi particles interacting with a gauge field. To make this connection, write U(x,x′) in terms of a line integral of a vector potential $A_\mu(x)$ as

$$U(x,x') = \exp\left[ig \int_x^{x'} d\vec{x}'' \cdot A(x'')\right] \qquad (2.30)$$

Here g will turn out to be the coupling constant of the gauge field theory. Since x and x′ will be separated by one lattice constant, one can expand the exponential in Eq. (2.30) in a power series in a_0. When this expansion is applied to the right-hand side of Eq. (2.25) one finds a part of the action

$$A_U = - \sum_{\mu\nu} \sum_x a_0^4\ F_{\mu\nu}\ F_{\mu\nu}\ g^2 J_{\mu\nu}$$

with

$$F_{\mu\nu} = \partial_\mu A_\nu(x) - \partial_\nu A_\mu(x) - ig[A_\mu(x), A_\nu(x)] \tag{2.31}$$

The standard coupling term [14] is then recovered if we replace the sum over x by an integral and choose $J_{\mu\nu}$ as

$$J_{\mu\nu} = 1/2g^2 \tag{2.32}$$

Then, A_U becomes simply

$$A_U = -\frac{1}{2}\sum_{\mu\nu}\int dx\, F_{\mu\nu}F_{\mu\nu} \tag{2.33}$$

Notice one very important point. To the statistical mechanic weak coupling means $J \rightarrow 0$; to the field theorist weak coupling means $g \rightarrow 0$. Therefore, according to Eq. (2.32) the field theorist and statistical mechanic have exactly opposite views of weak and strong coupling.

The fermion term A_ψ can be handled in a similar fashion. Take the coupling in Eq. (2.26) to be of the form [14]

$$K_{\mu f} = \frac{1}{2d}(1 - m_f a_0/d) \tag{2.34}$$

where m_f is physically the mass of a particle with flavor f. Then, an expansion of Eq. (2.26) in a_0 yields

$$A_\psi = \frac{1}{a_0^{\;3}}\int dx \sum_f \bar{\Psi}_f [m_f - \gamma_\mu(\partial_\mu - igA_\mu)]\Psi_f \tag{2.35}$$

This is then the standard structure [14] of a fermion term in a gauge field theory.

III. Renormalization Theory: A Point of View and a Calculational Method

3.1 Why Renormalization

The quark-string theory involves a formulation in which the variables $\psi(1)$ and $U(x,x')$ appear on a lattice with lattice constant much smaller than one fermi. Clearly, this lattice is only a formulational and calculational tool. It must disappear from all the final results of the theory.

The renormalization theory can be viewed as a description of how we can simultaneously change the lattice constant and the basic action but nonetheless leave all physical results of the theory entirely unchanged. This approach then permits us to visualize how it might be possible that the original lattice drops out of any physical end results of the theory.

At the same time as the renormalization method provides an important insight into the formulation of the problem, it can also provide a very useful calculational tool. In statistical physics, a variety of problems which do not yield to perturbation theory or any other "classical" analytic tool, were attacked with considerable success by approximate renormalization methods.

In this chapter, we describe the general formulation of these methods and their particular application to one-dimensional problems. In distinction to statistical physics, we are not especially interested in the partition function, Z. Instead, our interest will mainly focus upon the Green's functions.

3.2 Formulation

Given a set of statistical variables σ on a lattice with lattice constant a_o and an action $A[K,\sigma]$ which depends upon a set of coupling constants or coupling functions, K, we can, in principle, compute the partition function and all the correlation functions via Eqs. (1.1) - (1.6). Now imagine that we have another set of variables, μ, on a lattice embedded in the original lattice. This new lattice has lattice constant λa_o. We view these new variables μ as providing an alternative description of the original problem.

To make the conversion from one description to the other, we define a function $T(\mu, \sigma)$. At the start, T is arbitrary; in the end, we shall make very specific choices of T to achieve calculational convenience. This T is used to define a new effective action via

$$e^{A'(\mu)} = \mathrm{Tr}_{\sigma}\, e^{T(\mu,\sigma) + A[K,\sigma]} . \tag{3.1}$$

We demand the one condition that the transformation leave the partition function invariant, i.e., that

$$\mathrm{Tr}_{\mu}\, e^{T(\mu,\sigma)} = 1 . \tag{3.2}$$

Then, the partition function generated by $A'(\mu)$, i.e.,

$$Z' = \mathrm{Tr}_{\mu}\; e^{A'(\mu)}$$

will be identical with the partition function generated by $A[K,\sigma]$.

The new action $A'(\mu)$ will, in general, be very complex. If the transformation obeys translational (Galilian) invariance, $A'(\mu)$ can be characterized by the same kind of coupling functions which might appear in a very general $A[K,\sigma]$, i.e., coupling functions K describing two body, three body, ... interactions. Thus, we can write $A'(\mu)$ as $A[K',\mu]$ where K' is the set of new couplings generated by the transformation.

In summary the change of description $\sigma \rightarrow \mu$ has two effects. The lattice constant changes from a_0 to λa_0; the couplings change from K to K'. If we could calculate the sum in Eq. (3.1), we could find out how the new couplings K' depend upon the old. In general, we write this dependence as

$$K' = R^{\lambda}[K] . \tag{3.3}$$

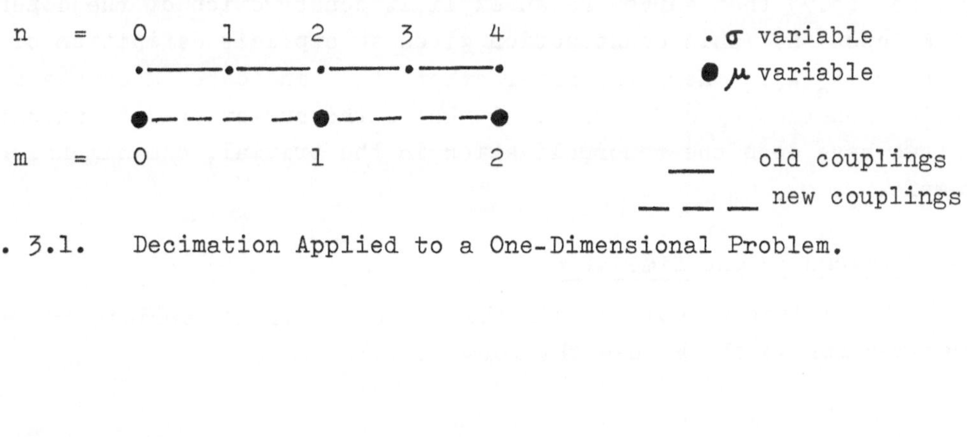

Fig. 3.1. Decimation Applied to a One-Dimensional Problem.

3.3 Example: Decimation in One-Dimension

All this can be made very explicit in a one-dimensional example [15,16] . Let the x in $\sigma(x)$ be simply $n\, a_0$, where n is an integer. Let the basic action be

$$A[\sigma, K] = \sum_n K(\sigma(na_0), \sigma((n+1)a_0)) \quad . \tag{3.4}$$

Let the new variables, $\mu(X)$, be defined at the points

$$X_m = \lambda\, m\, a_0 = 2\, m\, a_0 \, .$$

Hence we have a change in lattice constant by a factor of 2. We choose these new variables to be essentially identical with the old ones at the same point by writing

$$e^{T(\mu,\sigma)} = \prod_m \delta(\mu(X_m) - b\, \sigma(X_m)) \tag{3.5}$$

where δ is the delta function. In Eq. (3.5), b is a parameter which we shall choose for our eventual calculational convenience. The net result of the transformation function (3.5) is to decimate or thin out the statistical variables leaving us with half as many variables in the new problem as in the old. Each new variable is essentially equal to b times an old variable. But half of the old variables are pure summation variables and therefore disappear. In all the applications of this chapter, b will equal 1.

In one dimension the new action $A'(\mu)$ is easy to calculate from Eq. (3.1). The result is exactly of the nearest neighbor form (3.4),

$$A'(\mu) = \sum_m K'(\mu(ma_0\lambda), \mu((m+1)a_0\lambda)) \tag{3.6}$$

and

$$e^{K'(\mu,\mu')} = \mathrm{tr}_\sigma\; e^{K(\mu,\sigma) + K(\sigma,\mu')} \tag{3.7}$$

Equation (3.7) then serves as an explicit construction of the dependence of K′ upon K. This construction gives an explicit definition of the function $R_0^2(K)$. Here the superscript "2" indicates a change in lattice constant by a factor of 2 while the subscript "o" indicates that we have done the renormalization in the trivial, one-dimensional, example.

3.4 Fermions in One Dimension

The simplest example of all this is the case of fermions in one-dimension for which K has the form (1.13), i.e.,

$$K(\psi, \psi') = -\tfrac{1}{2}\bar{\Psi}\Psi - \tfrac{1}{2}\bar{\Psi}'\Psi' + K\bar{\Psi}(1-\gamma_1)\Psi' + K\bar{\Psi}'(1+\gamma_1)\Psi \quad . \tag{3.8}$$

On the right-hand side of Eq. (3.8), K represents the hopping interaction between neighboring sites. The exact Green's function for this problem is given by Eq. (1.22) which implies

$$G(x,x') = \int_{-\pi}^{\pi} \frac{d\,p\,a_0}{2\pi} \frac{e^{i\,p(x-x')}}{1 - 2Ke^{-ip\gamma_1 a_0}} \quad .$$

If $2K \ll 1$, this Green's function may be evaluated in the form

$$G(x,x') = \frac{1+\gamma_1\ \mathrm{sgn}(x-x')}{2}\ e^{-m|x-x'|} \tag{3.9}$$

with

$$m = -(\ln 2\,K)/a_0 \quad . \tag{3.10}$$

Now let us see how to get to this result via a renormalization technique. Calculate the sum (3.7). The sum turns out to be exactly of the form (3.8) except that in the last two terms the parameter K is replaced by a new parameter K', given by

$$K' = 2K^2 \tag{3.11a}$$

The change $K \to K' = 2K^2$ is the coupling renormalization produced by the lattice constant redefinition $a_0 \to 2a_0$. In this way we have evaluated the recursion function as

$$R_0{}^2(K) = 2K^2 \quad . \tag{3.11b}$$

The first consequence of Eq. (3.11b) is that there are two special values of the coupling, K^*, which obey

$$K^* = R_0{}^2(K^*) \quad .$$

These special values are $K^* = \tfrac{1}{2}$ and $K^* = 0$. These fixed points of the recursion relation are of significance because, at these values of the coupling, the scale affects no physical results. These special, scale-invariant points, are termed fixed points.

In the case outlined here, the fixed point $K^* = 0$ is just the case in which no particle can move, $m = \infty$; while the point $K^* = \tfrac{1}{2}$

has scale invariance because $m = 0$.

The next consequence is most easily drawn if we write the recursion relation which appears after n-iterations of Eq. (3.11). In that case we find

$$a_0 \longrightarrow 2^n a_0 = \lambda a_0$$

$$K \longrightarrow K' = \tfrac{1}{2}(2K)^{\lambda} \tag{3.12a}$$

so that

$$R_0^{\lambda}(K) = \tfrac{1}{2}(2K)^{\lambda} . \tag{3.12b}$$

We then argue that our result will be most conveniently understood if we introduce, following Wegner [17], a variable $h(K)$ with a simple recursion relation in the form that a change

$$a_0 \longrightarrow \lambda a_0 \tag{3.13a}$$

will produce a new h

$$h_\alpha' = \lambda^{y_\alpha} h_\alpha . \tag{3.13b}$$

Here y_α is termed a critical index for the problem.

In our case, the appropriate Wegner variable is

$$h(K) = -\ln 2K \tag{3.14}$$

since then

$$h' = h(K') = -\ln 2K'$$
$$= -\lambda \ln 2K = \lambda h . \tag{3.15}$$

Hence the critical index is

$$y = 1 . \tag{3.16}$$

Notice then that h/a_0 has a scale-invariant meaning, since both numerator and denominator change by a factor of λ as the result of a recursion. If we refer back to Eq. (3.10), we see that this scale-invariant is exactly the mass.

At this point, one can easily see the whole recursion scheme. Start from a strong coupling problem in which $K \approx \tfrac{1}{2}$ and thus $h \approx 0$. Do many recursions. Then, according to Eq. (3.15), h continually increases until finally $h \gg 1$ and thus $K \gg 1$. This resulting weak coupling problem has then been related to the original strong-coupling problem.

Specifically examine the Green's function at lattice constant a_0, and coupling K, $G(x-x', a_0, K)$. If x and x′ are points which are unsummed in the recursion, the new variables and the old are exactly the same. Hence, in this case

$$G(x-x', 2a_0, K') = G(x-x', a_0, K)$$

for

$$K' = R_0^2(K) .$$

After $\lambda = 2^n$ iterations of this same argument, one finds the identity

$$G(x-x', a_0, K) = G(x-x', \lambda a_0, R_0^{\lambda}(K)) . \tag{3.17}$$

Apply Eq. (3.17) to the particular case in which

$$x-x' = [\mathrm{sgn}(x-x')] \lambda a_0 \tag{3.18}$$

and $R_0^{\lambda}(K)$ is very small. Then, the right-hand side of Eq. (3.17) may be evaluated by first order perturbation theory as applied to the nearest-neighbor Green's function in a weakly interacting problem. The result of this first order perturbation theory is

$$G(x-x', \lambda a_0, K') = (1 + \mathrm{sgn}(x-x') \gamma_1) K'$$

so that Eq. (3.17) implies

$$G(\lambda a_0 \,\mathrm{sgn}(x-x'), a_0, K) = (1 + \mathrm{sgn}(x-x')\gamma_1) K'$$

$$= \frac{1 + \mathrm{sgn}(x-x')\gamma_1}{2} (2K)^{\lambda} . \tag{3.19}$$

The final line of Eq. (3.19) follows from the evaluation (3.12b) of the recursion formula. Now use Eq. (3.18) to eliminate λ from Eq. (3.19). The final result is

$$G(x-x', a_0, K) = \frac{1+\mathrm{sgn}(x-x')\gamma_1}{2} \exp\left[(\ln 2K) \frac{|x-x'|}{a_0}\right] \tag{3.20}$$

This is exactly the same as Eqs. (3.9) and (3.10). Hence the Green's function has been evaluated via the renormalization scheme. This example clearly demonstrates how one can use the renormalization technique to solve a "strong coupling problem" if one possesses a solution to the problem in the weak coupling limit.

3.5 Renormalization and Phase Diagrams

Consider a renormalization which changes the length scale from a_0 to λa_0 and the couplings from $\underline{K}$ to $\underline{K}' = R^{\lambda}(\underline{K})$. To understand

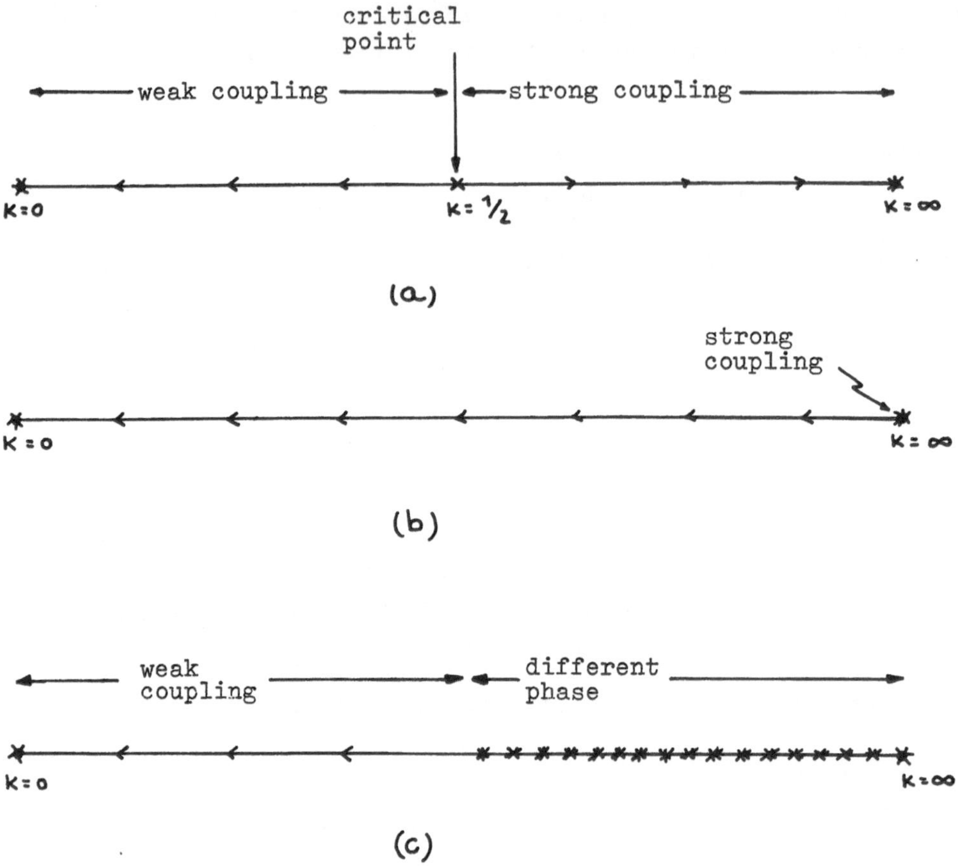

Fig. 3.2 Different phase diagrams.
The stars are fixed points. The arrows-heads diagrams show the directions of flow of couplings as the lattice constant m increased.

the qualitative importance of the renormalization, we must recognize that a renormalization is a charge of variables which cannot change any of the qualitative properties of the system [27]. A system with long range correlations will retain long-range correlations under the renormalization. A system with short-range correlations will remain short-ranged after a renormalization. A statistical mechanic describes this situation by saying that a renormalization is a change in coupling constants which leaves the system in exactly the same thermodynamic phase as it was originally. We then state that:

The couplings $\underline{K}_1$ and $\underline{K}_2$ describe precisely the same thermodynamic phase if there is some renormalization transform such that

$$\underline{K}_1 = R^{\lambda}(\underline{K}_2)$$

or

$$\underline{K}_2 = R^{\lambda}(\underline{K}_1)$$

for some value of λ.

Consider, for example, the fermion case in one dimension, which has according to Eqs. (3.12) - the recursion relation

$$K' = \frac{1}{2}(2K)^{\lambda} \quad .$$

Notice that in this case there are three phases for real $K > 0$, i.e.

The weak coupling phase:	$0 \leq 2K < 1$
The strong coupling phase:	$2K > 1$
The critical phase:	$2K = 1$.

The phase diagram for this situation is depicted in Fig. 3.2a. According to Eq. (3.20) these phases are really different in that the first has exponentially decreasing correlations, the second has correlations which grow for large separations, and the third has correlations independent of distance. A point separating two phases, e.g. $2K = 1$, is termed a critical point. It is also a fixed point since it is a point which remains unchanged under a recursion [18].

The arrows on the phase diagram 3.2a indicate the direction of $\underline{K}' = R_o^{\lambda}(\underline{K})$ as λ is increased infinitesimally from unity. An increase in λ increases the couplings in the strong coupling phase, decreases them in the weak coupling phase, and leaves the critical point unchanged.

An alternative phase diagram is shown in Fig. 3.2b. This kind of diagram shows no phase transition except an exceptional one at $K = \infty$. This kind of phase diagram might be produced, for example, by the recursion relation

$$\tanh K' = (\tanh K)^{\lambda} .$$

As we shall see this recursion actually occurs in the one-dimensional Ising model.

More complex situations are conceivable. For example, let

$$K' = 1 + (K-1) \left[1 - \mathrm{Re}\sqrt{1-K}\right]^{\lambda-1} .$$

Then all points for $K < 1$ are in a weak coupling phase while each point for $K > 1$ is a fixed point. Hence each of these $K > 1$ points is a separate phase. We believe that such a possible existence of an infinite number of different phases is characteristic of the Baxter [20,21] model for $d = 2$. It is also probably the kind of behaviour shown by the XY model at $d = 2$. This phase diagram is shown in Fig. 3.2c.

3.6 Transitive Representations

The one-dimensional renormalization can also be carried out in a very explicit fashion in the case in which $\sigma(x)$ is a transitive representation of some symmetry group and in which $K(\sigma,\sigma')$ is invariant under the operations of that group. In this situation, we can use two different schemes for describing $K(\sigma,\sigma')$, a class description and a representation description.

The class description arises from the fact that for any variable σ the group element G^{α} will generate a new variable σ' according to

$$\sigma \rightarrow \sigma' = G^{\alpha}\sigma .$$

Thence, for any pair of elements (σ_1,σ_2) we will generate a new pair according to

$$(\sigma_1,\sigma_2) \rightarrow (\sigma_1,\sigma_2)' = (G^{\alpha}\sigma_1, G^{\alpha}\sigma_2) .$$

We then state that the two pairs (σ_1,σ_2) and (σ_1',σ_2') will be in the same class if and only if there exists a G^{α} such that

$$\sigma_1' = G^{\alpha}\sigma_1$$

$$\sigma_2' = G^{\alpha}\sigma_2 . \qquad (3.21)$$

In general, then, $K(\sigma_1,\sigma_2)$ will depend upon the class into which (σ_1,σ_2) falls. For example, the Ising model $(\sigma = \pm 1)$ is connected with a group with two elements, which can be represented by $G^0 = 1$ and $G^1 = -1$. Each group element is in a class by itself. There are then two classes for (σ_1, σ_2); the class in which $\sigma_1 = \sigma_2$ and the class in which $\sigma_1 = -\sigma_2$.

In general, we denote the different classes by the index r and say that

$$K(\sigma,\sigma') = \sum_r K^r \delta_r(\sigma,\sigma') \quad . \tag{3.22}$$

Here $\delta_r(\sigma,\sigma')$ is a delta symbol defined by

$$\delta_r(\sigma,\sigma') = \begin{cases} 1 & \text{if } (\sigma,\sigma') \text{ falls in class } r \\ 0 & \text{otherwise} \quad . \end{cases} \tag{3.23}$$

In the Ising case, Eq. (3.22) reduces to

$$K(\sigma,\sigma') = K^0 \frac{1+\sigma\sigma'}{2} + K^1 \frac{1-\sigma\sigma'}{2} . \tag{3.24}$$

The other natural description of $K(\sigma,\sigma')$ is the representation description. To discuss this, one writes down the basis functions $u_{pm}(\sigma)$ for all the different irreducible representations of the group which can be constructed from σ. Here p labels the different representations and m labels the different basis functions for a given representation. Here the index m runs from 1 to $d_p \times M_p$, where d_p is the dimensionality of the irreducible representation and M_p is the number of times it appears. In general, it is possible to expand any class function, like $K(\sigma,\sigma')$, in terms of the functions

$$\chi_p(\sigma,\sigma') = \sum_m u_{pm}(\sigma) u^*_{pm}(\sigma') \quad . \tag{3.25}$$

Thus, an alternative to the expansion (3.22) is the representation description of $K(\sigma,\sigma')$, namely

$$K(\sigma,\sigma') = \sum_p K_p \chi_p(\sigma,\sigma') \; . \tag{3.26}$$

In the Ising case the two representations each have the dimension 1 and have $u_0(\sigma) = 1$ and $u_1(\sigma) = \sigma$. Thence $\chi_0(\sigma,\sigma') = 1$ and $\chi_1(\sigma,\sigma') = \sigma\sigma'$, so that (3.26) becomes

$$K(\sigma,\sigma') = (K_0 + K_1 \sigma\sigma') \; , \tag{3.27}$$

so that $K_0 = (K^0 + K^1)/2$ and $K_1 = (K^0 - K^1)/2$.

One can freely move back and forth between the two descriptions because of the orthogonality of the characters $\chi_p(\sigma,\sigma')$. Write

$\chi_p(\sigma,\sigma')$ in a class expansion as

$$\chi_p(\sigma,\sigma') = \sum_r \chi_p^r \, \delta_r(\sigma,\sigma') M_p \tag{3.28}$$

and find the orthogonality relations

$$\sum_r n_r \chi_p^r (\chi_{p'}^r)^* = d_p \, \delta_{p,p'} g/M_p$$

$$\sum_p \frac{\chi_p^r (\chi_p^{r'})^* M_p}{d_p} = \frac{\delta_{r,r'}}{n_r} g \tag{3.29}$$

Here n_r is the number of elements in class r, while g is the number of basis states $g = \sum_p d_p$. In general, M_p and n_r depend upon the nature of the representation σ. For the regular representation, $M_p = d_p$ and χ_p^r is exactly the character table.

From Eq. (3.29), we see that the "Fourier transforms" are connected by

$$K^r = \sum_p \chi_p^r K_p M_p$$

$$K_p = \sum_r K^r (\chi_p^r)^* n_r g/d_p \, . \tag{3.30}$$

3.7 One Dimensional Recursions

This point of view proves its usefulness when it is applied to the recursion relation

$$e^{K'(\mu,\mu')} = \mathrm{tr}_\sigma \, e^{K(\mu,\sigma) + K(\sigma,\mu')} \, .$$

Now take the quantity

$$Q(\sigma, \sigma') = e^{K(\sigma,\sigma')} \tag{3.31}$$

and apply our expansion to it. On the one hand

$$Q(\sigma, \sigma') = \sum_r \delta_r(\sigma,\sigma') \, e^{K^r} \, .$$

On the other hand, we can use a representation expansion

$$Q(\sigma,\sigma') = \sum_p \chi_p(\sigma,\sigma') \, e^{\widetilde{K}_p}/g \, . \tag{3.32}$$

Here $\widetilde{K}_p$ is not the same as K_p, but is instead given by

$$e^{\tilde{K}_p} = \sum_r e^{K^r} \, (\chi_p{}^r)^* \, n_r/d_p$$

$$e^{K^r} = \sum_p \left[\chi_p{}^r \; e^{\tilde{K}_p}/g\right] M_p \; . \tag{3.33}$$

We summarize the relations given by Eq. (3.33) by writing

$$\underline{\tilde{K}} = D\,(\underline{K}) \; ; \quad \underline{K} = D^{-1}\,(\underline{\tilde{K}}) \tag{3.34}$$

where $\underline{K}$ and $\underline{\tilde{K}}$ are abbreviations for the vectors K^r and $\tilde{K}_p$

The usefulness of Eq. (3.32) is demonstrated by substituting Eq. (3.32) into the recursion relation and calculating the sum over σ. The result is

$$e^{K'(\mu,\mu')} = \sum_{pp'} e^{\tilde{K}_p + \tilde{K}_{p'}} \, \mathrm{tr}_\sigma \, \chi_p(\mu,\sigma')\, \chi_{p'}(\sigma',\mu').$$

Since the characters obey

$$\mathrm{tr}_\sigma \, \chi_p(\mu,\sigma')\, \chi_{p'}(\sigma',\mu') = \delta_{p,p'}\, \chi_p(\mu,\mu')\, g$$

we find that the recursion relation states

$$Q'(\mu,\mu') = e^{K'(\mu,\mu')} = \sum_p \chi_p(\mu,\mu') \, \frac{e^{2\tilde{K}_p}}{g} \; .$$

If then Q' is expanded in the same manner as Eq. (3.32),

$$Q'(\mu,\mu') = \sum_p \chi_p(\mu,\mu') \, \frac{e^{\tilde{K}_p'}}{g}$$

we see the simple result

$$\tilde{K}_p' = 2\,\tilde{K}_p$$

as the effect of a change in lattice constant $a_o \rightarrow 2\,a_o$. More generally, the recursion formula states that if $a_o \rightarrow \lambda\, a_o$, $\tilde{K}_p$ obeys

$$\tilde{K}_p(\lambda\, a_o) = \lambda\, \tilde{K}_p(a_o) \; . \tag{3.35}$$

By using Eqs. (3.33) or (3.34) we can relate the recursion relations (3.35) back to the more immediately understandable variables $K^r(a_o)$, $K_p(a_o)$, $K^r(\lambda a_o)$ and $K_p(\lambda a_o)$. The "dual" variables $\tilde{K}_p$ are, for this one-dimensional case, Wegner variables which obey the simple recursion relations of the form (3.13). These Wegner variables form a convenient tool for solving the problem. After the solution is written down, then one must transform the solution back into the K_p

or K^r language.

To illustrate this point, we shall simply write out some of the more significant physical elements of the solution and express these in the most explicit form for the Ising model. The Ising model ($\sigma = \pm 1$) is the regular representation of Z_2, the permutation group of two objects. It has two classes (each with $n_r = 1$), two irreducible representations (each with $d_p = 1$) and each representation appears once in $K(\sigma,\sigma')$ (i.e. $M_p = 1$). The character table takes the form

$$\chi_p^r = \begin{pmatrix} 1 & 1 \\ 1 & -1 \end{pmatrix} \tag{3.36}$$

so the relationship between K_p and $\tilde{K}_p$ is

$$e^{K_0 + K_1} \frac{1+\sigma\sigma'}{2} + e^{K_0 - K_1} \frac{1-\sigma\sigma'}{2} = \frac{e^{\tilde{K}_0}}{2} + \frac{e^{\tilde{K}_1}\sigma\sigma'}{2} .$$

Thus

$$\begin{aligned} \tilde{K}_0 &= K_0 + \ln 2 \cosh K_1 \\ \tilde{K}_1 &= K_0 + \ln 2 \sinh K_1 \end{aligned} \tag{3.37}$$

It is convenient to define two particularly important couplings $K = K_1$ and $\tilde{K} = 2(\tilde{K}_0 - \tilde{K}_1)$. These are then connected by the dual relations

$$\tilde{K} = -\frac{1}{2} \ln \tanh K$$

$$K = -\frac{1}{2} \ln \tanh \tilde{K} .$$

Notice that in the strong coupling limit $(K \to \infty)$ $\tilde{K} \to 0$ while in the weak coupling limit $(K \to 0) \tilde{K} \to \infty$.

There are two fixed points, $K = 0$ and $K = \infty$. The general recursion relation is

$$\tilde{K}(\lambda a_0) = \lambda \tilde{K}(a_0)$$

so that the recursion relation for K is

$$\tanh K(\lambda a_0) = [\tanh K(a_0)]^{\lambda} \tag{3.38}$$

Let us consider another simple example which illustrates the general result (3.35). Consider the XY model when $\sigma = (\cos\theta, \sin\theta)$ is a two-dimensional vector. The tr_σ is then replaced by an integration over angles so that our basic recursion relation may be written

$$e^{K'(\theta - \theta')} = \int_0^{2\pi} \frac{dx}{2\pi} e^{K(\theta - x)} e^{K(x - \theta')} .$$

Let us expand $e^{K(\theta - \theta')}$ in a Fourier series

$$e^{K(\theta - \theta')} = \sum_{p=-\infty}^{\infty} e^{ip(\theta - \theta')} e^{\bar{K}_p}$$

and similarly for $e^{K'(\theta - \theta')}$. On substituting the Fourier series into the recursion relation, one easily finds that $\bar{K}_p' = 2\bar{K}_p$ for a change in lattice constant $a_0 \rightarrow 2a_0$. In the more general case $a_0 \rightarrow \lambda a_0$, we obtain (3.35). On recalling (3.32) we see that the characters in our simple example are given by $\chi_p(\theta,\theta') = e^{ip(\theta - \theta')}$. It is trivial to show that the character orthogonality relation is obeyed. The analogue of (3.33) in this example is the Fourier series expansion for $e^{K(\theta - \theta')}$ $(r = \theta - \theta')$. Note that if $e^{\bar{K}p} = 1$ for all p, we can sum the Fourier series and obtain $e^{K(\theta - \theta')} =$ $= 2\pi\delta(\theta - \theta')$ corresponding to strong coupling. Conversely, weak coupling corresponds to $e^{\bar{K}p} = \delta_{p0}$ which gives $K(\theta - \theta') = 0$.

In the general case, the transitive representation has a nearest neighbor correlation function of $u_{pm}(\sigma)$ and $u^*_{p'm'}(\sigma)$ in the weak coupling limit,

$$\langle u_{pm}(\sigma) \; u^*_{p'm'}(\sigma') \rangle = \delta_{p,p'} \, \delta_{m,m'} \, e^{\bar{K}_p - \bar{K}_0} .$$

From this result we can calculate the Green's function

$$G_{pm;p'm'}(x-x',a_0,K) = \langle u_{pm}(\sigma(x)) \; u^*_{p'm'}(\sigma(x')) \rangle$$

by exactly the same line of argument which led to Eq. (3.20). The net result is

$$G_{pm;p'm'}(x-x',a_0,K) = \delta_{p,p'} \, \delta_{m,m'} \, e^{-|x-x'| \, m_p} \qquad (3.39a)$$

with

$$m_p = \frac{1}{a_0} (\bar{K}_0 - \bar{K}_p) . \qquad (3.39b)$$

In the case of the Ising model, the only interesting correlation function formed this way is the spin-spin correlation function, which has a "mass"

$$m = - \frac{1}{a_0} \ln \tanh K_1 \qquad (3.40)$$

of course, the "mass" defined by Eq. (3.40) remains invariant under all recursions.

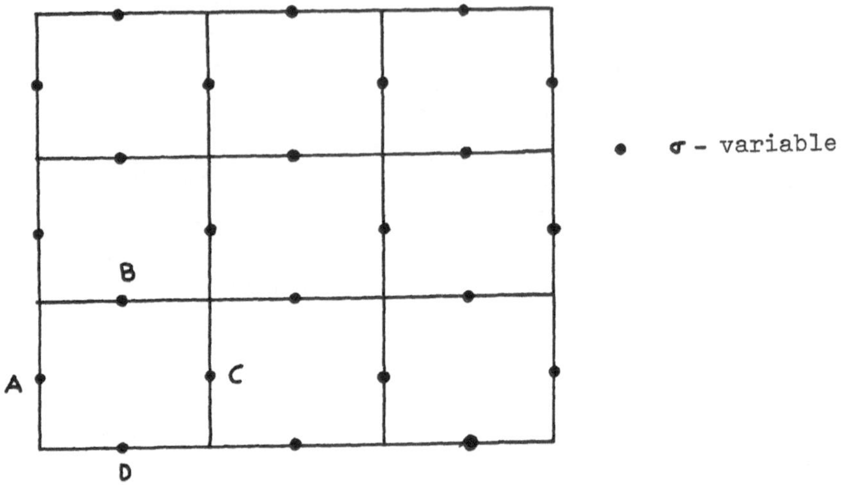

Figure 3.3 A Two Dimensional Lattice with Gauge Symmetries.

3.8 Two-Dimensional Problems with Gauge Symmetries [13,11,3,7]

Consider a two-dimensional problem on the square lattice drawn in Fig. 3.3. An elementary square is shown. For this elementary square, the coupling is

$$Q(\sigma_A\ \sigma_B\ \sigma_C\ \sigma_D) \quad = \quad e^{K(\sigma_A\ \sigma_B\ \sigma_C\ \sigma_D)} \quad . \tag{3.41}$$

Here each σ is a matrix, $\sigma_A = (\sigma_A)_{ij}$. We consider the case in which σ_{ij} is a basis for a transitive representation of the basic symmetry. This can be attained by simply making σ_{ij} the defining representation of the group. Thus, for example, for SU_n, σ_{ij} is a $n \times n$ unitary matrix with unit determinant. On the other hand, for the Ising model, σ_{ij} has no matrix indices and is just the Ising variable $\sigma = \pm 1$.

When we had a $K(\sigma,\sigma')$ which depended upon the scalar product $\sum_i \sigma_i(\sigma_i')^*$ we expanded $Q = e^K$ as

$$Q(\sigma,\sigma') \quad = \quad \sum_p \chi_p(\sigma,\sigma') \frac{e^{K_p}}{g} \quad .$$

Now, we have a K which can be written in terms of several different scalar products in the form

$$\sum_{ij} a_{ij} (b^*)_{ij}$$

with, for example,

$$a = \sigma_A \sigma_B \sigma_C \sigma_D \qquad b = 1$$

$$a = \sigma_A \sigma_B \sigma_C \qquad b = \sigma_D^+$$

$$a = \sigma_A \sigma_B \qquad b = \sigma_D^+ \sigma_C^+$$

$$a = \sigma_C \sigma_D \sigma_A \qquad b = \sigma_B^+$$

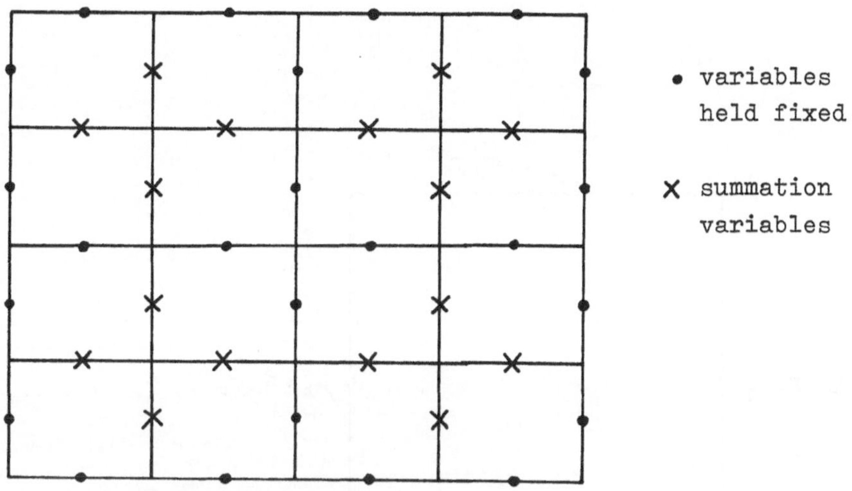

Fig. 3.4a. The Recursion Calculation

For any one of these we can expand the Q defined by Eq. (3.41) in the form

$$Q(\sigma_A \sigma_B \sigma_C \sigma_D) = \sum_p \chi_p(a,b) \frac{e^{\bar{K}_p}}{g} . \tag{3.42}$$

Since these different forms of writing are connected by a unitary transformation, the coefficients exp $\bar{K}$ are the same in all different forms of writing. The expansion (3.42), together with the orthogonality condition

$$\mathrm{tr}_\sigma \chi_p(\mu,\sigma) \chi_{p'}(\sigma,\mu') = g \, \delta_{pp'} \chi_p(\mu,\mu') \tag{3.43}$$
$$g = \mathrm{tr}_\sigma 1$$

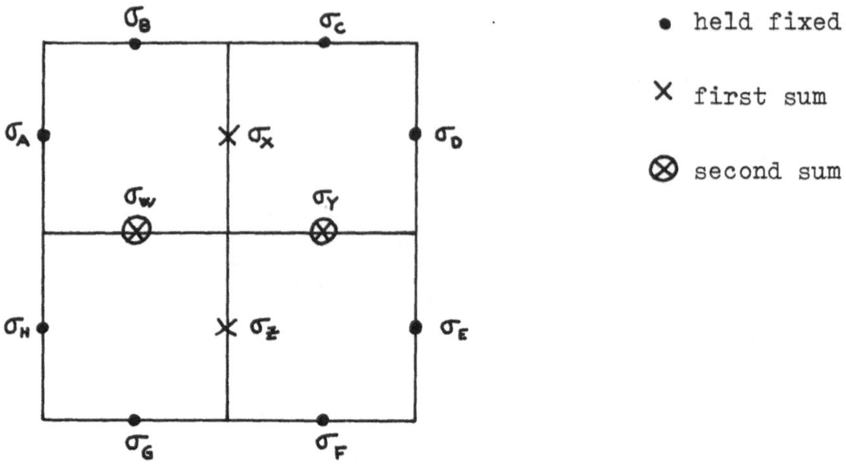

Fig. 3.4b The Elementary Square before the Sum

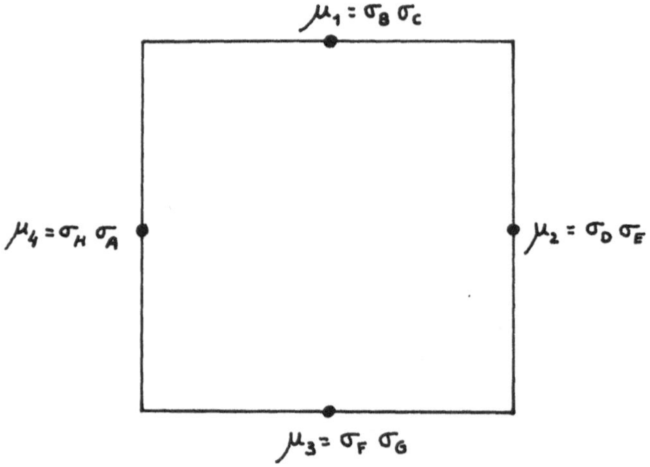

Fig. 3.4c After the Sum

enables us to carry out a recursion calculation for the configuration shown in Fig. 3.3. In Fig. 3.4a, we have indicated by x's the σ' variables which are to be summed. The remaining variables are held fixed. In Fig. 3.4b, we show an elementary square for the recursion calculation. The summation is divided into two steps: In the first step we sum over 2 variables and hold the other two summation variables fixed. The summation for the top pair of squares is, for example,

$$Q_{top} = \mathrm{tr}_{\sigma_x} \chi_p(\sigma_w \sigma_A \sigma_B, \sigma_x^{+}) e^{\bar{K}_p}$$

$$\chi_p(\sigma_x(\sigma_C \sigma_D \sigma_y)^{+}) e^{\bar{K}_p'}$$

$$= \sum_p \chi_p(\sigma_w \sigma_A \sigma_B, (\sigma_C \sigma_D \sigma_y)^{+}) \frac{e^{2\bar{K}_p}}{g} .$$

This result can then be rearranged in the form

$$Q_{top} = \sum_p \chi_p(\sigma_A \sigma_B \sigma_C \sigma_D, (\sigma_y \sigma_w)^{-1}) \frac{e^{2\bar{K}_p}}{g} .$$

Notice that this summation has generated a result of the same form as before, that is, Q_{top} depends upon the product of all the σ's on a path around the basic figure. The sum over σ_z generates a Q_{bottom} which is of exactly the same form. The only real effect of the summations is to change $e^{\bar{K}_p}$ into $g\, e^{2\bar{K}_p}$. The summation over σ_w can be carried out in exactly the same manner, with the result that the total summation generates a new Q which looks like

$$Q' = \mathrm{tr}_{\sigma_x \sigma_y \sigma_w} Q(\ ) Q(\ ) Q(\ ) Q(\ )$$

$$= \sum_p \frac{e^{4\bar{K}_p}}{g} \chi_p(\sigma_A \sigma_B \sigma_C \sigma_D \sigma_E \sigma_F \sigma_G \sigma_H, 1) .$$

Now relabel the variables as shown in Fig. 3.4c. Write

$$\begin{aligned} \mu(x_1,x_2) &= \sigma_B \sigma_C = \mu_1 \\ \mu(x_2,x_3) &= \sigma_D \sigma_E = \mu_2 \\ \mu(x_3,x_4) &= \sigma_F \sigma_G = \mu_3 \\ \mu(x_4,x_1) &= \sigma_H \sigma_A = \mu_4 \quad . \end{aligned} \tag{3.43}$$

The net result is a Q' of the form

$$Q' = e^{K'}(\mu_1\mu_2\mu_3\mu_4) = \sum_p \frac{e^{\bar{K}'_p}}{g} \chi_p(\mu_1\mu_2\mu_3\mu_4, 1)$$

with the new expansion coefficients being related to the old ones by

$$\bar{K}'_p = 4\,\bar{K}_p \ . \tag{3.44}$$

Equation (3.44) is the exact recursion relation for the gauge system in two dimensions for a change in lattice constant $a_o \longrightarrow 2a_o$. In the more general case

$$a_o \longrightarrow \lambda\, a_o \tag{3.45a}$$

the recursion is

$$\bar{K}_p(\lambda\, a_o) = \lambda^2\ \bar{K}_p(a_o) \ . \tag{3.45b}$$

A comparison between Eqns. (3.45) and (3.35) shows that the recursion scheme for the two-dimensional gauge system is exactly similar to that for the one-dimensional system except that the critical index, y, as defined by Eq. (3.13) has changed from 1 to 2.

IV. Properties of the Quark-String Model

This chapter is devoted to describing how the quark-string theory can lead to two different pictures of elementary particle phenomena, both of which seem to have a good experimental basis. These pictures are:

1. Asymptotic freedom. For high energy phenomena elementary particles appear to be composed of weakly interacting quarks.
2. The infra-red trap. No free quarks have ever been observed.

The renormalization group point of view suggests a way out of this dilemma presented by the apparent incompatibility of these two statements. How can quarks be at once weakly interacting and unobservable? They can be so if the qualitative nature of the couplings charge in the different energy ranges (length scales) so that a free quark picture, which is asymptotically correct for small distances, becomes vastly wrong for large distances.

To get to these two different pictures, we shall describe two different ways of approaching the Wilson action. First, we shall eliminate the quark variables and calculate Green's functions which manifestly do not permit the propagation of free quarks. Secondly, we shall calculate the properties of the system in the limit of strong string-string interactions and show how this limit gives physical results which are fully equivalent to a picture of free quarks.

4.1 Elimination of Quark Variables

The quark portion of the action, A_ψ given by Eq. (2.26), is exactly of the form of the non-interacting quark Hamiltonian, (1.13). The self-energy is

$$\begin{aligned}\Sigma(1,1') &= \Sigma_{if;jf'}(x_1, x_{1'}) \\ &= \delta_{ff'}\, K_f\, U_{ij}\,(x_1, x_{1'}) \\ &= \sum_\mu \big[\delta(x_{1'} - x_1 - \hat{e}_\mu a_0)(1 - \hat{e}_\mu \cdot \gamma) \\ &\qquad + \delta(x_{1'} - x_1 + \hat{e}_\mu a_0)(1 + \hat{e}_\mu \cdot \gamma)\end{aligned} \quad . \quad (4.1)$$

The analysis of section 1.3 can be carried out once more to find expressions for physical qualities which are independent of all quark variables. For example, the partition function can be evaluated as

$$Z = \mathrm{Tr}_U \; e^{A_U} \; \mathrm{Tr}_\Psi \; e^{A_\Psi}$$
$$= \mathrm{Tr}_U \; e^{A_U + \mathrm{trace}\, \ell n(1 - \Sigma)} \tag{4.2}$$

Thus, for the U variables, there is an effective action

$$A^{\mathrm{eff}}[U] = A_U + \mathrm{trace}\, \ell n(1 - \Sigma) \tag{4.3}$$

Similarly, one can define a set of Green's functions for the Ψ variables. These functions will come in two kinds. The functions $G(1, 1')$, $G_2(12; 1'2')$, ... will be, as before, averages over both U and Ψ. However, $g(1, 1')$, $g_2(12; 1'2')$, ... will be averages over Ψ and $\bar{\Psi}$, with U held fixed. The latter quantities are relatively simple. For example,

$$g(1, 1') = \frac{\mathrm{Tr}_\Psi \Psi(1) \; \Psi(1') \; e^{A_\Psi}}{\mathrm{Tr}_\Psi \; e^{A_\Psi}} \tag{4.4}$$

is

$$g = \frac{1}{1 - \Sigma} \tag{4.5a}$$

while g_2, which is a similar average of $\Psi(1)\Psi(2)\bar{\Psi}(2')\bar{\Psi}(1')$, is

$$g_2(12; 1'2') = g(1, 1')g(2, 2') - g(1, 2')g(2, 1') \tag{4.5b}$$

and in general

$$g_n(12...n; 1'2'...n') = \sum_{\substack{\text{permutations}\\ \text{of } 1'2'...n'}} (-1)^P \; g(1,1')g(2,2')...g(n,n') \tag{4.5c}$$

with $(-1)^P$ being the order of the permutation.

On the other hand, the true Green's functions are averages of these functions of U over the string variables. For example,

$$G(1,1') = \frac{\mathrm{Tr}_U \; g(1,1') \; e^{A_{\mathrm{eff}}[U]}}{Z} \tag{4.6}$$

$$G_n(12...n; 1'2'...n') = \mathrm{Tr}_U \; g_n(12...n; 1'2'...n') \; e^{A_{\mathrm{eff}}[U]}/Z \; .$$

Of course, the true Green's functions G are observable, while the other functions g are just an intermediate construct.

To study the trapping of free quarks, we look at the effect of colour symmetry upon G_n. According to this symmetry everything is left unchanged if we make the transformations

$$U(x,x') \rightarrow u(x)U(x,x')[u(x')]^{-1} \tag{4.7a}$$

$$\Psi(1) \rightarrow u(x_1)\Psi(1) \tag{4.7b}$$

$$\bar{\Psi}(1') \rightarrow \bar{\Psi}(1')[u(x_{1'})]^{\dagger}$$

Specifically this means that $A_{eff}[U]$ is left invariant under the transformation (4.7a) while g_n transforms according to

$$g(1,1') \rightarrow \sum_{i_1 \bar{i}_{1'}} u_{i_{1'}\bar{i}_1}(x_1)\, g(x_1\bar{i}_1 f_1;\, x_{1'}\bar{i}_{1'} f_{1'})[u(x_{1'})]^{\dagger}_{\bar{i}_{1'} i_1} \tag{4.8}$$

More generally, the transformation properties of the g_n's follow directly from Eq. (4.8) and the fact that g_n is a product of g's.

Now we reach the crucial step in the argument. The average over $U(x, x')$ includes an average over $u(x)$ for each x. In the calculation of the Green's functions, G, $u(x)$ only appears in explicit factors like that in Eq. (4.8). If there is no breaking of the colour symmetry -- either in the action or through a spontaneous symmetry breakdown (a phase transition) then in the process of going from g_n to G_n, we have the right to average over each individual $u(x)$ as it appears in Eq. (4.8). This averaging is, in fact, the projection of only colour symmetrical combinations in the final answer.

Now calculate $G(1,1')$ with the aid of Eq. (4.6). Include in the averaging the replacement (3.8) and average over all $U(x)$. We then have

$$G(1,1') = \sum_{\bar{i}_1 \bar{i}_{1'}} \langle u_{i_{1'}\bar{i}_1}(x_1)[u^{\dagger}(x_{1'})]_{\bar{i}_{1'} i_{1'}} \rangle \; \mathrm{Tr}_U\, g(x_1\bar{i}_1 f_1;\, x_{1'}\bar{i}_{1'} f_{1'})\, e^{A_{eff}[U]}/Z \; . \tag{4.9a}$$

The average of a single u must vanish (see Eq. (2.18)). Therefore the right hand side of Eq. (4.9) will only be non-zero if $x_1 = x_1$. The quark "propagator" $G(1, 1')$ then asserts that the quarks literally go nowhere by themselves. From Eq. (2.18) we see that this result may be expressed in symbols as

$$G(1, 1') = \delta(x_1, x_{1'})\, \delta_{i_{1'}, i_{1'}} F(x_1 f_1 f_{1'}) \; .$$

The first non-trivial propagation is $G_2(12;\ 1'2')$. If x_1 is different from x_2, this propagator can describe the motion through space and time of a quark-antiquark pair. But, by exactly the same argument as before, we cannot have a non-zero G_2 for arbitrary values of the spatial arguments. In fact, the quark and anti-quark must be at precisely the same place for a non-zero propagation to occur. In symbols, the result is, for $x_1 \neq x_{2'}$,

$$G_2(12;1'2') = \delta(x_1,x_{1'})\,\delta(x_2,x_{2'})\,\delta_{i_1,i_{1'}}\,\delta_{i_2,i_{2'}}\,F_2(x_1 f_1 f_{1'};\ x_2 f_2 f_{2'}) \qquad (4.9b)$$

The net result is $F_2(x_1;\ x_2)$ which will describe the propagation of mesons made of a bound quark anti-quark pair but neither G nor G_2 will give direct evidence of the free quark propagation.

Because there is a third order combination of u's which has a non-zero average, something new happens in $G_3(123;1'2'3')$. If $x_1 = x_2 = x_3$ and $x_1 = x_2 = x_3$, G_3 can be of the form

$$G_3(123;\ 1'2'3') = \epsilon_{i_1 i_2 i_3}\,\epsilon_{i_1' i_2' i_3'}\,F_3(x_1 f_1 f_2 f_3; x_{1'} f_{1'} f_{2'} f_{3'}) \qquad (4.9c)$$

Thus, three quarks can propagate together as a fermion excitation. Since F_3 must be symmetrical in its flavor and spinor indices, this new propagator can include such symmetrical combinations as the 56 dimensional representation of SU_6.

Therefore, according to this argument, we can only see colour singlets, and since free quarks are not singlets they cannot be seen.

4.2 A More Careful Discussion

The above argument is, however, quite superficial. A more careful argument would have to be built upon the possibility of observing the energy carried by a free quark with some local detector even if its colour was not observable. The necessity for constructing this more careful argument becomes obvious if we notice that the same argument which gives the vanishing of $G(x_1,x_{1'})$ for $x_1 \neq x_{1'}$ can also be applied to electrodynamics for which G is the electron propagator. Hence if we take the above argument really literally, we would conclude that electrons were not observable either!

To make a more careful analysis, we consider a process in which a (quark)-(anti-quark) pair is produced at the space-time point $x_o = (0,0,0,0)$ annihilated at $x = (0,0,0,t)$. In the meantime a quark is observed at the spatial point $r = (0,0,\frac{z}{2})$ and an anti-quark at $r = (0,0,-\frac{z}{2})$. A heuristic picture of such a process is shown in

Fig. 4.1.

The picture suggests that it is not impossible to observe a separated quark and anti-quark. But is it likely? We shall follow Wilson [1] and others and use perturbation theory in J to argue that such an observation is so unlikely for large values of z and t so that we could not possibly interpret the quark and anti-quark as unbound.

First of all, notice that the probability for the process shown in Fig. 4.1 is, to lower order in K of the form:

$$\text{Prob} \sim \frac{(\mathrm{Tr}_U \, e^{A_U} \, X_U)}{(\mathrm{Tr}_U \, e^{A_U})} \, K^{(2z + 2t)/a_o} \tag{4.10}$$

Here X_U is a product over the closed loop shown in Fig. 4.1 of the product of U′s which generate the motion of the quark

$$X_U = U(x_o,x_1)U(x_1,x_2)\ldots\ U(\ldots,\ldots)\ \ldots\ U(x_{n-1},x_n)U(x_n,x_o) \tag{4.11}$$

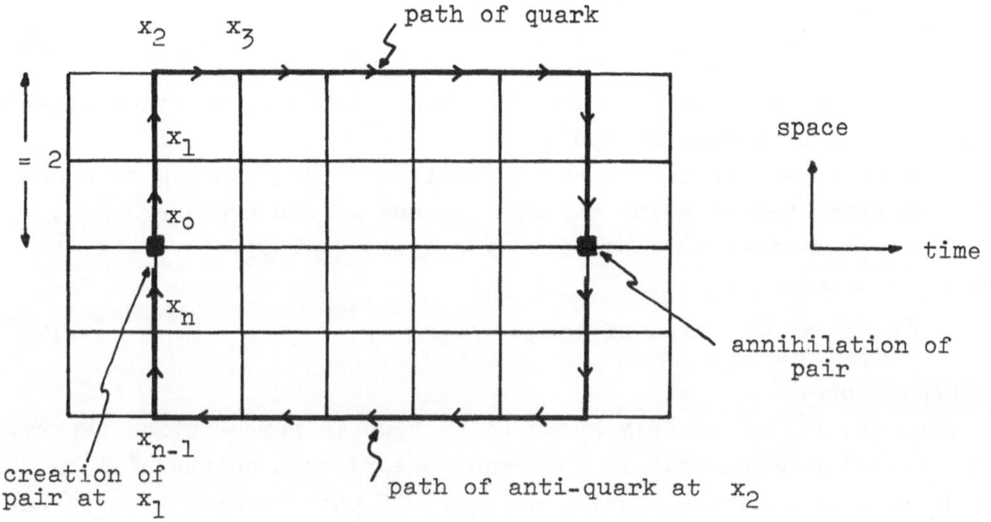

Fig. 4.1 Separation of a Quark - Anti-quark Pair

Since the average of each $U(x,x')$ is zero, to get a non-zero result we must use the interaction term A_U in Eq. (4.10). We wish to find terms like

$$U(x_i,\ x_{i+1})\ U^+(x_i,\ x_{i+1})$$

which indeed have a non-zero average.

It is easy to see the first non-zero term of this structure. It arises when one takes a term in perturbation theory

$$J\ \ U^+(1,2)\ U^+(2,3)\ U^+(3,4)\ U^+(4,1)$$

for each and every square contained in the path of Fig. 4.1. Thence, for small J we estimate the probability of this kind of process as

$$\text{Prob} \sim J^{zt/a_o^2} = e^{-(zt/a_o^2)\,|\ell n\ J|} . \tag{4.12}$$

To interpret Eq. (4.12), remember that in the Euclidian space the probability of a process is proportional to the exponential of $-\Delta E \times \Delta t$, where ΔE is the energy of the process and Δt is the time during which that energy is available. We then see from Eq. (4.12) that the energy of separating two quarks by a distance z is proportional to the separation distance

$$\Delta E \sim z\ |\ell n\ J|/a_o^2 \tag{4.13}$$

If the energy grows linearly with the separation distance then clearly the quarks cannot become unbound.

Thus we have established the infinitely strong binding of quarks in lowest order perturbation theory. It has arisen because $\langle X_U \rangle$, where X_U is defined as a product over a closed loop, is of the form

$$\ell n \langle X_U \rangle \sim -\ \text{Area of loop} \tag{4.14}$$

for large loops.

But, Eq. (4.14) is only known to be true in perturbation theory. If Eq. (4.14) remains true in the exact theory then quarks will be trapped with an infinite binding energy. If this result of perturbation theory disappears in the exact theory quarks can become unbound. What will happen?

Our experience in statistical mechanics indicates that qualitative results of perturbation theory, like Eq. (4.14), will remain true for some range of couplings J, whenever J is too weak to produce a phase transition. Thus, if J is lower than some critical value J_c, we might expect Eq. (4.14) to remain true so that free quarks will be unobservable. We follow W. Bardeen [28] and call this situation a baryon phase. In particular, in the baryon phase for large loops of linear dimension L [11],

$$\ell n \ \langle X_U \rangle \sim - L^2 \qquad \text{(baryon phase)} \tag{4.15a}$$

In the opposite limit of large J one can do an expansion in 1/J and find, to lowest order, that [11]

$$\ell n \ \langle X_U \rangle \sim - L \qquad \text{(quark phase)} \tag{4.15b}$$

As we shall see, in this situation, free quarks are definitely observable. More generally, one can imagine that

$$\ell n \ \langle X_U \rangle \sim - L^{\delta(J)} \qquad \text{(complex phase)} \tag{4.15c}$$

where perhaps δ depends continuously upon J.

To study the observability of quarks, we are then impelled to understand the phase transitions of the system as a function of J [3,4].

4.3 Asymptotic Freedom

There is a host of theoretical work in which the behavior of elementary particles at high energy is described by assuming that these particles are made up of non-interacting or weakly interacting quarks. The renormalization group point of view and the quark-string model provide a beautiful description of how this might occur.

In the renormalization group picture, one can have different forms of the action to describe different energy ranges. In particular, many renormalizations are required to move from high energy phenomena to lower energy regions. Assume that the action which describes high energy phenomena includes a very strong four-string interaction J. This strong interaction will tend to suppress fluctuations in $U(x,x')$. If U cannot fluctuate, then the quark interaction A_ψ is a pure two-body term. There is no higher order quark interaction. Hence the quarks behave as free particles.

Wilson [1] has shown how to make this conceptual framework more explicit. A general form of $U(x,x')$ which will maximize the action A_U is

$$U(x, x') = u(x) \ u^{-1}(x') \tag{4.16}$$

for arbitrary special unitary matrices $u(x)$. When J is large, this form of U might be expected to describe all-short ranged correlation phenomena reasonably well. However, we cannot expect to apply

(4.17) to long-range correlations, e.g., to products like

$$P = U(x,x + \hat{e}_1 a_0)\; U(x + \hat{e}_1 a_0, x + 2\hat{e}_1 a_0) \;\ldots$$
$$\ldots\; U(x + (n-1)\hat{e}_1 a_0, x + n\hat{e}_1 a_0) \tag{4.17}$$

for large n. According to (4.16), $P = u(x)\, u^{+}(x + n\hat{e}_1 a_0)$. But, in each step of the product (4.17), there will be some error arising from the imperfection in the approximation (4.16). If there is no phase transition, (that is, if we are not in the quark phase) these errors will accumulate after a large number of steps. Thus, we expect to find that there is a characteristic distance ξ_0 and a characteristic number of steps na_0 such that the

$$\begin{aligned} P \;&=\; u(x)\, u^{+}(x + na_0\hat{e}_1) \quad \text{for} \quad na_o \ll \xi_o \\ &\approx\; 0 \quad \text{for} \quad na_o \gg \xi_o \,. \end{aligned} \tag{4.18}$$

For larger distances than ξ_0, fluctuations in U will be very important; for smaller distances they will be unimportant. For the small distances, we can expect to use a theory based upon Eq. (4.16). As we shall see, this theory automatically has the quarks behave as free particles. Hence, we do have asymptotic freedom at short distances.

To follow this idea through in a mathematical fashion, combine Eq. (4.16) and (4.1) to get

$$\Sigma(1, 1') \;=\; u(x_1)\; \Sigma_0(1, 1')\; u^{-1}(x_{1'})$$

where Σ_0 is defined by Eq. (1.46). Then we find

$$g(1, 1') \;=\; u(x_1)\, g_0(1,1')\, u^{-1}(x_{1'}) \tag{4.19}$$

and

$$g_2(12,1'2') \;=\; u(x_1)\, u(x_2) \left[g_0(1,1')\;\; g_0(2,2') - g_0(1,2')\; g_0(2,1') \right] u^{-1}(x_1{'})\; u^{-1}(x_2{'}). \tag{4.20}$$

In Eqs. (4.19) and (4.20), g_0 is the propagator discussed in section 1.3. Now, calculate the averages, (4.6) of g and g_2. Since A[U] has already been picked to be independent of u and since trace $\ell n(1 - \Sigma)$ is also independent of u, the only u's are shown explicitly in Eqs, (4.19) and (4.20). Furthermore g_0 is independent of any other part of U, so that within the context of this approximation the only average to be done is the average over u. Thus, in this strong-coupling limit we can explicitly calculate the functions F, F_2, F_3, ... They are all exactly free quark results, i.e. a product of free quark propagators. In particular,

$$F(x_1 f_1 f_{1'}) \quad = \quad \delta_{f_1 f_{1'}} g_0(x_1 f_1;\ x_1 f_1) \tag{4.21a}$$

$$F_2(x_1 f_1 f_{1'};x_2 f_2 f_{2'}) = -\delta_{f_1 f_{2'}} \delta_{f_2 f_{1'}}\ g_0(x_1 f_1;x_2 f_{2'})\ g_0(x_1 f_2;\ x_1 f_{1'}) \tag{4.22b}$$

$$F_3(x_1 f_1 f_2 f_3;x_x\ f_{1'} f_{2'} f_{3'}) \sim g_0(x_1 f_1;x_{x'} f_1)\ g_0(x_1 f_2;x_{x'} f_{2'})\ g_0(x_1 f_3;x_{1'} f_{3'}). \tag{4.22c}$$

From these results we can get the asymptotic fall-off of the propagators F_2 and F_3 as being of the form $e^{-m|x - x'|}$, where m is the sum of the masses of the constituent quarks. Thus, in all respects the $J \rightarrow \infty$ theory is equivalent to a theory of free quarks.

4.4 Summary of Chapter 4

This chapter made two essential arguments:

1. If J is sufficiently small, i.e. J is smaller than a critical coupling at which a phase transition occurs, then quarks will be bound together with an infinite binding energy. Call this critical value of the coupling J^*.

2. In the limit $J \rightarrow \infty$, the quarks will show almost free particle behavior, except when they are separated by a very large distance. The larger the value of J, the greater the distance (measured in lattice constants) over which free particle behavior will be seen.

In short, trapping is characteristic of a theory with sufficiently weak couplings; freedom is characteristic of a theory with very strong couplings, J. In nature, freedom and trapping are both observed - but in different regions of energy, i.e. on different distance scales.

These contrasting observations can be made to agree within the context of the very simplest renormalization group point of view. Imagine that under successive increases of lattice constant $a_0 \rightarrow \lambda a_0$, J continually decreases, and that after many renormalizations J approaches zero. Thus, no matter how large J is naturally, a sufficient number of renormalizations will being it close to zero. This kind of behavior is characteristic of systems which show no phase transition. (See Fig. 3.2b.) Then, for small distance scales we can have an action with very large J - i.e. asymptotic freedom - while we always remain in the baryon "phase" there is no quark "phase" so unbound quarks cannot be observed.

Hence our contrasting observations of freedom and trapping will be consistent if there is no phase transition in the four-dimensional system of quarks and strings, no matter how large J might be.

These observations can be expressed graphically by redoing Fig. 3.2 as in Fig. 4.2. In the latter, we have indicated the physical nature of the different phases which arise. The arrows on the lines show the directions of charge of the couplings when a_0 decreases. An arrow going toward $g = 0$ (or $J = \infty$) shows asymptotic freedom [32,33] . Only the diagram without a phase transition (4.1b) is consistent with asymptotic freedom. In this diagram all values of the coupling (save $g = 0$) put the system in the baryon phase, and therefore show quark trapping. In the other two cases, the existence of quark trapping is a function of coupling and only occurs for sufficiently small J.

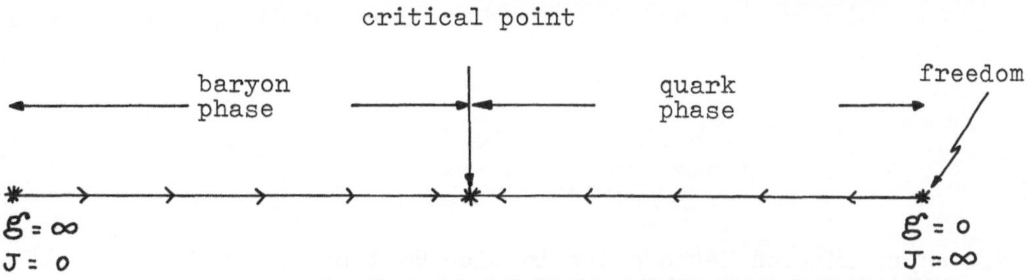

(a)

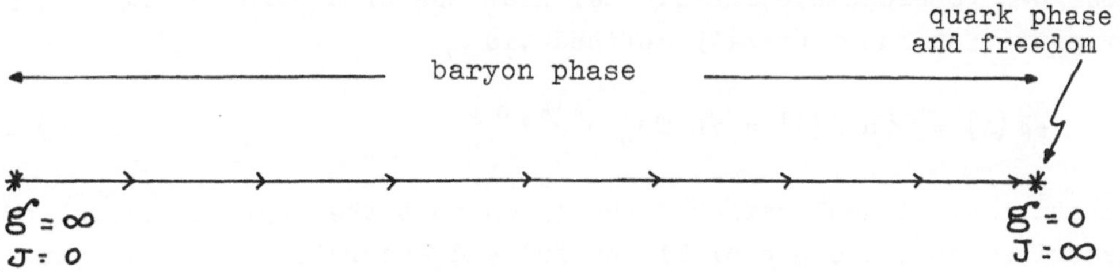

(b)

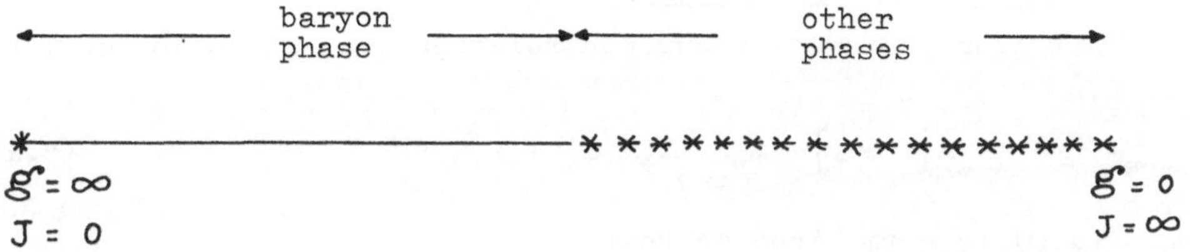

(c)

Fig. 4.2 Possible structures of phases for the string system. The arrows show the flow of couplings for decreasing lattice constant. We hope case (b) appears in the quark-string theory.

V. Approximation Methods for Lattice Systems

To make further progress, one needs approximation techniques. In this section, we will describe some approximation techniques borrowed from statistical mechanics [5], [6], [29], [30]. They are all designed to produce effective calculations of a "free energy" $F[A]$ or partition function $Z[A]$ defined via

$$-F[A] = \ln Z[A] = \ln \mathrm{Tr}_{\sigma}\, e^{A[K,\sigma]} \tag{5.1}$$

In statistical applications, one finds that the approximations described here are especially useful and accurate in calculations of critical indices. They have not yet been extensively employed for correlation functions. Hence we have no experience which would inform us about how accurate mass calculations might be.

5.1 Lower Bound Approximations [5] [6]

Start from the exact recursion relation for the action defined by

$$e^{A'(\mu)} = e^{A[K',\mu]} = \mathrm{Tr}_{\sigma}\, e^{T(\mu,\sigma) + A[K,\sigma]} \tag{5.2}$$

Here, $T(\mu,\sigma)$ is normalized so that

$$\mathrm{Tr}_{\mu}\, e^{T(\mu,\sigma)} = 1 \, . \tag{5.3}$$

Hence the free energy or partition function defined from A' is identical to that defined from A, i.e.,

$$F[A'] = F[A] \tag{5.4}$$

Unfortunately, one cannot calculate the sum in Eq. (5.2). To circumvent this difficulty, we define an approximate calculation that we can indeed perform. We add to the exponent in (5.2) an error term $\Delta(\mu,\sigma)$ which makes the sum calculable. Then, we find an approximate recursion equation

$$e^{A'_a(\mu)} = \mathrm{Tr}_\sigma\, e^{T(\mu,\sigma) + A[K,\sigma]}\, e^{\Delta(\mu,\sigma)} \tag{5.5}$$

This approximate action can be described in terms of some new coupling functions K′, which are some functional of the coupling functions. This form of the approximate recursion relation is then written

$$K' = R_a\,[K] \tag{5.6}$$

How do we choose a good approximation of this nature? More specifically, given several possible choices of $\Delta(\mu,\sigma)$, how can we choose the one which is "smallest" and thus generates the smallest possible error?

One guide comes from a variational principle -- or rather an inequality. This inequality requires the following conditions:

a. $\mathrm{Tr}_\sigma\, e^{T(\mu,\sigma) + A[K,\sigma]}$ is a sum with positive semi-definite weights.

b. $\Delta(\mu,\sigma)$ is real.

c. The average of Δ is zero, i.e.,

$$\mathrm{Tr}_\mu\, \mathrm{Tr}_\sigma\, e^{T(\mu,\sigma) + A[K,\sigma]}\Delta(\mu,\sigma) = 0 \tag{5.7}$$

Under these conditions, the free energy generated from A'_a is smaller than the true free energy, i.e., instead of Eq. (5.4) we have

$$F[A_a] < F[A'] = F[A] \tag{5.8}$$

To prove Eq. (5.8), one defines

$$F(\lambda) = -\ln \mathrm{Tr}_\mu\, \mathrm{Tr}_\sigma\, e^{T(\mu,\sigma) + A[K,\sigma] + \lambda\Delta(\mu,\sigma)} \tag{5.9}$$

Then F(0) is the exact free energy, $F(1) = F[A'_a]$. In virtue of Eq. (5.7)

$$\left.\frac{dF(\lambda)}{d\lambda}\right|_{\lambda=0} = 0 \tag{5.10}$$

Also,

$$\frac{d^2F}{d\lambda^2} = -\langle(\Delta - \langle\Delta\rangle_\lambda)^2\rangle_\lambda \tag{5.11}$$

where $\langle\ \rangle_\lambda$ is an average with weight $\exp(T + A + \lambda\Delta)$. If the weight is positive, the second derivative is negative and the theorem is proved.

Thus, from all possible Δ's, we choose that Δ which maximizes the approximately calculated free energy and we then get the "best" possible answer. In the meantime, the average of the squared fluctuations in Δ, as defined by integrating Eq. (5.11) over λ between 0 and 1 has been minimized.

The first problem is to find a Δ which obeys Eq. (5.7). To do this imagine any set of local variables $a_i(x)$. For example, $a_i(x)$ might be $\sigma(x)\sigma(x+\hat{e}_1 a_0)$. The labels i on $a_i(x)$ distinguish among different kinds of inequivalent variables; the labels x describe equivalent variables at different positions. Then if

$$\Delta(\mu,\sigma) = \sum_i c_i(x) \quad a_i(x) \tag{5.12}$$

where $c_i(x)$ is some set of coefficients independent of μ and σ which obeys

$$\sum_x c_i \ (x) = 0 \quad \text{for all } i \tag{5.13}$$

then, Eq. (5.10) will certainly be satisfied because, at $\lambda = 0$, the average of $a_i(x)$ is independent of x.

If we consider $a_i(x)$ to be in effect bits and pieces of the action, $A[K,\sigma]$, then the net effect of $\Delta(\mu,\sigma)$ is to add something to the action at some points and subtract something at others. The condition (5.13) says that we are allowed to add and subtract such couplings within the variational constraint if we just demand that for every bit of strength we add at one set of points we make sure we subtract an equivalent total strength at other points.

More simply stated: the variational principle allows us to move potential terms from one set of bonds in the lattice to equivalent bonds but not to increase or decrease the total amount of any type of bond.

This potential moving method will permit us to conduct approximate recursion calculations in a controllable fashion. The basic technique is to use the potential moving to move hard-to-handle bonds into a location where their effect may be taken into account.

5.2 Migdal Approximation [3], [4], [7]

To see this approximation technique in its simplest form, we consider the derivation of an approximation similar to that employed by Migdal. We start with variables $\sigma(x)$ and nearest neighbor bonds in the x,y,z,... directions. Thus, the action is

$$A[K,\sigma] = \sum_{x\alpha} K_\alpha (\sigma(x),\ \sigma(x + \hat{e}_\alpha a_0))$$

The label α on K_α distinguishes the different bonds in the different directions.

Now we employ a recursion calculation in which the new variables $\mu(x)$ are defined to be exactly the same as the old variables $\sigma(x)$ on a fraction $1/\lambda$ of the lattice sites ·, i.e.,

$$\mu(x) = \sigma(x) \text{ for } x = (\lambda n_1, n_2, n_3, \ldots)a_0$$

The remaining σ's are summation variables. (See Figure 5.1)

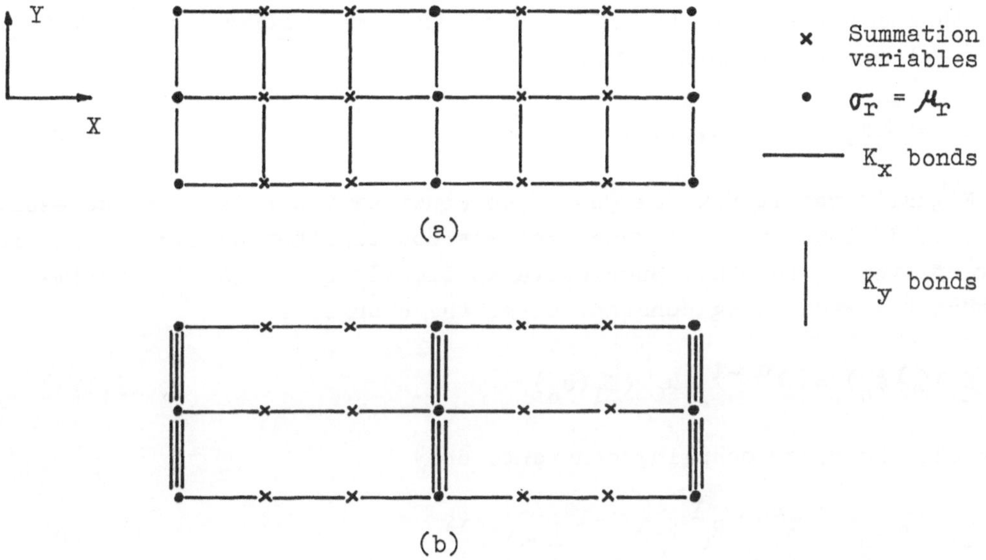

Fig. 5.1 Potential moving in the Migdal approximation depicted for d=2 and $\lambda = 3$. Part (a), before the potential moving; part (b), afterward.

$$\Delta(\sigma) = \sum_{x} a_{\alpha}(x) \quad K(\sigma(x), \sigma(x + \hat{e}_{\alpha} a_0 1) \tag{5.14}$$

and take a_α to be exactly zero if $\alpha = 1$. For the remaining bonds, we choose

$$a_{\alpha}(x) = -1 \tag{5.15a}$$

when $\sigma(x)$ and $\sigma(x + \hat{e}_{\alpha} a_0)$ are summation variables and

$$a_{\alpha}(x) = (\lambda - 1) \tag{5.15b}$$

for the bonds which connect two μ-variables. The net effect of (5.15) is to move $(\lambda-1)K_\alpha$ bonds ($\alpha = 2,3,\ldots$) from summation bonds to the bonds between two μ-variables.

Now the summation over σ is easy to perform. We sum as before and find a new x-coupling

$$\underset{\sim}{K}_1' = R_0^{\lambda}(\underset{\sim}{K}_1) \tag{5.16a}$$

The other bonds are the sum of the bond that was always present and the $\lambda - 1$ bonds that were moved:

$$\underset{\sim}{K}_\alpha' = \lambda K_\alpha \qquad \alpha = 2,3,\ldots \tag{5.16b}$$

Migdal's result now emerges if we consider the effect of successive x,y,z,... decimations. All these decimations together change the lattice constant from a_0 to λa_0. Successive applications of Eqs. (5.16) imply that the x-coupling constant after the change is:

$$\underset{\sim}{K}_1(\lambda a_0) = \lambda^{d-1} R_0^{\lambda}(\underset{\sim}{K}_1(a_0)) \tag{5.17a}$$

while all the other coupling constants obey

$$\underset{\sim}{K}_\alpha(\lambda a_0) = \lambda^{d-\alpha} R_0^{\lambda}(\lambda^{\alpha-1} \underset{\sim}{K}_\alpha(a_0)) \tag{5.17b}$$

Here $\alpha = 2,3,4\ldots$ is the index which describes the coupling constant in the y,z,t,... direction.

Eq. (5.17a) is exactly the same as Migdal's result. We shall discuss the consequences of Eq. (5.17) below.

5.3 "Fourier Transform" Representation [7], [31]

The potential moving described above moves potentials defined as functions of σ and μ. This "coordinate" representation of the system is not the only possibility. In Chapter II, we described a "Fourier Transform" representation in which functions of random variables were expanded in irreducible representations in the form

$$e^{K(\sigma,\sigma')} = \sum_p e^{\widetilde{K}_p} \chi_p(\sigma,\sigma') /g.$$

In strong coupling problems, the potential is so strong that the elimination of unwanted bonds via $\Delta(\mu,\sigma)$ requires a very large Δ and hence produces a large error. However, in these strong coupling problems the fourier transformed potentials $\widetilde{K}_p$ tend to be almost independent of p so that <u>their</u> motion will produce small errors. Can we phrase the potential-moving scheme in the Fourier transform language?

To do this, imagine that we partition the variables σ and μ into sets labeled by the index r. For example, all the variables which lie within a given square on the lattice might be assigned to a set labeled by an r denoting the center of that square. More simply, the set might just consist of the two points at the ends of a nearest neighbor bond. The sets need not be exclusive: A given $\sigma(x)$ might fall into several sets. Let us label the set variables by s_r and assume that the action may be expanded as a sum over these sets.

$$T(\mu,\sigma) + A[K,\sigma] = \sum_i \sum_r K_i(s_r) \tag{5.18}$$

Here the sum over r is essentially a spatial sum. Different kinds of terms in the action are distinguished by different indices i. For a given i, $K_i(s_r)$ and $K_i(s_{r'})$ only differ by a spatial translation which takes r' into r. For example, if we are doing a recursion like the one in the last section s might be all pairs of nearest neighbor variables, and $K_i(s)$ might be the nearest neighbor bonds $K_\alpha(\sigma,\sigma')$, with i being the label α which describes the direction of the bonds.

Now, start from Eq. (5.18) and imagine that we wish to derive a variational technique for attacking

$$Z[A] = \mathrm{Tr}_\mu \mathrm{Tr}_\sigma \prod_{ri} e^{K_i(s_r)} \tag{5.19}$$

To derive a Fourier transform representation of this sum take each factor

$$Q_i(s_r) = e^{K_i(s_r)}$$

and imagine that it can be expanded in a set of basis functions as

$$Q_i(s_r) = \sum_p e^{\tilde{K}_{pi}} \Phi_p(s_r) \tag{5.20}$$

This expansion is, of course, always possible. However, we must carefully choose the signs of the basis functions to ensure that $\tilde{K}_{p,i}$ is real.

Now write the partition function (5.19) in terms of the expansion (5.20). We have

$$Z[A] = \prod_i (\prod_r \sum_{p_r} e^{\tilde{K}_{p_r,i}}) \ wt[p] \tag{5.21}$$

Here wt[p] is a weight which depends upon all of the expansion variables p_r. It is defined by

$$wt\ [p] = Tr_\mu Tr_\sigma \prod_{ri} \Phi_{p_r}(s_r) \tag{5.22}$$

If this weight is positive semi-definite and if $\tilde{K}_{p,i}$ is real, we describe our problem as being ferromagnetic. The reason for this terminology is that we can prove the conditions of reality of $\tilde{K}_p$ and positiveness of wt[p] for Ising-type problems with positive (ferromagnetic) couplings

Let us describe one such problem: An Ising model with nearest neighbor positive couplings. As before, the μ's are equal to some of the σ's. The set σ_r is the set of all pairs of nearest neighbor variables ($\sigma(x), \sigma(x + e_\alpha a_0)$). The coupling is of the form

$$e^{K_0+K_1\sigma\sigma'} = e^{K_0} \cosh K_1(1 + \tanh K_1 \sigma\sigma')$$

Hence if we choose $\Phi_0 = 1$, $\Phi_1 = \sigma\sigma'$, and take K_1 to be greater than zero, the coefficients $\tilde{K}$ in Eq. (5.20) are certainly positive. The weight in Eq. (5.22) is either 0 or 2^N, where N is the number of spins. The conditions we need are then satisfied.

Now we can apply potential moving techniques once more, but now apply this potential moving to the "potentials" $\tilde{K}_{p,i}$ at the different sites r. For ferromagnetic problems these "potentials" may also be freely moved from site to site, if we are willing to accept a decrease in the approximately calculated free energy. Once more the "best" potential moving is the one which produces the maximum calculated free energy for a given set of coupling constants.

This kind of potential moving, is much more appropriate for strongly interacting systems than the motion of coordinate-space potentials described in Section 5.1.

5.4 Migdal Revisited [7], [3], [4]

To see how all this works, return to the decimations of nearest neighbor potentials described in Section 5.2. Let σ be an irreducible representation of the symmetry of $K_\mu(\sigma, \sigma')$ and expand, as in Chapter III,

$$Q_\alpha(\sigma,\sigma') = e^{K_\alpha(\sigma,\sigma')} = \sum_p \chi_p(\sigma,\sigma')\, e^{\overline{K}_{p\alpha}}/g \qquad (5.23)$$

These bonds are distributed on the lattice, as shown in Fig. (5.2). To produce a workable recursion calculation, divide the x-direction bonds ($\mu=1$) into two categories: ($\lambda-1$) "donor" bonds for each "recipient" bond. Move every bit of $\overline{K}_{p,i}$ from the bonds on the recipient bond. Then, on the donor bonds we have

$$Q_1(\sigma,\sigma') = \sum_p \chi_p(\sigma,\sigma') = \delta(\sigma,\sigma') \qquad (5.24a)$$

while on the recipient bond

$$Q_1(\sigma,\sigma') = \sum_p \chi_p(\sigma,\sigma')\, e^{\lambda\overline{K}_p}/g \qquad (5.24b)$$

The delta symbols permit us to sum over $\lambda-1$ of the σ-variables as shown in Fig. (5.2b). The unsummed σ-variables are the renamed μ-variables. The result is that the new Q_1 is given by

$$Q_1'(\mu,\mu') = \sum_p \chi_p(\mu,\mu')\, e^{\lambda K_p}/g \qquad (5.25a)$$

The bonds Q in the other directions are multiplied by one another because of the delta functions. The results for these bonds is

$$Q_\alpha'(\mu,\mu') = Q_\alpha[(\mu,\mu')]^\lambda \quad \text{for } \alpha > 1 \qquad (5.25b)$$

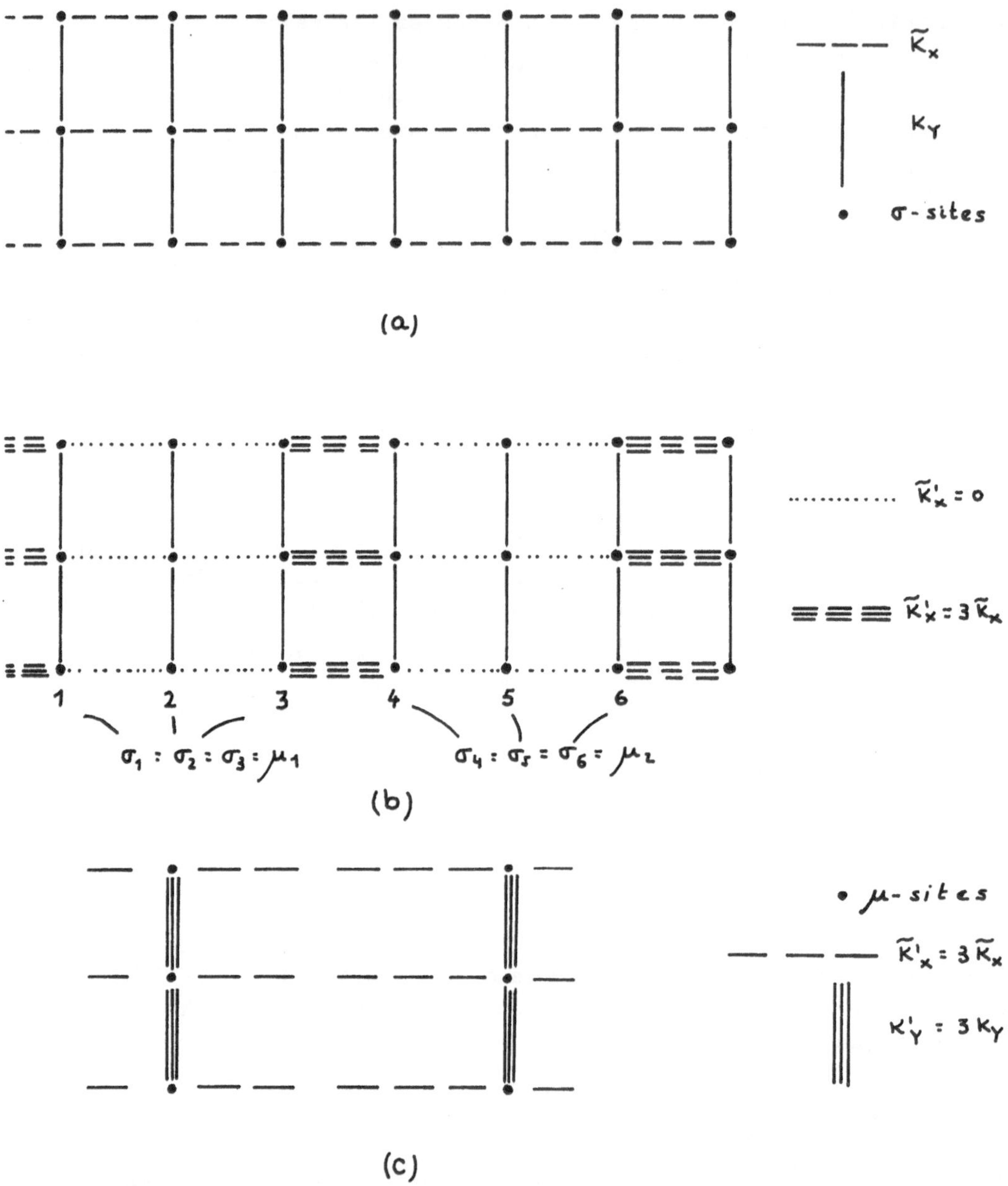

Fig. 5.2 Fourier Transform version of Migdal approximation

Now compare these results with our earlier calculations. According to Eq. (5.25b)

$$K_\alpha'(\mu,\mu') = \lambda K_\alpha(\mu,\mu') \quad \text{for } \alpha > 1 \tag{5.26a}$$

which is exactly the same as Eq. (5.16b). Moreover the statement $\bar{K}_p' = \lambda \bar{K}_p$ is exactly the one dimensional recursion relation

$$K_1' = R_o^\lambda [K_1] \tag{5.26b}$$

which is exactly the earlier result (5.16a), Thus the fourier transform potential moving in this case leads us directly back to our earlier "Migdal" recursion relation.

The earlier derivation indicated that the Migdal result was appropriate for weak potentials $K_1(\sigma,\sigma')$; this derivation points out that for transitive representations the Migdal result also applies to strong ferromagnetic interactions.

5.5 Gauge Systems

In this section, we carry through the Migdal conception for systems with string variables linked together in a U^4 coupling on squares. The interaction is assumed to be strong so that the Migdal scheme can work.

Start from the two dimensional case -- which we know to be exactly soluble. A basic lattice cell for the recursion is shown in Fig. 5.3a. Once again, we sum over the σ-variable being between the two squares and define new μ variables by

$$\begin{aligned} \mu_{AA'} &= \sigma_{AA'} \\ \mu_{BB'} &= \sigma_{BB'} \\ \mu_{AB} &= \sigma_{A1}\,\sigma_{1B} \\ \mu_{A'B'} &= \sigma_{A'1'}\,\sigma_{1'B'} \end{aligned} \tag{5.27}$$

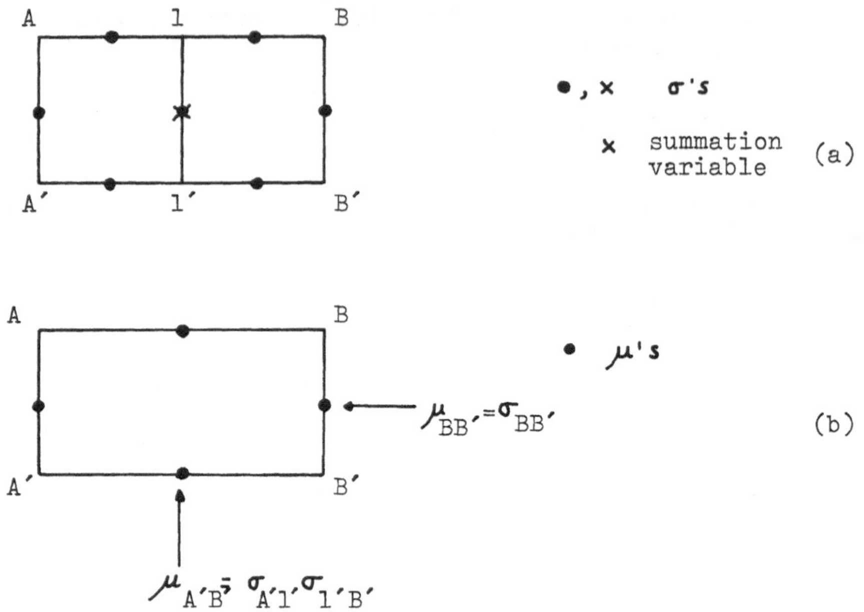

Fig. 5.3. The basic cell for the d=2 recursion.
Part a, before the recursion. Part b, afterward.

But now, instead of doing the sum exactly, we move the interaction (in the form $\tilde{K}_p$) from the left hand cell to the right hand one. Then the interaction in the left hand cell becomes

$$Q(\sigma_{A1}\ \sigma_{11'}\ \sigma_{r'A'}\ \sigma_{AA'}) = \delta_{\sigma_{11'}\ \sigma_{1A}\ \sigma_{AA'}\ \sigma_{A'1'},1} \tag{5.28a}$$

while the interaction in the right-hand cell becomes

$$Q(\sigma_{11'}\ \sigma_{1'B'}\ \sigma_{B'B}\ \sigma_B) = \sum_p e^{2\tilde{K}_p}\ \chi_p(\sigma_{11'}\sigma_{1'B'}\ \sigma_{B'B}\ \sigma_{B'1})/g \tag{5.28b}$$

With these new interactions the sum over $\sigma_{11'}$ can easily be performed to give a new interaction

$$Q' = \sum_p e^{2\tilde{K}_p}\ \chi_p(\sigma_{1A}\,\mu_{AA'}\mu_{A'B'}\mu_{B'B}\,\sigma_{B1'}1)\ /g$$

$$= \sum_b e^{2\tilde{K}_p}\,\chi_p(\mu_{BA}\,\mu_{AA'}\ \mu_{A'B'}\ \mu_{B'B'}1)\ /g \tag{5.29}$$

Thus, a change in lattice constant in x-direction from a_o to $2a_o$

gives a recursion in which the interaction $\tilde{K}_p$ increases by multiplication by a factor of two. When we return to the coordinate space representation, this result reads $K' = R_o^2\,[K]$. More generally a change by a factor of λ in a_o yields $K' = R_o^{\lambda}[K]$.

Now imagine a similar calculation in higher dimensionality. In Fig. 5.4 we have depicted a three dimensional version of the problem. Notice that there are three different couplings K_{xy}, K_{yz}, K_{xz}. Let us change the lattice constant in the x-direction by summung over variables indicated by crosses in the Figure. The new variables, μ, are defined in Fig. 5.4 . If $x_1 - x_1'$ points in the x-direction $\mu(x,x')$ is a product of λ σ-variables; if $x - x'$ points in the other directions, $\mu(x,x')$ is just a single unsummed σ .

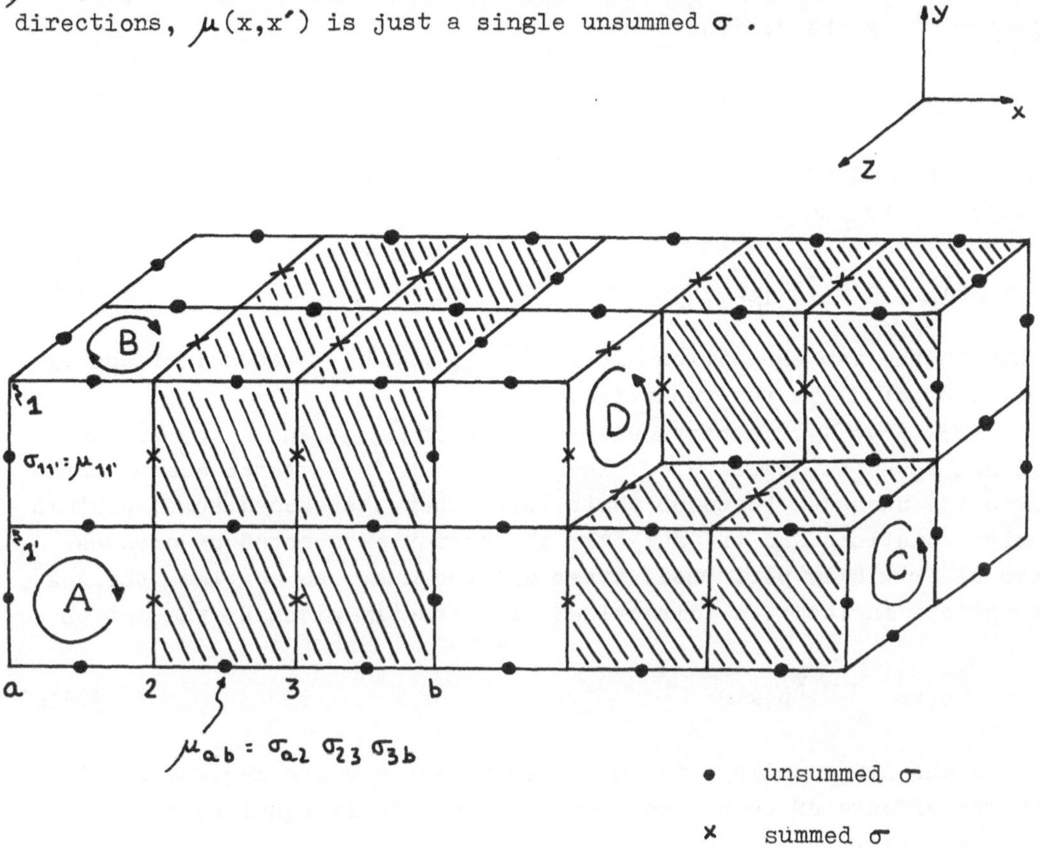

Fig. 5.4 The three dimensional Migdal recursion for a gauge system.

In the case of nearest neighbor interactions, we saw that there were two different ways of deriving an approximate "Migdal" recursion formula. The first was to move K; the second was to move $\widetilde{K}$. These two methods can also be applied here, and once again give equivalent answers.

First, move K. We could sum up the effect of all interactions in the x-y and x-z planes (e.g., couplings A and B on Figure 5.4) but we cannot handle coupling like D in the yz plane, which couple together four summation variables. Hence, move all couplings like D to sites like C where they couple together four unsummed variables. The net result is a calculable sum. The y-z couplings are increased by a factor λ by the motion

$$K_{\alpha\beta} = \lambda K_{\alpha\beta} \quad \text{if } \alpha \neq 1 \text{ and } \beta \neq 1 \tag{5.30a}$$

while the summations over the variables labeled by crosses can be performed to yield

$$K'_{1\beta} = R_o^{\lambda}[K_{1\beta}] \tag{5.30b}$$

Thus in the representation picture $\widetilde{K}_{p,xy}$ and $\widetilde{K}_{p,xz}$ each increase by a factor of λ in the recursion.

Now move $\widetilde{K}_p$. In particular focus upon $\widetilde{K}_{p,xy}$ and $\widetilde{K}_{p,xz}$ in the shaded boxes in Figure 5.4. Move all the $\widetilde{K}_p$'s in these shaded boxes into the unshaded boxes on their left. Then the shaded boxes contain delta functions and the sum over the variables denoted by crosses can be trivially performed. After all the dust has settled, the new couplings in the x, α directions are directly given by the motion

$$\widetilde{K}'_{p,x\alpha} = \lambda \widetilde{K}_{p,x\alpha} \tag{5.31a}$$

while the new couplings in the other directions are derived by adding up the effects of couplings like D to be exactly equal to those in couplings like C so the net result is, in configuration space

$$K'_{\alpha\beta} = \lambda K_{\alpha\beta} \quad \text{for} \quad \begin{matrix} \alpha \neq x \\ \beta \neq x \end{matrix} \tag{5.31b}$$

Of course, the results 5.31 are exactly the same as the results (5.30). We thus obtain a relatively simple result (essentially similar to Migdal's) for the case of gauge interactions with irreducible variables.

To derive the full consequences of this approximation, consider the effect of successive decimations in the directions 1,2,...,d upon $K_{\alpha\beta}$. Take the spatial indices α β so that $\alpha < \beta$ and $\beta = 2,3,\ldots$ d. Then Eqs. (5.31) imply the Migdal style recursion relations

$$K_{\alpha\beta}(\lambda a_0) = \lambda^{\alpha-\beta} R_0^{\lambda}[\lambda^{\beta-\alpha-1} R_0^{\lambda}[\lambda^{\alpha-1} K_{\alpha\beta}(a_0)]] \tag{5.32}$$

An especially interesting example of this recursion occurs if we take $\alpha=1$, $\beta=d/2+1$. In that case, the recursion (5-32) is a composition of two identical steps. Each step can be described by an effective recursion,

$$K \rightarrow K' = R_1^{\lambda}[K] = \lambda^{d/2-1} R_0^{\lambda}[k] \tag{5.33}$$

In terms of this effective recursion function R_1^{λ}, Eq. (5.32) may be written as

$$K_{\alpha\beta}(\lambda a_0) = R_1^{\lambda}[R_1^{\lambda}[K_{\alpha\beta}(a_0)]] \tag{5.34}$$

But, if we view the change in lattice constant $a_0 \rightarrow \lambda a_0$ as taking place in two steps

$$a_0 \rightarrow \sqrt{\lambda} a_0 \rightarrow \sqrt{\lambda}(\sqrt{\lambda}\, a_0)$$

then we can consider R_1^{λ} to be the recursion function for a single step. In this way, we reinterpret (5.34) as

$$\begin{aligned} K_{\alpha\beta}(\sqrt{\lambda}\, a_0) &= R_1^{\lambda}[K_{\alpha\beta}(a_0)] \\ &= \lambda^{d/2-1} R_0^{\lambda}[K_{\alpha\beta}(a_0)] \end{aligned} \tag{5.35}$$

The net result of this argument is that a particular coupling function $K_{1,d/2+1}$ in the gauge case obeys exactly the same style of recursion as Eq. (5.17a) - which describes the nearest neighbor case. Thus, for these particular couplings, the recursion in the d-dimensional gauge case is just the same as the recursion in the d/2-dimensional nearest neighbor situation [4] . For example, the four dimensional gauge case has some recursions which are exactly the same (in the Migdal approximation) as the two-dimensional nearest neighbor situation. Thus, if the Migdal approximation is accurate, the four-dimensional gauge-case can be understood in terms of the much simpler problem of nearest neighbor interactions in two dimensions.

But the Migdal approximation is accurate when the couplings are strong. And, it is exactly this strong-interaction limit which is significant for discussing whether or not the quark-string system has a phase transition. Therefore, the Migdal approach can be very useful for determining whether the Wilson model does in fact give both asymptotic freedom and quark trapping.

5.6 The Ising Example

To illustrate the considerations of this chapter, consider a nearest-neighbor Ising model [4]. In this case, the couplings

$$\underline{K} = (K_0, K_1) = (K_0, K)$$

obey the one-dimensional recursion relation

$$(K_0', K') = R_0^{\lambda}(K_0, K)$$

where, according to Eq. (3.38)

$$\tanh K' = (\tanh K)^{\lambda}$$

Therefore, from Eq. (5.17a), the x-direction coupling for the system with nearest neighbor interactions obeys:

$$K_x(\lambda a_0) = \lambda^{d-1} \tanh^{-1} [\tanh K_x(a_0)]^{\lambda} \tag{5.36}$$

Eq. (5.36) expresses the result of one recursion in which the lattice constant increases by a factor of λ.

The case $d \to 1$ is especially interesting. In this situation, there is a fixed point for large values of K_x. When $K_x \gg 1$, Eq. (5.36) implies

$$K_x(\lambda a_0) = \lambda^{d-1} [K_x(a_0) - \ln \lambda/2] \tag{5.37}$$

Then, $d \to 1$, there is a fixed point at $K_x = K^*$, where K^* goes to infinity as $d \to 1$, in the form

$$2K^* = \frac{1}{d-1} \tag{5.38}$$

Notice also that the recursion relation (5.37) has a critical index

$$y = d-1 \tag{5.39}$$

which goes to zero.

Clearly one dimension is a very special limit of the Ising model. In this limit, the critical couplings go to infinity. For $d < 1$, the critical point disappears entirely. We describe a value of the dimension at which the critical couplings go to infinity and then the phase transition disappears as a <u>lower critical dimension</u>, d_L^*. For the nearest-neighbor Ising model $d_L^* = 1$.

A very similar analysis can be applied to the gauge-style coupling. In this case, for Ising variables $U(x,x') = \pm 1$, the basic coupling on a plaquette takes the form (2.25). For each pair of spatial indices (α, β) there is a single coupling $J_{\alpha\beta}$, which is directly analogous to the K_α described above.

Apply Eq. (5.32) to the case in which $J_{\alpha\beta}$ is very large. Then by using the same calculation which led from Eq. (5.36) to Eq. (5.37), we find a recursion relation for the gauge case

$$J_{\alpha\beta}(\lambda a_0) = \lambda^{d-2} J_{\alpha\beta}(a_0) - \frac{1}{2} \ln (\lambda^{d-\beta} + \lambda^{d-\alpha-1}) \tag{5.40}$$

This recursion relation shows a lower critical dimensionality, d_L^*, equal to 2. Near the lower critical dimensionality there is a fixed point at very strong values of the coupling.

$$J^*_{\alpha\beta} = \frac{\lambda^{2-\beta} + \lambda^{1-\alpha}}{2(d-2)} \tag{5.41}$$

and a critical index

$$y = d-2 \tag{5.42}$$

We now make an analogy between these results for the group Z_2 near d=2 and the desired results for the group SU_3 at d=4. Assume for a moment that nature had one space and one time dimension and had a "color" symmetry Z_2. For $d \simeq 2$, and strong coupling, the Migdal relations of the section would be reliable. They would show a phase transition, i.e. the structure of Fig. 4.2a, for $d=2+\epsilon$, with $\epsilon > 0$. However at $\epsilon = 0$, there would be no phase transition and the phase diagram would look like Fig. 4.2b. Thence the theory would show both quark trapping and asymptotic freedom. The theory, however, would be far from trivial since a perturbation theory in g would only work in the quark phase. Since this phase only exists at g=0, it is very likely that the radius of convergence of this perturbation theory would be zero.

Two dimensions is special for a Z_2 gauge theory because $d_L^*=2$ for this theory. According to the Migdal approximation, this value of the lower critical dimensionality is in turn derivable from the fact that $d_L^*=1$ for the Z_2 nearest neighbor coupling. In fact, the general result is that for a given representation of a given symmetry, the lower critical dimensionality of the gauge theory, (d_L^*) gauge, is related to the lower critical dimensionality of the corresponding nearest neighbor theory, $(d_L^*)_{\text{global}}$

$$(d_L^*)_{\text{gauge}} = 2\,(d_L^*)_{\text{global}} \tag{5.43}$$

Now turn to the consideration of the interesting case, one in which $U(x,x')$ is the fundamental representation of a particular Lie group. According to Migdal [4] and to the more accurate calculations of Brezin and Zinn-Justin [34] and of Polyakov [35] the lower critical dimension for the global symmetry in this situation is $(d_L^*)_{\text{global}}=2$. Therefore, the gauge case shows a lower critical dimensionality at $d=4$!

In fact, one can make a slightly (but crucially) stronger statement. For all the regular representations of compact semi-simple Lie groups (e.g. SU_3), according to calculational methods of reference [34] there will be no phase transition in the d=2 nearest neighbor case and consequently no phase transition in the d=4 gauge case. Therefore, if the string-string interactions dominated the quark-string interaction, the strings would show no phase transition. In this case, we would indeed obtain the desired phase diagram, i.e. Fig. 4.2b. We would then have a theory containing as we wish both asymptotic freedom and trapping . It would be near-critical (since $d_L^*=4$) so it might even be Lorentz invariant. What could be more satisfying!

But there is more. Quantum electrodynamics can be expressed in this same language, with a symmetry group U_1, an abelian group. The corresponding nearest-neighbor problem is called the XY model. For this special case, all proofs of the non-existence [3],[4],[34],[35] of a phase transition at d_L^* fail. In fact, there are plausible [22 - 26],[35] arguments which suggest the existence of a phase transition at d_L^*, perhaps [22],[35] of the nasty nature shown by the phase diagram 4.2c. But, this phase transition is quite desirable from the point of view of experiment. It permits the observation of electrons and positrons and also permits the theory to not be asymptotically free.

However, the reader should be aware that all of the analysis of this section depends on the idea that the strings determine their own interactions with no help from the quarks. In the Migdal approximation, this idea is valid. Different couplings can be moved independently of one another in this approximation. But, in the real world, the quarks may enter into the string recursion relations in an essential way and thereby invalidate the reasoning outlined here.

ACKNOWLEDGEMENTS

This material is a revised version of a set of lecture notes for a series of talks delivered at the University of Chicago. Several members of the staff, including the entire many body group helped with constructing and criticizing these presentations. I owe special thanks to G. Mazenko, who organized most of the show.

The work described here started at the IBM Research Laboratory in Zurich and was continued at Brown and Harvard. Particularly helpful criticism came to me during seminars delivered at Harvard, NAL, Princeton, and Urbana. In addition, I wish to thank G. Mazenko, S. Coleman, B. Lee, S. Weinberg, K. Wilson, W. Bardeen, R. Ditzian, and A.A. Migdal for their helpful discussions and private communications.

REFERENCES

1) K.G. WILSON, "Quarks and Strings on a Lattice", Erice Lecture Notes, 1975.

2) K.G. WILSON, Phys. Rev. D10, 2445 (1974).

3) A.A. MIGDAL, Zh. E.T.F. 69, 810 (1975).

4) A.D. MIGDAL, Zh. E.T.F. 69, 1457 (1975).

5) L. KADANOFF, A. HOUGHTON, and M.C. YALABIK, J. Stat. Phys. Mar.1976.

6) L. KADANOFF, Phys. Rev. Lett. 34, 1005 (1975).

7) L. KADANOFF, Annals of Phys. (N.Y.) to be published.

8) BEREZIN, "The Method of Second Quantization", Acad. Press (N.Y.) (1966). The particular formulation given here was described to me by K. Wilson.

9) R. HERMANN, Lie Groups for Physicists, W.A. Benjamin, N.Y. (1966) p. 3.

10) R. BALIAN, J. DROFFE and C. ITZYKSON, Phys. Rev. D10, 3364 (1974).

11) R. BALIAN, J. DROFFE and C. ITZYKSON, Phys. Rev. D11, 395 (1975).

12) R. BALIAN, J. DROFFE and C. ITZYKSON, Phys. Rev.

13) F. WEGNER, J. Math. Phys. 12, 2259 (1971).

14) E.S. ALBERS and B.W. LEE, Phys. Reports 9C (1973).

15) A. HOUGHTON and L. KADANOFF, in Renormalization Group in Critical Phenomena and Quantum Field Theory : Proceedings of a Conference, 1973, (ed. J.D. Gunton and M.S. Green), Dept. of Physics, Temple University, Philadelphia, 1973.

16) D. NELSON and M. FISHER, Annals of Phys. 91, 226 (1975).

17) F. WEGNER Lecture Notes in Physics 37, 171.

18) The relation between critical points and fixed points forms the basis of the modern theory of critical phenomena. See K.G. WILSON, Phys. Rev. B4, 3174, 3184 (1971), K.G. WILSON and M.E. FISHER, Phys. Rev. Lett. 28, 240 (1972), K.G. WILSON, Phys. Rev. Lett. 28, 548 (1972) and also R. BALESCU, "Equilibrium and Non Equilibrium Statistical Mechanics", John Wiley N.Y. 1975, Chapters 9 and 10.

19) F.D. MURNAGHAN, The Theory of Group Representations (1938), Chapter 8.

20) R. BAXTER, Phys. Rev. Lett. 26, 832 (1971), Annals of Phys. (N.Y.) 70, 193 (1972).

21) L.P. KADANOFF and F. WEGNER, Phys. Rev. B4, 3989 (1971).

22) J. ZITTARTZ, Z. Phys. B23, 55,63 (1976).

23) V.L. BEREZINSKI, Sov. Phys. JETP 32, 493 (1971).

24) J.M. KOSTERLITZ and D.J. THOULESS, J. Phys. C6, 1181 (1973).

25) J.M. KOSTERLITZ, J. Phys. C7, 1046 (1974).

26) A. LUTHER, Private Communication.

27) K. WILSON, "Relativistically Invariant Lattice Theories", presented at Coral Gables Conference, Jan. 1976.

28) Private Communication. See also W. BARDEEN and R.B. PEARSON, "Local Gauge Invariance the Bound State Nature of Hadrons", Fermilab preprint 76/24-THY 1976.

29) THE NIEMEIJER and J.M.J. VAN LEEUWEN, Phys. Rev. Lett. 31, 1411 (1973) and Physica (Utrecht) 71, 17 (1974).

30) M. NAUENBERG and B. NIENHUIS, Phys. Rev. Lett. 33, 944, 1598 (1974) and Phys. Rev. B11, 4152 (1975).

31) It is useful to notice that this "Fourier Transformation" is the first step in constructing a dual transformation (H.A. KRAMERS and G.H. WANNIER, Phys. Rev. 60, 252 (1941)). See references 11 and 13.

32) D. GROSS and F. WILCHEK, Phys. Rev. Lett. 30, 1343 (1973).

33) H.D. POLITZER, Phys. Rev. Lett. 30, 1346 (1973).

34) E. BREZIN and J. ZINN-JUSTIN, Phys. Rev. Lett. 36, 691 (1975).

35) A.M. POLYAKOV, Phys. Lett. 59B, 79 (1975).

THE ROLE OF SPONTANEOUS BROKEN SYMMETRY

R. BROUT

Pool de Physique
Université Libre de Bruxelles

The Role of Spontaneous Broken Symmetry

R. BROUT

Pool de Physique
Université Libre de Bruxelles

Within the past decade or two, the application of spontaneous broken symmetry (s.b.s) to elementary particle theory has found widespread use, so much so that most particle theorists now consider that s.b.s. is almost surely going to be one of the building blocks of the ultimate theory. Essentially all the concepts used in this program are borrowed from the theory of phase transitions. It is our purpose in these lectures to show how this is done. Applications of these ideas to various phenomena will be discussed and the successes and hopes of the entire enterprise will be assessed.

I. Spontaneous Broken Symmetry and the Nambu-Goldstone Theorem

Our starting point, chosen because it offers a convenient bridge between the theory of phase transitions and quantum field theory is the Landau-Ginzburg formulation of phase transitions. The partition function is written as a functional integral over a set of (scalar) fields φ_i. The φ_i form a representation of the symmetry group of the problem.

One writes in d-dimensions

$$
\begin{aligned}
Z &= \int \prod_i \mathcal{D}\{\phi_i\} \exp - \int d^d x \, L \\
L &= \frac{1}{2} \sum |\nabla \phi_i|^2 + P(\{C_n(\phi)\}) + \sum J_i \, \phi_i
\end{aligned}
\qquad (1)
$$

$\int \mathcal{D}\{\phi_i\}$ is the functional integral over all configurations of the field ϕ_i . P is a polynomial which is a local function of the group invariants $C_n(\phi)$ and J_i are external sources. Limitation of time does not permit me to derive from fundamental molecular models the form (1) which is the relevant phenomenological expression for the partition function in the critical region. Rather I shall indicate the significance of the different terms in L in the context of a few interesting cases.

1. The Ising model : Here ϕ has only one component which simulates the two-valued spin on each lattice site; lattice sums have become integrals so that the representation is good only for phenomena occurring on a length scale large compared to the lattice distance- hence critical phenomena. The origin of the term $|\nabla\phi|^2$ is the exchange energy. If two spins differ in orientation, this costs energy - a phenomenon simulated by $|\nabla\phi|^2$, a term which is non-negative and which vanishes when ϕ is the same all over. The polynomial $P(C(\phi))$ must be a function of ϕ^2 since the Ising model has reflection symmetry between ϕ and $-\phi$. A popular version is to take

$$
P(\phi) = \frac{\mu_0^2}{2} \phi^2 + \lambda \phi^4 \qquad (2)
$$

We now know from renormalization group arguments that for $d=3$, it could be "irrelevant" to consider higher-order terms. The factor μ_0^2 is of the form $\mu_0^2 = A(1 - \alpha/kT)$. The form of this mass term arises from an entropy term and an energy term, the latter being α/kT , which comes from the exchange energy of a given particle with its neighbours. For kT such that μ_0^2 is small (in units of the range of the potential) one is then in the "critical region". In molecular field approximation, the critical point is when μ_0^2 vanishes. $\lambda\phi^4$ is another contribution to the entropy. Finally the source J is an external magnetic field.

2. Heisenberg model : Here ϕ_i form the components of a three vector, representing a spin on a lattice, exchange coupled to its neighbours through the scalar interaction $(\vec{S}_i \cdot \vec{S}_j)$.Scalarity then tells us that P is a function of $|\vec{\phi}|^2$. The group of symmetry is O_3. Otherwise everything said for the Ising model applies.

3. Liquid helium and superconductivity : For liquid helium ϕ is the field function for helium atoms in second quantization. Then $\frac{|\nabla\phi|^2}{2}$ represents the kinetic energy. The term in $\lambda\phi^4$ represents the repulsive interaction between atoms. (It is necessary that $\lambda > 0$ in the phenomenology. Of course there is attraction as well; in that case λ is to be interpreted as the s-wave scattering matrix and this must be positive i.e. repulsion must dominate attraction in s-waves in order to have a meaningful phenomenology (and presumably a stable medium)). In this case, Z means the grand partition function so that the term in $\mu_0^2/2|\phi|^2$ comes from the chemical potential. At the critical point of the ideal Bose gas, μ_0 vanishes. Thus the ideal Bose gas plays the role of a zeroth order approximation in the same way that molecular field theory does for the Ising or Heisenberg model. In both cases the bare mass, μ_0^2 , vanishes in the zeroth order approximation.

A superconductor is described by the same phenomenological Lagrangian. Here ϕ is the Fourier transform of the bound electron pair. $\phi(r) = \int dk dq \langle a_k a_{-k+q} \rangle e^{iqr}$ where $\langle a_k a_{-k} \rangle$ are the BCS order parameters in the quiescent superconductor i.e. for $\langle\phi\rangle$ constant. The integral over k is over the relative momentum of the pair. This integral is non vanishing over a range related to the correlation length. The latter parameter appears in the theory in the following way. $P(C(\phi))$ is taken to have a non trivial minimum at some ϕ_c $(\neq 0)$. The curvature at this minimum is the square of the inverse of this correlation length.

In both instances the field ϕ is a Schrodinger field, hence complex. Conservation of the number of particles requires that

$C(\phi)$ be $\phi^*\phi$. Writing ϕ as $\phi_1 + i\phi_2$, the invariant is $\phi_1^2 + \phi_2^2$, so the symmetry group is O_2, often referred to as the gauge group. If the particles are charged then conservation of particles implies conservation of charge and this version is the one used in particle theory. Gauge then is the familiar electromagnetic gauge. One of the important applications of s.b.s. is to consider what happens due to the coupling of the charged field ϕ with the gauge field A_μ (the electromagnetic field). This is discussed in detail in Section III together with the Meissner effect, and massive vector mesons in elementary particle theory.

4. We mention some examples drawn from particle theory which have no counterparts in many-body physics. If SU_3 is taken to be a good symmetry then the invariants constructed from an octet of scalar fields ϕ_i need no longer be quadratic. The cubic $\sum d_{ijk}\,\phi_i\phi_j\phi_k$ is also invariant and $P(\phi)$ involves this term as well. Another favourite group is $SU_2 \times SU_2$ or $SU_3 \times SU_3$. These will be discussed in more detail in Section II.

We now explore the consequences of s.b.s. in the context of the models presented. We begin with the example of a broken gauge symmetry. Here $C(\phi) = \phi_1^2 + \phi_2^2$. We first present the problem in molecular field theory. This is the theory where the functional integral over all configurations is approximated by neglecting all configurations but the one which maximizes the integral i.e. minimizes L. In the absence of inhomogeneous sources the minimum of L clearly corresponds to $\nabla\phi_i = 0$ or ϕ_i = const. The constants ϕ_i are then chosen to minimize $P(C(\phi)$

$$\frac{\partial P}{\partial \phi_i} = 0 \qquad i = 1,2 \tag{3}$$

or

$$\frac{\partial P}{\partial C(\phi)}\,\phi_i = 0 \tag{4}$$

We suppose a non trivial minimum exists, corresponding to Fig. 1

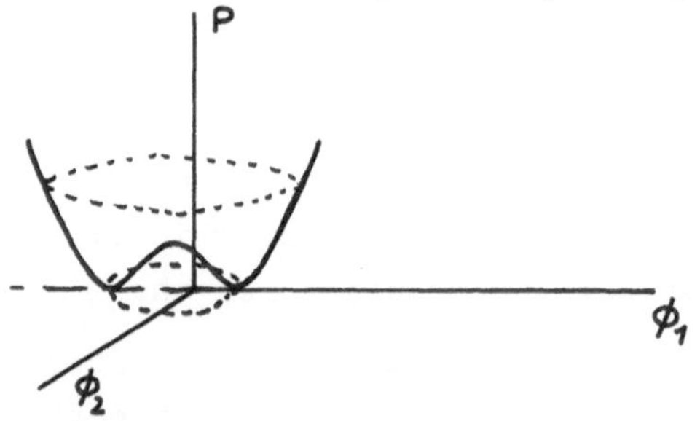

There is a locus of minima lying on a circle. s.b.s corresponds to picking one point on this circle, say $\langle\phi_1\rangle = \phi_c \neq 0$ and $\langle\phi_2\rangle = 0$. The free energy log Z is then given by

$$\log Z = V\, P(\phi_c^{\,2}) \tag{5}$$

where V is the volume. Had we included all configurations lying on the minimal circle, then Z would have contained an extra factor of (2π) corresponding to the phase integral over these configurations and we would have

$$\log Z = V\, P(\phi_c^{\,2}) + \log 2\pi \quad .$$

The thermodynamic limit is obtained by taking $V \to \infty$. Hence $\log(2\pi)$ is negligible. It suffices to take one configuration. This is the crux of s.b.s. The free energy is effectively evaluated by taking one of a number of configurations which are degenerate due to symmetry, since inclusion of the degenerate set is negligible in $O(1/V)$.

There are two ways to understand this physically. In the first instance take a source $\vec{J}$ (the vector direction is in the gauge plane). Then the maximal configuration is obtained from

$$\frac{\partial(P + \sum J_i\phi_i)}{\partial\phi_i} = 0 \tag{6}$$

or

$$\left.\frac{\partial P}{\partial C(\phi)}\right|_{\phi=\phi_c} \phi_i + J_i = 0 \tag{7}$$

Since $\partial P/\partial C(\phi)$ is a function of $C(\phi)$ only, we get $\vec{\phi} \parallel \vec{J}$. For every $\vec{J}$ no matter how small the configuration direction is given and all other configurations are then negligible. Now go to the limit $V \to \infty$ and then $J \to 0$. All other configurations cut out and the originally imposed direction is adhered to.

Another way to understand s.b.s. is to orient $\vec{\phi}(x)$ at one point (with the finger of an angel). Then the $|\nabla\vec{\phi}|^2$ term forces $\vec{\phi}$ to orient in the same way at all points. Thus an infinitesimal local orientation fixes the global orientation.

To go from molecular field to the rigorous theory is an easy step in so far as the kinematic consequences of broken symmetry are concerned. One introduces the "effective potential" in the following way. The functional integral in the presence of a spatially homogeneous source leads to a form

$$-\log Z = V\left[F(|\phi_c|^2) + \vec{J}\cdot\vec{\phi}_c\right] \tag{8}$$

where $\vec{\phi}$ is determined by minimization

$$\partial \log Z / \partial \vec{\phi}_c = 0 \tag{9}$$

F, the effective potential, is a function of the invariants $C(\phi)$ by symmetry arguments to be sketched below (in the gauge case $C(\phi)=\phi_1^2+\phi_2^2$)). Note in molecular field theory that $F = P$. The form of F, at least its relevant part, is the same as that of P. Namely

$$F = \frac{\mu^2}{2} \vec{\phi}^2 + \lambda' (\vec{\phi}^2)^2$$

where μ^2 will depend on the temperature. This parameter vanishes at the true critical temperature, in most models somewhat lower than the molecular field value, where μ_0^2 vanishes. It is clear that once s.b.s. can occur the previous discussion obtains

We sketch a simple argument for the forms 8) and 9). Write eq.(1) in the presence of a homogeneous source as

$$Z = \int d\phi_c \, e^{-V[F(\phi_c) + J\phi_c]} \tag{10}$$

where

$$V\,F(\phi_c) = -\log \int \mathcal{D}\{\phi\} \, \delta\Big(\int \vec{\phi}(x)\, d^dx - V\vec{\phi}_c\Big)\, e^{-L} \tag{11}$$

In (10) we have used : 1/ $\int J\phi(x) = J\int \phi(x) = V J\phi_c$ for each member of the integrand in (1) which contributes to the "slice" selected by $\delta\,[\int \vec{\phi}(x)dx - V\phi_c]$ 2/ the integral over all configurations in such a slice is exponential in the volume, since L is proportional to the volume.

In the limit $V \to \infty$, only the minimum of $F(\phi_c) + J\phi_c$ contributes to (10). Configurations characterized by values of ϕ_c which are close to the minimum = [quantitatively $(\phi_c - (\phi_c)\text{min})^2 = O(1/V)$] contribute a small term to (log Z/V) of O(log V/V) . We therefore have demonstrated the forms 8) and 9) but for the proof that F is only a function of $C(\phi)$ (= $|\vec{\phi}|^2$ in the gauge case).

To prove this invariance, one need but evaluate (11) for two orientations of $\vec{\phi}_c$ which differ by a group transormation R e.g. a rotation about the 3 axis for the gauge group. That F is invriant is seen by making a change of integration variables from $\{\phi_c\}$ to $\{R\phi_c\}$. L is invariant under the change and the Jacobian is unity. Therefore we have

$$\begin{aligned}
&\int \mathcal{D}\{\phi\} \, \delta\Big(\int \vec{\phi}(x)d^dx - VR\vec{\phi}_c\Big)\, e^{-L(\phi)} \\
&= \int \mathcal{D}\{R\phi\} \, \delta\Big(\int R\vec{\phi}(x)d^dx - VR\vec{\phi}_c\Big)\, e^{-L(R\phi)} \\
&= \int \mathcal{D}\{\phi\} \, \delta\Big(\int \vec{\phi}(x)d^dx - V\vec{\phi}_c\Big)\, e^{-L}
\end{aligned} \tag{12}$$

in $F(\vec{\phi}_c) = F(R\phi_c)$. Thus $F(\phi_c) = F(R\phi_c)$ can depend only on the invariants $C(\phi)$.

We are now prepared to discuss the Nambu-Goldstone theorem in phase transition theory. In fact the germs of the theorem go back to the beginning of the century when Debye and Born and von Karman recognized that longitudinal phonons (i.e. quantities whose eigenfrequencies vanish with their wave number) are implied by the broken translational symmetry of crystalline lattices. The idea reappeared once more in the spin wave theory of Bloch in the thirties and again in Anderson's analysis of broken gauge theories. Application to particle physics was introduced by Nambu and further formalized by Goldstone and successive authors following them.

The most direct way to proceed is as follows. (For simplicity we continue with the example of gauge theory). Suppose we have s.b.s. where $\vec{\phi}_c$ points in the 1 direction. Now turn on an infinitesimal source pointed in the 2 direction. As we have explained the minimal configuration now has $\vec{\phi}_c$ pointed in the 2 direction. We conclude

$$\lim_{J_2 \to 0} \frac{\partial \langle \phi_2 \rangle_1}{\partial J_2} = \infty \qquad (13)$$

where the subscript 1 means average in the state where ϕ_c points in the 1 direction. All we are expressing is that such a state is unstable under any disturbance which has a 2 component. Now when a system responds infinitely strongly to an infinitesimal driving force one is said to have a resonance phenomenon. One is "exciting" modes of the frequency and wave length of the perturbing force. Such modes must be eigenmodes of the system in order to respond in this resonant way. Conclusion : there is an eigenmode in an s.b.s state oriented in the 1 direction which has zero frequency and infinite wave length and which has orientation in the 2 direction. To make this last statement more explicit we remark that the definition of Z, (eq. 1) gives us

$$\partial \log Z/\partial J_i = - \langle \phi_i \rangle \; ; \; \partial^2 \log Z/\partial J_i \, \partial J_j \, , = \langle \phi_i \phi_j \rangle - \langle \phi_i \rangle \langle \phi_j \rangle = \partial \langle \phi_i \rangle / \partial J_j \qquad (14)$$

Thus from (13) we get

$$\left\langle \int \phi_2(x)\, dx \int \phi_2(x)\, dx \right\rangle = \infty \qquad (15)$$

It is interesting to dwell somewhat further on the formal structure of the theory. We first establish the theorem :

$$\frac{\partial^2 \log Z}{\partial J_i \, \partial J_j} \quad \text{and} \quad \frac{\partial^2 \log Z}{\partial \langle \phi_i \rangle \, \partial \langle \phi_j \rangle} \quad \text{are reciprocal matrices} \tag{16}$$

To this end we express the fact that the $\langle \phi_i \rangle$ minimize log Z for a set of J_k i.e. the system responds to a small change in the J_k in such a way as to keep $\partial \log Z / \partial \langle \phi_k \rangle = 0$. (This is the equation of state which gives $\langle \phi_i \rangle$ as a function of J_k) . Thus we have

$$\partial^2 V / \partial \langle \phi_i \rangle \, \partial \langle \phi_k \rangle + (\partial^2 V / \partial \langle \phi_i \rangle \, \partial J_\ell)(\partial J_\ell / \partial \langle \phi_k \rangle) = 0 \tag{17}$$

using (14), the second term of (17) becomes $-\delta_{\ell i}(\partial J_\ell / \partial \langle \phi_k \rangle)$. Reference once more to the second equation in (14) establishes the theorem (true for all J_i , but we are interested in the case $J_i = 0$ only).

The Nambu-Goldstone theorem is $\partial^2 \log Z / \partial J_i \, \partial J_j$ has some infinite matrix elements at $q = 0$. Alternatively in the eigenfunction basis, this matrix shows up some infinite eigenvalues. From (16), the quest for these eigenvalues and eigenfunctions is equivalent to the search of zero eigenvalues of $\partial^2 \log Z / \partial \langle \phi_i \rangle \partial \langle \phi_j \rangle$, (or $\partial^2 \log Z / \partial \langle \phi_i \rangle^2$ in the principal axes frame; the latter is obviously a frame which is lined up so that some of the "axes" will coincide with the "axes of broken symmetry"). But this is then a simple job in elementary geometry in most cases. We only need ask : for a given set of $\langle \phi_i \rangle$, what changes in the $\langle \phi_j \rangle$ do not provoke a change in log Z but rather simply cause the initial point in $\langle \phi_i \rangle$ space to slip along the locus of the minimum of log Z ? Thus in the gauge case we confirm that in the vacuum $\langle \phi_1 \rangle \neq 0$, $\langle \phi_2 \rangle = 0$, then ϕ_2 is the Goldstone boson. The reader is requested to confirm this from eq. (10) by writing $\chi_1 = \phi_1 - \langle \phi_1 \rangle$ and $\chi_2 = \phi_2$ and then checking that the term in χ_2^2 vanishes - thereby establishing the zero eigenvalue as well as the corresponding eigenfunction. Similarly if the group is SU_2 and $\langle \phi_i \rangle$ is a triplet with the state chosen so that $\langle \phi_i \rangle = \phi_c \, \delta_{i3}$, we will check that ϕ_1 and ϕ_2 are the Goldstone mesons.

We now give a somewhat more formal prescription on how to find these Goldstone modes. Namely we use the fact that the F of eq. (8) is a function of the group invariants constructed from the $\langle \phi_i \rangle$ where $\langle \phi_i \rangle$ is in some representation of the group. Thus under an infinitesimal group transformation we have

$$\delta \langle \phi_i \rangle = \epsilon_a \, T_{a\,ij} \, \langle \phi_j \rangle \tag{19}$$

the T_a's being the group generators in the $\langle \phi_i \rangle$ representation and the ϵ_a the infinitesimal group parameters. Expressing the invariance of F under (19) gives us

$$\delta F = 0 = \partial F/\partial \langle \phi_i \rangle \; \delta \langle \phi_i \rangle + \frac{1}{2} \partial^2 F/\partial \langle \phi_i \rangle \partial \langle \phi_j \rangle \; \delta \langle \phi_i \rangle \; \delta \langle \phi_j \rangle \tag{20}$$

Because the ϵ_a of (19) are arbitrary each term in (20) must vanish separately. The first already does so by the minimization condition. Let us condense the notation a bit to

$\langle \vec{\phi} \rangle$ = vector whose components are $\langle \phi_j \rangle$

$\mathcal{F}$ = matrix whose components are $\partial^2 F/\partial \langle \phi_i \rangle \partial \langle \phi_j \rangle$

Eq. (20) then gives us

$$\overrightarrow{T_a \langle \phi \rangle} \; \mathcal{F} \; \overrightarrow{T_a \langle \phi \rangle} = 0$$

i.e. the zero eigenvalues of $\mathcal{F}$ have eigenfunctions along the directions $\overrightarrow{T_a \langle \phi \rangle}$ for a state in which the direction of broken symmetry is given by $\langle \vec{\phi} \rangle$. The Goldstone modes are those which are obtained by the group generators operating on the direction $\langle \vec{\phi} \rangle$.

The next important question to face is whether this kinematical theorem of s.b.s. is the terminus at $q = 0$ of a physical spectrum of finite frequency at finite q. We address ourselves here to this point without yet complicating the situation by introducing gauge vector mesons.

The answer to our question is in the affirmative provided there is a length scale for which eq. (1) is truly valid, i.e. for which the term in $|\nabla \phi_i|^2$ makes sense. Note that this term was never set into play in a crucial way in our demonstration since we looked at the $q = 0$ response only. Now, in point of fact, the term $|\nabla \phi|^2$ in eq. (1) applied to spin problems should have been multiplied by a constant $= \frac{1}{J}\left[d^2 J/dr^2 \big|_{r=0} \right] \equiv R^2$ where J is the exchange potential and R its range. The phenomenological representation 1) is good only for $q R << 1$. Hence as $R \rightarrow \infty$, the kinetic energy term gets lost and the Goldstone mode becomes an isolated pole. For R finite the spectrum of $\omega(q)$ vs. q rises from zero to some value at $q = O(R^{-1})$ and then flattens out. We will now make this clear by both physical and formal argument.

Consider the Heisenberg model constructed from spin ½ particles. The ground state has all spins aligned, say in the Z direction. This ground state is N-fold degenerate corresponding to the eigenstates of S_Z ($\equiv \frac{1}{2} \Sigma \sigma_i^Z$) with all spins parallel ($S = N/2$). These are obtained by rigid rotations of the ground state i.e. by successive operations of S^- ($\equiv \frac{1}{2}\sum \sigma_i^-$). We see that the Goldstone excitation operator is thus S^- since $dS^-/dt = [H, S^-] = 0$ i.e. this is the zero frequency, zero wave number eigenmode for the present problem.

Now take half the spins and rotate them a little bit (say all the spins on the left are aligned along the z-axis, but on the right are turned by the operator $\sum_{i \in \text{right side spins.}} \sigma_i^-$). This will cost energy in the region where the spins on the left side see the spins on the right side i.e. within a distance R of the dividing line. We will say that this costs one node of energy (obviously only a little bit because the angle between the unaligned spins is $O(1/N)$). We will call this one node's worth of energy. Now do this repeatedly by dividing the system up into thirds, fourths, etc. We will get 2,3,4... nodes of energy. The energy it costs to make these excitations thus rises with wave number, being zero at zero wave number. The Goldstone mode is the terminus of a spectrum. The spectrum continues to rise with the number of nodes until the interval between the nodes is $O(R)$ the range of the force where upon turning spins within such intervals causes no further rise in energy. For example for $R \to \infty$, turning half the spins one against the other causes a disalignment of $(N/2)$ spins with respect to a given spin so even if the angle of disalignment is $O(1/N)$ the energy involved in this excitation is $O(1)$. The spectrum then has a gap.

Quantitatively this goes as follows. Define $\sigma_q^- = \frac{1}{\sqrt{N}} \sum e^{i\vec{q}\cdot\vec{R}_i} \sigma_i^-$ where $\vec{R}_i$ is the site of the ith spin and $\vec{q}$ is in the reciprocal lattice. The Hamiltonian is $\frac{1}{2}\sum v_{ij}\, \vec{\sigma}_i \cdot \vec{\sigma}_j = \frac{1}{2}\sum v(q)\, \vec{\sigma}_q \cdot \vec{\sigma}_{-q}$ where

$$v(q) = \sum_{R_j} v(R_i - R_j) e^{iq \cdot (R_i - R_j)} . \quad \text{Using : (1)}$$

$$[\sigma_q^+, \sigma_{-q'}^-] = \sigma_{q-q'}^z \quad ; \quad [\sigma_q^z, \sigma_{-q'}^+] = \sigma_{q-q'}^+$$

and $\sigma_q^z |0\rangle = \delta_{oq} \sqrt{N}$ where $|0\rangle$ is the ground state with all spins up, one readily deduces

$$i \frac{d}{dt} \sigma_q^- |0\rangle = [H, \sigma_q^-] |0\rangle = [v(0) - v(q)]\, \sigma_q^- |0\rangle \qquad (20)$$

Thus the energy of the state $\sigma_q^- |0\rangle$ is $(v(0)-v(q))$ higher than the ground state. If the range of v is R then $v(q) \to 0$ for $q \gtrsim R^{-1}$.

In this way we see that when supplemented by quite simple dynamical elements, the kinematic Goldstone theorem acquires dynamical significance, One easily confirms by analogous arguments the existence of the phonon spectrum in solids, gauge Goldstone modes in neutral superfluids etc. It is only required that the relevant length (range of force, correlation length in superfluids etc.) be finite.

II. Application of the Nambu-Goldstone Theorem to Particle Theory

A convenient starting point is once again the effective potential. The vacuum to vacuum S matrix of Lagrangian field theory for scalar fields is given by Z of eq. (1) with $d=4$. The time dimension is imaginary, but analytic properties of the S matrix permit an analytic continuation to real t (at least this can be established to every order in perturbation theory and as far as I know has never been known to fail in practical circumstances). Therefore quantum field theory may be likened to a 4 dimensional statistical mechanical problem. The whole formalism of Section I to describe s.b.s may then be taken over en masse.

The physical significance of log S (S here is the vacuum to vacuum S matrix) is important. As in statistical mechanics, log S is proportional to VT (the "volume of space-time"). The theorem is $\lim_{V,T\to\infty}(1/VT)\log S$ = energy density of vacuum (units are $\hbar = 1, c = 1$). In non s.b.s. situations the proof is evident by simply comparing the Feynman diagrams which contribute to the $\langle P(\phi)\rangle$ in the Hamiltonian (or Dyson) formalism and in the Lagrangian (functional integral) formalism. It is less straightforward to establish the identity of the kinetic energy terms due to ill defined integrals. This problem is of no importance however, since the relevant piece for s.b.s. is the potential energy $P(\phi)$. In the s.b.s. case, the same proof can be executed but one must replace the fields ϕ_i by $\chi_i \equiv [\phi_i - \langle\phi_i\rangle]$ in order to make Feynman graphs. $\langle\phi_i\rangle$ then appear as parameters in log S and they are fixed by requiring $\partial \log S/\partial\langle\phi_i\rangle = 0$ in accordance with the variational principle of quantum mechanics. These exercises will not be executed in these lectures, but are recommended as homework.

The important part of log S in s.b.s. is the dependence on $\langle\phi_i\rangle$ and to this end one defines the quantity

$$V\{\phi_i\} = \log\left[S(\langle\phi_i\rangle)\,/\,S(0)\right] \tag{21}$$

It is this quantity which is most often referred to as the effective potential.

Let us now turn to the phenomena of the zero mass scalar (again gauge fields are absent for the nonce). For a J_i or set of J_i's appropriately chosen, the response will be infinite for vanishing 4 vector q which we will designate by $(q_0, \vec{q})$. Relativistic invariance requires the response to be a function of q^2 and the simplest form permitted by analyticity is q^{-2}. Thus the response function is proportional to $1/q_0^2 + q^2$ in the Euclidean version and hence $1/q_0^2 - q^2 + i\epsilon$ in the correct analytic continuation to real time (Feynman causal propagation). We infer that there is a spectrum of poles at $q_0 = q$, and hence a true zero mass excitation.

We give two applications of these ideas. The first is the group $SU_2 \times SU_2$, thought by most physicists to be a weakly broken symmetry of the Nambu-Goldstone type (in fact this example is Nambu's inspiring inception of the corrupt of s.b.s. into elementary particle theory). The pion is the (almost) zero mass mode. We first wish to understand the significance of this all important fact.

Before the details we have first to acquire some notion of the role of symmetry breaking. A simple example is offered in the Heisenberg model. The Hamiltonian of eq. (15) now has an added term say $\lambda \Sigma \sigma_i^z$ so that the commutator reads

$$i \sigma_j^- |0\rangle = ([v(0) - v(q)] + \lambda) \sigma_q^- |0\rangle$$

As $q \to 0$ the energy of the lowest excitations is $O(\lambda)$. The zero-mass scalar now has acquired mass, linear in the symmetry breaking parameter. Though the linearity is not completely general, it is true in most cases, in particular in the case of chiral symmetry breaking a la Nambu-Jona-Lasinio.

We now turn to the essentials of the model. The free Dirac Lagrangian in the absence of isospin is

$$L_D = \bar{\Psi} \gamma_\mu \partial_\mu \Psi \tag{22}$$

This Lagrangian is invariant under the two transformations

$$\Psi \to e^{i\theta_V} \Psi \qquad \Psi \to e^{i \gamma_5 \theta_A} \Psi \tag{23}$$

and hence by Noether's theorem is possessed of two conserved currents

$$j_\mu = \bar{\Psi} \gamma_\mu \Psi \qquad ; \qquad j_{\mu 5} = \bar{\Psi} \gamma_\mu \gamma_5 \Psi \tag{24}$$

An instructive way to understand these conserved currents is to write the Dirac 4-spinor Ψ in the "chiral" representation (left handed and right handed 2-spinors u,v) such that

$$\sigma_\mu \partial_\mu u = 0 \qquad \sigma_\mu = 1, \vec{\sigma}$$
$$\bar{\sigma}_\mu \partial_\mu v = 0 \qquad \bar{\sigma}_\mu = 1, -\vec{\sigma} \tag{25}$$

One easily establishes that $u = [(1 + \gamma_5)/2]\Psi$, $v = (1 - \gamma_5)/2\, \Psi$. The parity operation shifts u into v and vice versa. In these terms we have

$$L_D = u^\dagger \sigma_\mu \partial_\mu u + v^\dagger \bar{\sigma}_\mu \partial_\mu v \tag{26}$$

The phases of u and v may be separately varied so there are two conserved currents $u^\dagger \sigma_\mu u$ and $v^\dagger \bar{\sigma}_\mu v$. The relationship with (24) is

$$j_\mu = u^\dagger \sigma_\mu u + v^\dagger \bar{\sigma}_\mu v \qquad \leftarrow \text{vector}$$
$$j_{\mu 5} = u^\dagger \sigma_\mu u - v^\dagger \bar{\sigma}_\mu v \qquad \leftarrow \text{axial vector} \tag{27}$$

If a mass term is present one must add $m_o \bar{\Psi}\Psi$ to L_D . This term is no longer invariant under $\Psi \to e^{i\gamma_5\theta_A}\Psi$; hence $j_{\mu 5}$ is no longer conserved. Rather

$$\partial_\mu j_{\mu 5} = 2\, m_o \bar{\Psi}\gamma_5\Psi \tag{28}$$

One understands this in u,v language from the fact that $\bar{\Psi}\Psi = u^\dagger v + v^\dagger u$. To keep L invariant then requires the same phase transformation for u and v , so only one current is conserved.

Nambu considered an example in which a massless Ψ field undergoes interactions in such a manner that the chiral symmetry is conserved in the Lagrangian but is spontaneously broken. A zero mass pseudoscalar meson ensues which is identified with the pion. (The details are presented below, but we wish first to give the qualitative reasoning). Let us now suppose that the chiral symmetry is not perfect, but is sullied by a small $m_o \bar{\Psi}\Psi$ term in L . Small means $m_o << m$, where m is the mass acquired by the Ψ field due to s.b.s. of chirality. Then the pion mass will not vanish. As we have mentioned, relativistic formulations will give us quite generally a propagator in q^2 so in the present case something in $(q^2 - m_\pi^2)^{-1}$ where m_π^2 will be linear in the symmetry breaker, m_o . The only expression one can make is $m_\pi^2 \sim m_o m$. If m is taken on the hadron mass scale (1 GeV), then m_o works out to be a about 10 MeV or i.e. something like a lepton mass. This estimate of m_o has been confirmed in a number of independent determinations from various hadronic phenomena so, at the least, the point of view is coherent. What is suggested is that the elementary hadronic constituents (quarks) acquire a mass , m_0 , by the same mechanism which endows leptons with their mass. Spontaneous broken chiral symmetry then boosts this up to m of O(1 GeV) for all hadrons but the pion which is an almost zero mass excitation.

This idea has been successfully tested quantitatively through what is called soft pion physics, one example of which is the Goldberger-Treiman relation. We sketch here some of the ideas behind this development. A simple method is to use the σ model. The first step is to extend matters to include i-spin. Then the Ψ , u and v of eqs (22) to (28) are i-spinors (i-spin ½). The interesting conserved currents are then

$$\vec{j}_\mu = \bar{\Psi}\gamma_\mu \frac{\vec{\tau}}{2}\Psi \qquad \vec{j}_{\mu 5} = \bar{\Psi}\gamma_\mu\gamma_5\frac{\vec{\tau}}{2}\Psi \tag{29}$$

The group of transformations which is responsible for the conservation of these 6 currents is defined by $\Psi \to e^{i\vec{\tau}\cdot\vec{\theta}/2}\Psi$ and $\Psi \to e^{i\frac{\vec{\tau}\cdot\vec{\theta}_A}{2}\gamma_5}\Psi$. This is the six parameter group $SU_2 \times SU_2$. The group generators are $\vec{V}$ and $\vec{A}$ where the $\vec{V}$ form an isospin subgroup

$$[V_i , V_j] = \epsilon_{ijk} V_k \tag{30}$$

and the $\vec{A}$'s transform like i-spin vectors. Their commutation properties are

$$\begin{aligned} [V_i , A_j] &= \epsilon_{ijk} A_k \\ [A_i , A_j] &= \epsilon_{ijk} V_k \end{aligned} \tag{31}$$

One checks that $(\vec{V} \pm \vec{A})/2$ are the generators responsible for conserving left and right handed i-spin currents $u^\dagger \sigma_\mu \vec{\tau}/2\, u$ and $v^\dagger \bar{\sigma}_\mu \vec{\tau}/2\, v$ respectively. If there is a breaking term $m_0\bar{\Psi}\Psi$ then L will no longer be invariant under the A transformations. Rather

$$\partial_\mu \vec{j}_{\mu 5} = 2m_0 \bar{\Psi}\gamma_5\frac{\vec{\tau}}{2}\Psi . \tag{32}$$

One may now check that the pseudoscalar quantities $\bar{\Psi}\gamma_5\vec{\tau}\Psi$ on the right hand side of (32) and the scalar quantity $\bar{\Psi}\Psi$ form a representation.

$$
\begin{aligned}
A_i \bar{\Psi}\gamma_5 \frac{\tau_j}{2} \Psi &= \delta_{ij} \bar{\Psi}\Psi \\
A_i \bar{\Psi}\Psi &= \Psi\gamma_5 \frac{\tau_i}{2} \Psi \\
V_i \bar{\Psi}\gamma_5 \frac{\tau_j}{2}\Psi &= \epsilon_{ijk} \bar{\Psi}\gamma_5 \frac{\tau_k}{2} \Psi \\
V_i \bar{\Psi}\Psi &= 0
\end{aligned}
\tag{33}
$$

We associate mesons to this representation: σ transforms like $\bar{\Psi}\Psi$ and $\vec{\pi}$ like $\bar{\Psi}\gamma_5\vec{\tau}/2\,\Psi$. A chiral invariant Lagrangian is

$$
\begin{aligned}
L = i\bar{\Psi}\gamma_\mu \partial_\mu \Psi &+ \frac{1}{2}\left((\partial_\mu \sigma)^2 + (\partial_\mu \vec{\pi})^2\right) + P(\sigma^2 + \vec{\pi}^2) + \\
&+ g\left[\left(\bar{\Psi}\frac{\vec{\tau}}{2}\gamma_5 \Psi\right)\cdot\vec{\pi} + \bar{\Psi}\Psi\sigma\right]
\end{aligned}
\tag{34}
$$

We introduce s.b.s. by requiring the polynomial P to have a non trivial minimum. External considerations (say weak and electromagnetic interactions) will force a direction of breakdown upon us, namely a choice of vacuum which is manifestly Lorentz and i-spin scalar. Thus; the vacuum has $\langle\sigma\rangle \neq 0$ and $\langle\vec{\pi}\rangle = 0$.

From our general theorem (eq. (20) et seq.), the π_i will be zero mass excitations as is seen by reference to the second of eqs. (33) [i.e. the T_a that operates on $\langle\sigma\rangle$ which gives a non vanishing result are the three axial generators]. Hence we predict an i-spin triplet of zero mass pseudoscalars. If chirality is broken slightly by adding an $m_0\bar{\Psi}\Psi$ term to (34) one checks the formula $m_\pi^2 \sim m_0 m$.

We now turn to the Goldberger-Treiman relation. Before we present it for chiral symmetry however, we wish to consider first a simpler example so the reader will appreciate the general content of the formulae.

Consider a group which is spontaneously broken and take a group representation of scalar mesons. Spontaneous breakdown will split the mass spectrum of this representation. To this group there will belong a set of conserved currents equal in numbers to the number of group generators. Take the matrix element of one of these currents between two such mesons $\langle i | j_\mu | j\rangle$. This is a Lorentz vector and since there are only two such available $(P_i)_\mu$ and $(P_j)_\mu$ or alternatively $P_\mu \equiv (P_i)_\mu + (P_j)_\mu$ and $q_\mu = (P_i)_\mu - (P_j)_\mu$, we may in all generality write

$$
\langle i | j_\mu | j\rangle = F_+(q^2)\, P_\mu + F_-(q^2)\, q_\mu \tag{35}
$$

The reason why $F_\pm$ depend only on q^2 is that there is only one free

Lorentz scalar in the problem since $P_i^2 = m_i^2$ and $P_j^2 = m_j^2$. We may choose this to be $q^2 = P_i^2 + P_j^2 - 2P_i \cdot P_j = m_i^2 + m_j^2 - 2P_i \cdot P_j$. Now multiply (35) by q_μ and use $\langle i | q_\mu j_\mu | j \rangle = \langle i | \partial_\mu j_\mu | j \rangle$. The latter is zero by the symmetry which implies current conservation. [We remind the reader of Noether's theorem. If a Lagrangian is invariant under an infinitesimal transformation of fields characterized by an infinitesimal parameter ϵ one then lets ϵ depend on x and forms the Lorentz vector $\partial L / \partial(\partial_\mu \epsilon(x))$. This quantity is given the symbol $j_\mu(x)$ and it is easily proved from the supposed symmetry and the equations of motion that $\partial_\mu j_\mu(x) = 0$. If L is not invariant under the transformation then $\partial_\mu j_\mu = \partial L / \partial \epsilon(x)$].

Eq. (35) then gives us

$$F_+(q^2)\, q_\mu \cdot P_\mu + q^2 F_-(q^2) = 0$$

or

$$F_+(q^2)\left[P_i^2 - P_j^2\right] + q^2 F_-(q^2) = 0 \qquad (36)$$

$$F_+(q^2)(m_i^2 - m_j^2) + q^2 F_-(q^2) = 0$$

There are two possibilities :

1) $F_- = 0$ and $m_i^2 = m_j^2$. This is Schur's lemma and is the realization of symmetry which is not spontaneously broken, or
2) $F_- \neq 0$, then $m_i^2 \neq m_j^2$ and F_- picks up a pole at $q^2 = 0$: $F_- = - F_+(m_i^2 - m_j^2)/q^2$.

Substituting into (35) we get

$$\langle i | j_\mu | j \rangle = F_+(q^2)\left[P_\mu - \left(\frac{m_i^2 - m_j^2}{q^2}\right) q_\mu\right] \qquad (37)$$

The pole at $q^2 = 0$ is of course the Nambu-Goldstone scalar excitation and our derivation here is tantamount to an alternative derivation of its existence. (A detailed analysis reveals the same group theoretical structure as previously derived). It is easy to understand the longitudinal coupling to this mode as it appears in (37) through the following consideration. Consider for simplicity the gauge group. The conserved current is $\phi_1^* \overleftrightarrow{\partial_\mu} \phi_2$. In a vacuum where $\langle \phi_1 \rangle \neq 0$ but $\langle \phi_2 \rangle = 0$ there is a term in the current equal to $\langle \phi_1 \rangle \partial_\mu \phi_2$ i.e. proportional to q_μ multiplied by the breaking parameter $\langle \phi_1 \rangle$, hence the second term in (37). This consideration is general, the ath current for a general group being $(T_a)_{ij} (\partial_\mu \phi_i) \phi_j$.

The second term in (37) is often diagrammed as follows

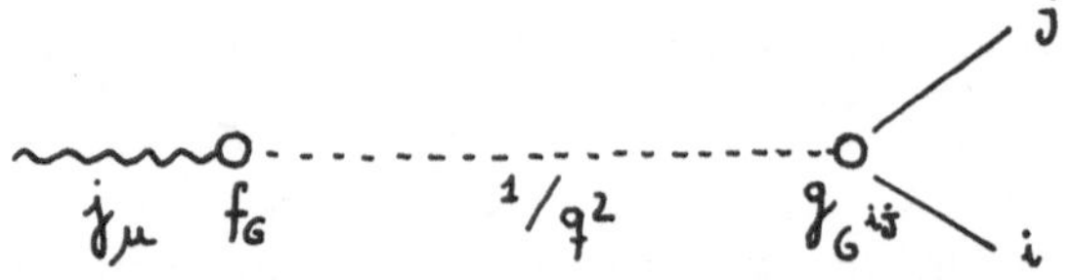

f_G is a matrix element for the current to create a Goldstone meson and g_G^{ij} is its coupling to mesons ij . From (37) we get the Goldberger-Treiman relation.

$$F_+(0)\ (m_i^2 - m_j^2) = f_G\, g_G^{ij} \tag{38}$$

F_+ is taken at $q^2 = 0$ since eq. (38) is a statement about residues of the pole at $q^2 = 0$.

This theorem was first presented in the present form by Nambu for the case of chiral symmetry where in place of mesons i,j were spin ½ fermions, the nucleons. The conserved current was $j_{\mu 5}$ of eq. (29). The analogy of (35) for the most general form factor decompositions

$$\langle N | j_{\mu 5} | N' \rangle = \bar{u}_N \left[F_A\, \gamma_\mu \gamma_5 \frac{\vec{\tau}}{2} + F_P q_\mu\, \gamma_5 \frac{\vec{\tau}}{2} \right] u_{N'} \tag{39}$$

u_N are Dirac spinors and isospinors. Multiply by q_μ and set the left hand side to zero. Use the Dirac equation as follows

$$\bar{u}_N\, q_\mu \gamma_\mu \gamma_5\, u_{N'} = \bar{u}_N\, (P_N - P_{N'})\, \gamma_\mu\, \gamma_5\, u_{N'}$$

$$= \bar{u}_N\, \not{P}_N \gamma_5\, u_{N'} + \bar{u}_N\, \gamma_5\, \not{P}_{N'}\, u_{N'} = (m_N + m_{N'})\, \bar{u}_N\, \gamma_5\, u_{N'} \tag{40}$$

where we use $\{\gamma_\mu, \gamma_5\} = 0$. We then get

$$0 = (m_N + m_{N'})\, F_A\, \bar{u}\, \gamma_5 \frac{\tau}{2} u + q^2\, F_P\, \bar{u}\, \gamma_5 \frac{\tau}{2} u = 0$$

$$F_P = - \frac{(m_N + m_{N'})\, F_A}{q^2} \tag{41}$$

the pole here represents pion propagation and the analogue of (38) is (in our convention)

$$F_A(0)\ \ (m_N + m_{N'}) = f_\pi\, g_{\pi N N'} \tag{42}$$

[Note here the chiral breaking is the non vanishing of m_N for fermions whereas in the example of eq. (38) symmetry breaking was manifested in the difference $m_i^2 - m_j^2$ for mesons].

Assuming that chirality is broken "softly" then Nambu changed (41) to a partial conservation law (PCAC). He postulated that $\partial_\mu j_{\mu 5}$

became conserved for $q^2 \gg m^2$. This postulate together with the existence of the pole in eq. (41) displaced to $q^2 = m_\pi^2$ can be realized by taking

$$F_P = - \frac{(m_N + m_{N'})\, F_A}{q^2 - m_\pi^2} \tag{43}$$

Eq. (42) still holds. PCAC can be confirmed in some Lagrangian models, but not all. Its viability is a function of the type of external breaking. We will not discuss this here.

All the constants in eq. (42) have been measured and agreement has been checked to about 10%. This is indeed a triumph that can hardly be accidental. Together with other successes of soft pion physics it has given most physicists confidence that the underlying ideas are correct.

We now turn to another, far more tenuous, attempt at applying s.b.s. to hadron physics. This is concerned with the breakdown of SU_3. Consider a Lagrangian constructed from an octet of scalar fields $\phi_1 \ldots \phi_8$. There are now two invariants

$$\begin{aligned} C_2 &= \Sigma\, \phi_i^2 \\ C_3 &= \Sigma\, d_{ijk}\, \phi_i\, \phi_j\, \phi_k \end{aligned} \tag{44}$$

The minimum equations for $\langle \phi_i \rangle$ are then

$$\frac{\partial F}{\partial C_2} \langle \phi_i \rangle + \frac{\partial F}{\partial C_3}\, d_{ijk} \langle \phi_j \rangle \langle \phi_k \rangle = 0 \tag{45}$$

The new kind of breaking consists in balancing the two terms of (50) one against the other, hence requiring far less stringent conditions. One easily verifies that one solution is $\langle \phi_8 \rangle \neq 0$ $\langle \phi_1 \rangle = \ldots = \langle \phi_7 \rangle = 0$. Now the 8 direction of SU_3 is a charge direction i.e. breaking along it leaves invariant an SU_2 subgroup, the isospin for the 8 direction. This characterization of the breaking is complete i.e. group transformations transform charge directions to charge directions. [The proof is most easily given in terms of the matrix representation of ϕ_i : $\Phi = \sum_{i=1}^{\infty} \lambda_i\, \phi_i$ when the λ_i are the 3x3 generators of the defining representation. λ_8 has the set of eigenvalues 1,1,-2 . These are invariant under group rotations : if the system spontaneously breaks along 8 , all vacua accessible from this one by group transformation will be characterized by a Φ with the same eigenvalues; but this is the unique characterization of a charge direction].

Now in fact SU_3 does break along a charge direction (8) so one is tempted to explain this fact by s.b.s. The conjecture is reinforced by the following remarkable fact. In the presence of a source term $J_i \phi_i$ one addends an inhomogeneity J_i to the right hand side of (45). When the term is $\partial F/\partial C_3$ is absent, one establishes the familiar law that the response is parallel to the external force. But when the $\partial F/\partial C_3$ is present this is not necesserily so. If we believe that the main perturbing force to SU_3 symmetry is electromagnetic we would be tempted to take $\vec{J}$ along the electromagnetic charge direction $\vec{Q}$ (i.e. at 120^0 to the 8 direction in the weight plane). One finds two solutions :

1) the expected breaking along $\vec{Q}$
2) a breaking along a set of hypercharge axes, a one parameter family corresponding to a rotated set of weight planes about the Q axis. One of these is the hypercharge.

Fixing this last angle is the project of predicting the Cabbibo angle, in response to the weak interactions. Though many ideas have been generated on this point, we still have no clear notion on how to solve this problem.

Nevertheless the fact that s.b.s. points up the special role of the hypercharge axis is impressive. The difficulty with the approach however, is the absence of the associated Goldstone bosons which by our general theorem should bear the quantum numbers 4,5,6,7. Nobody has ever seen a low mass strange scalar. Nor can gauge theory take care of the matter because this would require non-strange zero mass vectors. But this is another chapter to which we now turn our attention.

III. Spontaneous Broken Gauge Theory

A somewhat surprising turnabout in s.b.s. is what happens when gauge vector mesons are coupled to the theory. We explain below what the basic ideas of gauge theory are and why in systems without s.b.s., gauge vector mesons have zero mass. The surprising fact is that in the presence of s.b.s., the scalar Nambu-Goldstone bosons go away and a subset of the otherwise massless vectors acquires mass.

The physical applications of this idea in the physics of matter is in the domain of superconductivity where the theory explains the Meissner effect and superconductivity of the second kind. We remark that the idea goes back to London's work in the nineteen thirties. In particle theory, the idea was independently introduced by Higgs on one hand and by Englert and myself on the other. Within the past decade this development has profoundly affected our thinking on weak interactions. It was rapidly realized that the hypothetical intermediate vector meson W of weak interactions and the photon could be of the same nature, where the mass of the W's would be due to s.b.s. Also it was shown that the propagators of such mesons would remain transverse at all momenta so that their massiveness would not destroy the renormalizability of the theory. A concrete realization through a model was then proposed by Weinberg and by Salam who applied the gauge idea to an earlier phenomenological structure of Glashow. The program of renormalization could not be completed until an adequate quantum gauge formalism was found. Once this was delivered to us by Fadeev and Popov, the program was achieved by 't Hooft and Veltman and others.

We first review gauge invariance of the second kind and electromagnetism. (This is the Abelian case.) For a Lagrangian of type(1) generated by a complex field (i = 1,2) with $J_i = 0$, we have remarked the invariance group is the gauge transformation $\phi_1 + i\phi_2 \rightarrow e^{i\theta} (\phi_1 + i\phi_2)$. Now note that the term $|\nabla \phi|^2$ prevents the invariance of L from being extended to the case where θ is a function of the space-time point x. The generalization is nonetheless a natural and elegant requirement to demand. The way to do the job is to introduce the field $\vec{A}$ so that the $|\nabla\phi|^2$ term in L becomes $|(\vec{\nabla} - ie\,\vec{A})\phi|^2$ and to require the transformation property that when $\phi \to e^{i\theta(x)}\phi$, one transforms A to $\vec{A} - \frac{1}{e}\vec{\nabla}\theta(x)$. Since A is a function of x, it is a field. We must then build its Lagrangian. The latter is to be a rotational invariant (Lorentz invariant for d = 4) and a gauge invariant, i.e. invariant under $\vec{A} \to A - (^1/e)\nabla\theta$. To this end one builds the anti-symmetric tensor $F_{k\ell} = \partial_k A_\ell - \partial_\ell A_k$ hence $L_A = {}^1/4\, F_{k\ell} F^{k\ell}$. The $^1/4$ is required by canonical rules which I shall not discuss here. The full gauge problem then is stated by modifying 1) to

$$Z = \int \Pi \mathcal{D}\{\phi\}\,\mathcal{D}\{\phi^*\}\,\mathcal{D}\{\vec{A}\}\,\exp - \int d^d x\, L$$

$$L = \tfrac{1}{2}|(\vec{\nabla} - e\vec{A})\phi|^2 + \tfrac{1}{4} F_{k\ell} F^{k\ell} + P(|\phi|^2) \tag{46}$$

$$F_{k\ell} = \partial_k A_\ell - \partial_\ell A_k .$$

Actually Z defined by (46) is not quite right since one will get the same value of the functional integral for all values of $\vec{\nabla}.\vec{A}$. One must select some one of these. This is the Fadeev-Popov prescription. Before presenting the rule we introduce the relativistic version of (46)

$$\log S = \int \Pi \mathcal{D}\{\phi\}\mathcal{D}\{\phi^*\}\mathcal{D}\{A_\mu\}\ \exp - \int d^4 x\, L$$

$$L = \tfrac{1}{2}\,|(\partial_\mu - ie\,A_\mu)\phi|^2 + \tfrac{1}{4} F_{\mu\nu} F^{\mu\nu} + P(|\phi|^2) \tag{47}$$

The Fadeev prescription can be given in many ways. For our purposes, it is most convenient to state it as the rule to multiply the integrand of (47) by

$$\exp - {}^1/2\eta \int d^4x(\partial_\mu A_\mu)^2 \Big/ \int \mathcal{D}\{A_\mu\}\, e^{-{}^1/2\eta\int d^4x(\partial_\mu A_\mu)^2} . \tag{48}$$

It is easy to prove by the gauge invariance of L that log S is η independent, but the rule (48) is necessary. Otherwise the integral contains an infinite number of redundant configurations which causes it to diverge. We will often work in what is called Landau gauge: $-\eta = 0$. In this gauge the equations of motion of the A_μ field derived from L are

$$\partial_\mu F_{\mu\nu} = \Box A_\nu = e\,j_\nu$$
$$j_\nu = i\phi^* \overleftrightarrow{\partial_\nu} \phi + eA_\nu \phi^*\phi . \tag{49}$$

We see from (47) that gauge invariance has imposed the result that A_μ be a mass-less field. A term in $\mu^2 A_\mu{}^2$ is not invariant under $A_\mu \rightarrow A_\mu - \partial_\mu \epsilon$. It is clear that this must be the case. The A_μ field was introduced precisely to extend symmetry to situations where ϵ depends on space and time and hence configurations which vary with finite wave number. The symmetry with zero wave number variations of the fields is guaranteed without introducing the A_μ. Therefore there should be no manifestation of A_μ at $q = 0$. In particular there must be no energy associated with it; hence its mass is zero in the initial Lagrangian. The deep question is whether its mass can arise through interaction and the interesting answer is yes. An example is provided by s.b.s.

The Meissner effect follows from the existence of $\langle \phi \rangle \neq 0$ in Eq. (49), say real. In this case Eq. (49) becomes

$$\Box A_\nu + e^2 \langle \phi \rangle^2 A_\nu = i \langle \phi \rangle \partial_\nu \phi_2 + O(\delta) \tag{50}$$

The δ term in (50) contains only field fluctuation,ie. higher order terms in ϕ_2 and $[\phi_1 - \langle \phi \rangle]$. These are irrelevant to the discussion and will be dropped in what follows. In the more rigorous treatment of the problem their effect is included but the conclusions given below do not change. Observe in this approximation that ϕ_2 is the Goldstone boson field and hence of zero mass; in the present approximation - corresponding to the retention of only quadratic terms in ϕ_2 - its equation of motion is $\Box \phi_2 = 0$. Hence (50) is consistent with our gauge condition

$$\partial_\nu A_\nu = 0 \quad \left[\Box \partial_\nu A_\nu - e^2 \langle \phi \rangle^2 \partial_\nu A_\nu = i \langle \phi \rangle \Box \phi_2 = 0 \right].$$

To describe the Meissner effect one considers a static situation in which an external magnetic field is incident on the surface of a superconductor. Eq. (50) for $\vec{A}$ then reads

$$\Delta \vec{A} - \mu_v^2 \vec{A} = i \langle \phi \rangle \vec{\nabla} \phi_2 \tag{51}$$

$$\mu_v^2 = e^2 \langle \phi \rangle^2 . \tag{52}$$

Taking the curl of (51) gives us

$$\Delta \vec{B} - \mu_v^2 \vec{B} = 0 .$$

In a one dimensional geometry with a superconductor on the positive z axis and void on the negative side such that $B = B_0$ for $z < 0$, the solution of (53) is $\vec{B} = \vec{B}_0 e^{-\mu_v z}$. The field is excluded from superconductor within a penetration distance $= \mu_v^{-1}$, the London penetration depth. The role of s.b.s. is to supply this parameter through (52). This is London's very important theoretical discovery, some thirty years before it was rediscovered in the context of relativistic field theory.

The point of course is that the left-hand side of Eq. (50) describes the wave equation of a vector particle with mass equal to $e\langle\phi\rangle$ (the Klein-Gordon equation). This is the essence of the matter. An important bonus however arises from the study of the effect of the term in $\partial_\nu \phi_2$ in (50). We have mentioned that it is consistent with the gauge condition we have imposed, $\partial_\nu A_\nu = 0$. In fact one can show more. Namely its existence is precisely responsible for the maintenance of this transversality condition. Indeed this is the only role of the Goldstone field and in consequence the latter is no longer physically observable.

We now make these statements more transparent by showing that the A_μ which arises in response to an external current source J_μ, which is conserved, is in fact transverse - as is required by gauge invariance (i.e. no longitudinal component of A_μ should arise from a disturbance). In a general gauge, and in the approximation of retaining only linear terms in an A_μ, ϕ_2 and $(\phi_1 - \langle\phi_1\rangle)$ we have

$$\partial_\mu F_{\mu\nu} + e_v^2 \langle\phi_1\rangle^2 A_\nu - i\langle\phi_1\rangle \partial_\nu \phi_2 = J_\nu \tag{53}$$

$$\Box \phi_2 + i\partial_\mu A_\mu \langle\phi_1\rangle = 0 . \tag{54}$$

All the terms in (53) and (54) have been previously discussed save the second term in (54). This occurs in the ϕ_2 equation of motion by functional differentiation of the term $i\langle\phi_1\rangle(\partial_\mu \phi_2)A_\mu$ in L. An integration by parts is required to get the form (54). In Fourier transforms we get

$$-(q^2 g_{\mu\nu} - q_\mu q_\nu)A_\mu(q) + e^2\langle\phi_1\rangle^2 A_\nu(q) - eq_\nu\langle\phi_1\rangle\phi_2(q) = J_\nu(q) \tag{55}$$

$$q^2 \phi_2(q) - e\langle\phi_1\rangle q_\mu A_\mu(q) = 0 \tag{56}$$

Substituting the solution of (56) $\phi_2(q) = e(\langle\phi_1\rangle/q^2)q_\mu A_\mu$ we get (with $\mu_v^2 = e^2\langle\phi_1\rangle^2$)

$$(g_{\mu\nu} - q_\mu q_\nu/q^2)(q^2 - \mu_v^2)A_\mu(q) = -J_\nu(q)$$

$$A_\mu(q) = -\frac{g_{\mu\lambda} - q_\mu q_\lambda/q^2}{q^2 - \mu_v^2} J_\lambda(q) \tag{57}$$

Thus $A_\mu(q)$ does indeed remain transverse.

Thus calculated as a response function to conserved currents, the propagator of the A_μ field calculated from (14) is

$$\langle A_\mu(q)\, A_\nu(q)\rangle = \frac{g_{\mu\nu} - q_\mu q_\nu/q^2}{q^2 - \mu_v^2} \quad . \tag{58}$$

To this propagator one can add an arbitrary longitudinal part $\eta q_\mu q_\nu/q^4$, but this piece will be untouched by external currents and hence by interactions (the latter all being simulated by the term J_ν in Eqs. (53) - (57). Hence such gauge terms are without physical consequence. Furthermore the passage from Eqns. (55), (56) to (57) shows that the Goldstone field has been entirely absorbed into the A_μ field. The pole at $q^2 = 0$, now is in the $(q_\mu q_\nu/q^2)$ term of the A_μ propagator and this is the only place where the pole appears.

The all-important form (58) is to be compared to the propagator of an elementary vector meson, $(g_{\mu\nu} - q_\mu q_\nu/\mu^2)/q^2 - \mu^2$. The term $q_\mu q_\nu/\mu^2$, when used to calculate virtual processes (loops in Feynman graphs), causes ultraviolet divergences whereas (58) will give integrals the same ultraviolet behavior as in the unbroken theory. If the latter is renormalizable so is the former. This is the crux of renormalizability. The transversality of (58) has been proven in all generality as a consequence of gauge symmetry - as to be expected.

The generalization to the non-Abelian case can be induced without much effort and we shall not burden the reader with the often heavy formal structure of non-Abelian gauge theory. Suffice it to say that a group generated by n parameters Θ can be extended to have invariance of the second kind by introducing n gauge fields $A_{a\mu}$, each of which has zero mass in the unbroken theory. These gauge fields are coupled to the n matter currents, and again by our general argument have vanishing mass terms in L.

We may now induce from (58) which of these n gauge vector mesons acquire mass in the event of s.b.s. Recall that the term $q_\mu q_\nu/q^2$ in (58) arises from the absorption of the Goldstone pole. Therefore the quantum numbers of the A_μ fields which acquire mass are the same as those of the Goldstone mesons in the corresponding non gauge theory. These Goldstone modes are then physically unobservable in the gauge theory. Thus for s.b.s. in the 3 direction in SU_2 gauge symmetry, there is a zero mass gauge vector meson in the 3 direction and two (degenerate) massive vectors in the 1, 2 directions. In SU_3, 4,5,6,7 are massive gauge vectors and 1,2,3,8 are massless -

the symmetry being broken in the 8 direction.

The most interesting application of this idea is in the domain of the weak interactions. The interesting experimental fact is that the weak interactions are current-current short range interactions. For years a fundamental model was lacking in that such a short range interaction, either a direct 4 fermi coupling or a heavy elementary vector meson mediated interaction, is non renormalizable. The suggestion in 1965 that the heavy meson be an s.b.s. type object such as (58) then came as a welcome relief (at least to its authors). The model was put into a reasonable working form by Weinberg and by Salam in 1967, though it is still very far from the complete story.

Only the ideas will be discussed as we do not wish to become entangled with the intricacies of non-Abelian gauge theory. The experimental facts are that the weak interaction current found from leptons involves only the left-handed spinors, the u of Eq. (25). These spinors seem to come in pairs, the muon and its neutrino (μ, ν_μ) and the the electron and its neutrino (e, ν_e). One is tempted to identify these pairs with an i-spin doublet, with a current constructed as in Eqns. (29) - (32). The weak current is thus composed: $u^\dagger \sigma_\mu \frac{\vec{\tau}}{2} u$, u being a left-handed Lorentz spinor and i-spin spinor, the 2 components being either μ, ν_μ or e, ν_e or some unknown similar construction for hadrons - not to be discussed here. Were $m_\mu = \nu_\mu = m_e = \nu_e = 0$, one would have left-handed i-spin symmetry. It is assumed that spontaneous breakdown gives the charged leptons their mass. This is achieved as follows.

Consider, for definiteness the μ, ν_μ pair. Only the left-handed spinors are felt in weak interactions. But since the muon has a mass one must have a right-handed muon spinor as well, as discussed in conjunction with Eq. (28). The ν_μ (and ν_e) have zero mass and their right-handed spinors need not exist. And indeed they never have been seen. It is therefore compelling to assume that the right-handed muon, v is an i-spin singlet. A term that one can introduce into L which is an i-spin invariant, and which can generate mass by s.b.s, is then

$$G[\Phi u^\dagger v + \Phi^\dagger u v^+]. \qquad (59)$$

Φ is an i-spin spinor Lorentz scalar. Identifying the μ with $\tau_3 = +1$, we can then give the muon its mass by requiring $\langle\Phi_1\rangle \neq 0$ $\langle\Phi_2\rangle = 0$. (This can be brought about through a non trivial minimum in the usual type polynomial function $P(\Phi^\dagger\Phi)$). The muon mass is $G\langle\Phi_1\rangle$ and the left-handed i-spin group is spontaneously broken.

Observe now that there is a subgroup of symmetry of L left in this problem of a highly interesting structure and which is not concerned with the left-handed i-spin group. Namely one can transform v and $(\frac{1+\tau_3}{2})u$ (i.e. the muon component of u) by a common phase. This symmetry is not broken by $\langle \Phi_1 \rangle \neq 0$ in (59). We conclude that

$$v^\dagger \bar{\sigma} \; v \; + \; u^\dagger \left(\frac{1+\tau_3}{2}\right) \sigma_\nu \; u \; \equiv \; j_\nu \tag{60}$$

is a normally conserved current, untouched by s.b.s. generation of the muon mass. From (27) we see that this current is a true Lorentz vector (as opposed to axial vector or some mixture). Since it is generated by the charged lepton, we identify (60) as the muon's contribution to the electromagnetic current.

The gauge theory is built upon these elements. One introduces a triplet of $\vec{A}_\mu$ coupled to $u^\dagger \frac{\vec{\tau}}{2} \sigma_\mu u$ and a singlet B_μ coupled to that combination of currents, $(\frac{1}{2}u^\dagger \sigma_\mu u + v^\dagger \bar{\sigma}_\mu v)$, which in combinations with $u^\dagger \frac{\tau_3}{2} \sigma_\mu$ gives rise to the j_ν of (60). This latter is conserved in the standard way, untouched by s.b.s. Correspondingly the gauge boson coupled to it remains massless. It is a linear combination of $A_\mu{}^3$ and B_μ.

There are two charged gauge bosons $A_\mu^\pm$ coupled to $u^\dagger \tau^\pm \sigma_\mu u$. These are responsible for the usual weak interactions seen in β-decay. They acquire mass since $u^\dagger \vec{\tau} \sigma \; u = \vec{j}_\mu - \vec{j}_{\mu 5}$. The axial pieces are all spontaneously broken by the induced charged lepton mass through the analog of (32). (Lest there by any confusion, $\partial_\mu j_{\mu 5}$ still vanishes, but the form factor of $j_{\mu 5}$ has two pieces as in (39). Eq. (32) in the context of s.b.s. refers to matrix elements containing the $\gamma_\mu \gamma_5$ form factor only. Total conservation is then restored by the Goldstone pole in the second form factor.)

Finally there is one last neutral vector meson which is the linear combination of $A_\mu{}^3$ and B_μ orthogonal to the photon. It has mass. This is a new interesting prediction. There must be weak neutral currents for this approach to work. This prediction has now been experimentally confirmed and the results are in conformity with the allowed range of theoretical possibilities in the detailed construction of the various couplings.

One then has built a beautiful synthesis of weak interactions and electromagnetism which is renormalizable. This is no small advance and I believe this idea, like Nambu's spontaneous broken chiral symmetry on the hadron side, is here to stay. What is missing is an overall theory which incorporates both. Also there is no understanding of why hadrons have the complicated internal symmetry which they display (firstly SU_2, then SU_3 and now whatever the new Ψ resonances and di-lepton events are trying to tell us).

IV. Quantum Flux Lines: Type II Superconductivity: The Dual String

We have seen that s.b.s. in superconductors implies the Meissner effect. In fact this is not the only response possible to an external field. Already in the early work of London, it was seen how magnetic field could exist within superconductors provided the flux penetrated in appropriately quantized units. Subsequent work by Abrikosov and related considerations of Bohm and Aharanov have now made London's first suggestion entirely clear. We review these considerations and then sketch what is behind some recent activity along these lines in particle theory.

The equations of motion in Ginzburg-Landau theory are derived from the L of Eq. (46). We introduce the symbol D_k for co-variant derivative

$$D_k = \partial_k - ie A_k .$$

We then have

$$D_k F_{k\ell} = i \phi^* \overleftrightarrow{D_\ell} \phi$$

$$D_k D_k \phi = \partial P/\partial \phi^* \qquad (61)$$

Let us look for a solution of cylindrical symmetry within a superconductor penetrated by a tube of magnetic flux. We shall work with cylindrical coordinates, z = direction of flux tube, $\rho = \sqrt{x^2 + y^2}$, φ = azimuthal angle. The tube is centred on $\rho = 0$. For large ρ we shall see that this means $\rho \gg \mu_v^{-1}$), the superconductor should have the same energy that it has in the quiescent state. Thus

$$\lim_{\rho \to \infty} \begin{cases} \partial P/\partial \phi = 0 & \text{(minimum of potential energy} \quad (62a) \\ D_k \phi = 0 & \text{(minimum of kinetic energy)} \quad (62b) \end{cases}$$

(Recall that D_k is the velocity operator in the presence of an e.m. field.) The solution of (62a) is

$$\lim_{\rho \to \infty} \phi = |\phi| \, e^{i\chi(x)}$$

where $|\phi|$ characterizes the quiescent superconductor. Cylindrical symmetry requires that χ be a function of φ only and single-valuedness restricts the solution to be

$$\lim_{\rho \to \infty} \phi = |\phi| e^{in\varphi} \qquad n \text{ integer} . \tag{63}$$

Inserting into (62b) we get

$$n \vec{\nabla} \varphi - e \vec{A} = 0 , \qquad \vec{A} = \frac{n \vec{\nabla} \varphi}{e} . \tag{64}$$

Applying Stokes theorem we find

$$\Phi = \int B \, d\sigma = \oint A . d\ell = \frac{2\pi n}{e} . \tag{65}$$

The enclosed flux is quantized in units of $2\pi/e$. This is the essential point of penetrating quantum flux lines in superconductors. Type II superconductivity occurs when surface effects favour flux penetration so that rather than having a Meissner effect a lattice of penetrating flux lines arises within the superconductor in response to an external field.

The exact details of the solution of the non-linear equation (61) are hard to get at, but the following simple features emerge from a qualitative study. The solution of which (63) and (64) are the asymptotic limits has the properties that

1) $\lim_{\rho \to 0} \phi = A\rho^n e^{in\varphi}$ 2) $\lim_{\rho \to 0} A = 0$

3) $\lim_{\rho \to 0} B_z = \lim_{\rho \to 0} (\vec{\nabla} \times \vec{A})_z = \alpha - \beta \rho^{2n-1}$ (α, β are constants).

4) The bridge between $\rho = 0$ behavior and $\rho = \infty$ behavior occurs on the length scale μ_v^{-1} for $\vec{A}$ and $\vec{B}$. Thus the B_z field is constant inside the flux line up to $\rho = O(\mu_v^{-1})$, whereupon it falls off, first like a polynomial and then exponentially in $e^{-\mu_v \rho}$. The bridge between the $\rho = 0$ and $\rho = \infty$ behavior of ϕ occurs on a length scale $= \mu_s^{-1}$ where μ_s is the curvature of $P(|\phi|^2)$ around the minimum. ϕ rises from zero at $\rho = 0$, to $|\phi| e^{in\varphi}$ at $\rho = \infty$ where $\partial P/\partial |\phi| = 0$. The experimental side of flux lines is a science in itself and all the theoretical notions have been successfully confirmed.

Completely independently in 1972, Nielessen and Olessen discovered quantum flux lines in the relativistic version. They constructed the solution in order to obtain a field theoretic model for the hitherto mysterious dual string. The latter constitutes a phenomenological approach which explains many, but far from all of the features of Regge phenomenology and resonances and their dual behavior. The model identifies hadron resonances with the vibrational spectrum of a string. What string? A possible and appealling answer was delivered by Nielessen and Olessen - the magnetic flux line.

The next remarkable observation was due to Nambu and to Parisi who recalled that Dirac's monopole construction of 1948 required a quantization condition $g = {}^{n}/2e$, where g is the magnetic monopole charge. (This condition arises from a single-valuedness argument identical to that which leads to (65)). Thus the total flux emitted by Dirac monopoles is $2\pi n/e$ - exactly the value that can be channelled into quantum flux lines in superconductors.

What happens when a Dirac monopole is set into a superconductor? The lines of force first are emitted radially up to a radius μ_v^{-1}, whereupon they get channelled into a flux line which threads its way through the superconductor. If the universe is characterized by s.b.s., the length of this flux line is infinite. But a flux line is endowed with an energy per unit length. Therefore a single monopole will have infinite energy. (We are not concerned with the usual infinity at $r = 0$ due to self interactions. This must be handled by other considerations such as 't Hooft monopoles.) A single monopole then would not exist. It is identified with the unobservable quark. A finite energy structure is obtained by constructing a monopole-antimonopole pair. The flux line then is confined to a finite length between the two. This then is the string of phenomenological hadron physics. It terminates on quarks which themselves are unobservable.

Needless to say, much is required between this germ of an idea and the construction of viable phenomenology built upon it. In particular: how does one construct baryons (3 quarks bound together)? and what are the conditions of stability which leads to a stable structure of definite length? To this end, I believe that 't Hooft monopoles must be called into play. This is still another s.b.s. phenomenon, which will not be discussed here - and there are surely more to come.

V. Odds and Ends

These lectures have included most of the notions and applications of s.b.s. to many body and particle physics. We wish here to mention a few interesting chapters which were not touched upon.

A) Many Body Physics

1. Detailed structures. Crystalline structures, complicated sublattice structures of spins such as spirals, electric dipole structures of ferroelectrics - all these are the province of s.b.s. The main tool is the judicious use of the variational principle, which is the founding stone of the whole development of s.b.s.

2. Critical phenomena. Understanding the singular manner in which s.b.s. is approached as a function of temperature and external parameters has received an enormous boost in recent years due to the brilliant work of Wilson and others.

3. Bound state formation and s.b.s. The Cooper pair mechanism was the key to modern superconductivity theory. It is a mechanism of s.b.s. which is not based on scalar fields. Rather the latter are a phenomenological expression of the former. This is a very interesting subject in itself, and in fact each of the above three topics warrants a separate lecture series.

B) Particle Theory

1. Dynamical Breakdown. As in many body physics bound state mechanisms can be called upon to generate s.b.s. Here a new problem arises, due to convergence behavior in the ultraviolet region of integrals. This problem has been studied in chiral symmetry in the case where the binding glue is supplied by the gauge·mesons themselves. There are rapidly and slowly converging theories depending on the choice of model. What is amusing and perhaps important is that in the former class, the coupling constant is determined by the theory as well as all mass ratios. This is not the case in the slowly converging theory.

2. 't Hooft monopoles. Solutions of a monopole type occur in non-Abelian gauge theory in the presence of s.b.s. These monopoles are more appealing than Dirac's in that:

a) Their energy is finite; there is no $1/|r|$ term in the energy as $r \to 0$. s.b.s. plays a very important role in this in that it supplies a mass, μ_v, such that for $|r| < \mu_v^{-1}$, the energy density no longer behaves like $|B|^2 \sim 1/r^4$ as for a Dirac monopole.

b) Also, 't Hooft monopoles may be divested of their Dirac strings in well chosen gauges. Dirac's monopoles are accompanied by strings in all gauges and hence are not genuine point-like mathematical structures. One may build models of the dual string type using 't Hooft monopoles and confined non-Abelian flux lines.

3. Infinite Superconvergence. Some years ago, it was observed that exact duality could be made to work for a class of processes in which two members of a 4 point amplitude were pions. The processes considered were $\pi^+ A^- \to \pi^- B^+$, A, B are target particles of unit i-spin. The t channel in such processes is exotic and exact duality then requires that all moments $\int \nu^n / mT \, d\nu$ of such amplitudes vanish. In place of pions, if one used a conventionally conserved current, such as i-spin, these conditions solve to give back Schur's lemma. With pions of zero mass, one can look for other solutions and one finds a coherent set when the targets A, B are non strange mesons. The quantitative results are astounding: one finds that all resonances lie on lines in parallel Regge trajectories. In units where the slope is unity, one finds $m_\pi^2 = 0$ (input-observed is $m_\pi^2 = 0.017$)

$$m_\eta^2 = \frac{1}{4}, \quad m_\rho^2 = m_\omega^2 = \frac{1}{2}, \quad m_{\eta'}^2 = m_\delta^2 = \frac{3}{4}, \quad m_{A_1}^2 = m_H^2 = 1, \quad m_B^2 = m_D^2 = \frac{5}{4}$$

$$m_f^2 = m_{A_2}^2 = \frac{3}{2} \ldots\ldots .$$

Many of these results are obtainable from other related considerations, but the SU_3 breaking parameter $m_\eta^2 = \frac{1}{4}$ and $m_{\eta'}^2 = m_\delta^2 = \frac{3}{4}$ as far as I know can only be obtained through this scheme. None of the results is off by more than 2 or 3 % in the square of the mass.

It is tempting to identify this alternative solution to Schur's lemma, as a manifestation of spontaneous chiral breaking. The idea would be that the Goldstone particles are decoupled from exotic channels by some symmetry principle. This notion has never been carried farther, but the spectacular results which one finds warrant much more work on this point.

Spontaneous breakdown of symmetry in the context of local field theory clearly contains a wealth of physical phenomena and one should not be surprised if the list drawn up in these lectures doubles within the next few years.

REFERENCES

All of the material covered in these lectures is well known and there is little point in quoting detailed references. Rather, we will cite a few general sources each of which contains a good bibliography.

Section I

1. Role of s.b.s in phase transitions. R.Brout Phase Transitions WA Benjamin Inc. New York 1963

2. Nambu Goldstone Theorem in Field Theory: J. Goldstone, A. Salam, S. Wintery, Phys. Rev. 127 965 (1962)

Section II

1. Chiral symmetry and Goldberger Treiman relation: Y. Nambu and L. Jona Lasinio Phys. Rev. 122, 345 (1962)

2. SU_3 breakdown: R. Brout, Il Nuovo Cim. 47, 932 (1967)

Section III

See general reviews

E. S. Abers, B. W. Lee, Phys. Recpts 9C N° 1 (1973)
J. C. Taylor, Rutherford Laboratory RPP/T/29 (1972)

Section IV

Type II Superconductivity: A. L. Fetter and P. C. Hohenberg in Superconductivity edited by R. D. Parks (Marcel Dekker, New York 1969) vol. II p. 817

General review of flux lines and monopoles: R. Brout Lecture at Trieste Symposium on Dynamical Breakdown of Symmetry, Brussels preprint June 1976

Section V

1. Phase Transitions and Critical Phenomenon: The present volume

2. Superconductivity: R. D. Parks loc. cit.

3. Dynamical Breakdown: F. Englert, Weak and Electromagnetic Interactions at High Energies - Cargèse 1975, Plenum Press, New York and London (1976)

4. Infinite Superconvergence: R. Brout, F. Englert, C. Truffin Phys. Rev. 9D, 2694 (1974)

SPRINGER TRACTS IN MODERN PHYSICS

Ergebnisse der exakten Naturwissenschaften

Editor: G. Höhler

Associate Editor:
E.A. Niekisch

Editorial Board:
S. Flügge, J. Hamilton, F. Hund, H. Lehmann, G. Leibfried, W. Paul

Springer-Verlag
Berlin
Heidelberg
New York

Volume 66

30 figures. III, 173 pages. 1973
ISBN 3-540-06189-4

Quantum Statistics

in Optics and Solid-State Physics

R. Graham: Statistical Theory of Instabilities in Stationary Nonequilibrium Systems with Applications to Lasers and Nonlinear Optics.
F. Haake: Statistical Treatment of Open Systems by Generalized Master Equations.

Volume 67

III, 69 pages. 1973
ISBN 3-540-06216-5

S. Ferrara, R. Gatto, A. F. Grillo:

Conformal Algebra in Space-Time

and Operator Product Expansion

Introduction to the Conformal Group in Space-Time. Broken Conformal Symmetry. Restrictions from Conformal Covariance on Equal-Time Commutators. Manifestly Conformal Covariant Structure of Space-Time. Conformal Invariant Vacuum Expectation Values. Operator Products and Conformal Invariance on the Light-Cone. Consequences of Exact Conformal Symmetry on Operator Product Expansions. Conclusions and Outlook.

Volume 68

77 figures. 48 tables. III, 205 pages. 1973
ISBN 3-540-06341-2

Solid-State Physics

D. Schmid: Nuclear Magnetic Double Resonance — Principles and Applications in Solid-State Physics.
D. Bäuerle: Vibrational Spectra of Electron and Hydrogen Centers in Ionic Crystals.
J. Behringer: Factor Group Analysis Revisited and Unified.

Volume 69

13 figures. III, 121 pages. 1973
ISBN 3-540-06376-5

Astrophysics

G. Börner: On the Properties of Matter in Neutron Stars.
J. Stewart, M. Walker: Black Holes: the Outside Story.

Volume 70

II, 135 pages. 1974
ISBN 3-540-06630-6

Quantum Optics

G. S. Agarwal: Quantum Statistical Theories of Spontaneous Emission and their Relation to Other Approaches.

Volume 71

116 figures. III, 245 pages. 1974
ISBN 3-540-06641-1

Nuclear Physics

H. Überall: Study of Nuclear Structure by Muon Capture.
P. Singer: Emission of Particles Following Muon Capture in Intermediate and Heavy Nuclei.
J. S. Levinger: The Two and Three Body Problem.

Volume 72

32 figures. II, 145 pages. 1974
ISBN 3-540-06742-6

D. Langbein:

Theory of Van der Waals Attraction

Introduction. Pair Interactions. Multiplet Interactions. Macroscopic Particles. Retardation. Retarded Dispersion Energy. Schrödinger Formalism. Electrons and Photons.

Volume 73

110 figures. VI, 303 pages. 1975
ISBN 3-540-06943-7

Excitons at High Density

Editors: H. Haken, S. Nikitine
Biexcitons. Electron-Hole Droplets. Biexcitons and Droplets. Special Optical Properties of Excitons at High Density. Laser Action of Excitons. Excitonic Polaritons at Higher Densities.

Volume 74

75 figures. III, 153 pages. 1974
ISBN 3-540-06946-1

Solid-State Physics

G. Bauer: Determination of Electron Temperatures and of Hot Electron Distribution Functions in Semiconductors.
G. Borstel, H. J. Falge, A. Otto: Surface and Bulk Phonon-Polaritons Observed by Attenuated Total Reflection.

Selected Issues from

Lecture Notes in Mathematics

Vol. 507: M. C. Reed, Abstract Non-Linear Wave Equations. VI, 128 pages. 1976.

Vol. 501: Spline Functions, Karlsruhe 1975. Proceedings. Edited by K. Böhmer, G. Meinardus, and W. Schempp. VI, 421 pages. 1976.

Vol. 495: A. Kerber, Representations of Permutation Groups II. V, 175 pages. 1975.

Vol. 490: The Geometry of Metric and Linear Spaces. Proceedings 1974. Edited by L. M. Kelly. X, 244 pages. 1975.

Vol. 489: J. Bair and R. Fourneau, Etude Géométrique des Espaces Vectoriels. Une Introduction. VII, 185 pages. 1975.

Vol. 485: J. Diestel, Geometry of Banach Spaces – Selected Topics. XI, 282 pages. 1975.

Vol. 484: Differential Topology and Geometry. Proceedings 1974. Edited by G. P. Joubert, R. P. Moussu, and R. H. Roussarie. IX, 287 pages. 1975.

Vol. 481: M. de Guzman, Differentiation of Integrals in R^n. XII, 226 pages. 1975.

Vol. 480: X. M. Fernique, J. P. Conze et J. Gani, Ecole d'Eté de Probabilités de Saint-Flour IV–1974. Edité par P.-L. Hennequin. XI, 293 pages. 1975.

Vol. 477: Optimization and Optimal Control. Proceedings 1974. Edited by R. Bulirsch, W. Oettli, and J. Stoer. VII, 294 pages. 1975.

Vol. 474: Séminaire Pierre Lelong (Analyse) Année 1973/74. Edité par P. Lelong. VI, 182 pages. 1975.

Vol. 470: R. Bowen, Equilibrium States and the Ergodic Theory of Anosov Diffeomorphisms. III, 108 pages. 1975.

Vol. 468: Dynamical Systems – Warwick 1974. Proceedings 1973/74. Edited by A. Manning. X, 405 pages. 1975.

Vol. 464: C. Rockland, Hypoellipticity and Eigenvalue Asymptotics. III, 171 pages. 1975.

Vol. 463: H.-H. Kuo, Gaussian Measures in Banach Spaces. VI, 224 pages. 1975.

Vol. 461: Computational Mechanics. Proceedings 1974. Edited by J. T. Oden. VII, 328 pages. 1975.

Vol. 459: Fourier Integral Operators and Partial Differential Equations. Proceedings 1974. Edited by J. Chazarain. VI, 372 pages. 1975.

Vol. 458: P. Walters, Ergodic Theory – Introductory Lectures. VI, 198 pages. 1975.

Vol. 449: Hyperfunctions and Theoretical Physics. Proceedings 1973. Edited by F. Pham. IV, 218 pages. 1975.

Vol. 448: Spectral Theory and Differential Equations. Proceedings 1974. Edited by W. N. Everitt. XII, 321 pages. 1975.

Vol. 447: S. Toledo, Tableau Systems for First Order Number Theory and Certain Higher Order Theories. III, 339 pages. 1975.

Vol. 446: Partial Differential Equations and Related Topics. Proceedings 1974. Edited by J. A. Goldstein. IV, 389 pages. 1975.

Vol. 445: Model Theory and Topoi. Edited by F. W. Lawvere, C. Maurer, and G. C. Wraith. III, 354 pages. 1975.

Vol. 444: F. van Oystaeyen, Prime Spectra in Non-Commutative Algebra. V, 128 pages. 1975.

Vol. 443: M. Lazard, Commutative Formal Groups. II, 236 pages. 1975.

Vol. 442: C. H. Wilcox, Scattering Theory for the d'Alembert Equation in Exterior Domains. III, 184 pages. 1975.

Vol. 441: N. Jacobson, PI-Algebras. An Introduction. V, 115 pages. 1975.

Vol. 440: R. K. Getoor, Markov Processes: Ray Processes and Right Processes. V, 118 pages. 1975.

Vol. 439: K. Ueno, Classification Theory of Algebraic Varieties and Compact Complex Spaces. XIX, 278 pages. 1975.

Vol. 438: Geometric Topology. Proceedings 1974. Edited by L. C. Glaser and T. B. Rushing. X, 459 pages. 1975.

Vol. 437: D. W. Masser, Elliptic Functions and Transcendence. XIV, 143 pages. 1975.

Vol. 436: L. Auslander and R. Tolimieri, Abelian Harmonic Analysis, Theta Functions and Function Algebras on a Nilmanifold. V, 99 pages. 1975.

Vol. 435: C. F. Dunkl and D. E. Ramirez, Representations of Commutative Semitopological Semigroups. VI, 181 pages. 1975.

Vol. 434: P. Brenner, V. Thomée, and L. B. Wahlbin, Besov Spaces and Applications to Difference Methods for Initial Value Problems. II, 154 pages. 1975.

Vol. 433: W. G. Faris, Self-Adjoint Operators. VII, 115 pages. 1975.

Vol. 432: R. P. Pflug, Holomorphiegebiete, pseudokonvexe Gebiete und das Levi-Problem. VI, 210 Seiten. 1975.

Vol. 431: Séminaire Bourbaki – vol. 1973/74. Exposés 436–452. IV, 347 pages. 1975.

Vol. 430: Constructive and Computational Methods for Differential and Integral Equations. Proceedings 1974. Edited by D. L. Colton and R. P. Gilbert. VII, 476 pages. 1974.

Vol. 429: L. Cohn, Analytic Theory of the Harish-Chandra C-Function. III, 154 pages. 1974.

Vol. 428: Algebraic and Geometrical Methods in Topology, Proceedings 1973. Edited by L. F. McAuley. XI, 280 pages. 1974.

Vol. 427: H. Omori, Infinite Dimensional Lie Transformation Groups. XII, 149 pages. 1974.

Vol. 426: M. L. Silverstein, Symmetric Markov Processes. X, 287 pages. 1974.

Lecture Notes in Physics

Vol. 32: Particles, Quantum Fields and Statistical Mechanics. Proceedings 1973. Edited by M. Alexanian and A. Zepeda. V, 132 pages. 1975.

Vol. 33: Classical and Quantum Mechanical Aspects of Heavy Ion Collisions. Proceedings 1974. Edited by H. L. Harney, P. Braun-Munzinger, and C. K. Gelbke. VII, 311 pages. 1975.

Vol. 34: One-Dimensional Conductors GPS Summer School Proceedings, 1974. Edited by H. G. Schuster. VII, 371 pages. 1975.

Vol. 35: Proceedings of the Fourth International Conference on Numerical Methods in Fluid Dynamics, 1974. Edited by R. D. Richtmyer. V, 457 pages. 1975.

Vol. 36: R. Gatignol, Théorie Cinétique des Gaz à Répartition Discrète de Vitesses. II, 219 pages. 1975.

Vol. 37: Trends in Elementary Particle Theory. Proceedings 1974. Edited by H. Rollnik and K. Dietz. V, 472 pages. 1975.

Vol. 38: Dynamical Systems, Theory and Applications. Proceedings 1974. Edited by J. Moser. VI, 624 pages. 1975.

Vol. 39: International Symposium on Mathematical Problems in Theoretical Physics. Proceedings 1975. Edited by H. Araki. XII, 562 pages. 1975.

Vol. 40: Effective Interactions and Operators in Nuclei. Proceedings 1975. Edited by B. R. Barrett. XII, 339 pages. 1975.

Vol. 41: Progress in Numerical Fluid Dynamics. Proceedings 1974. Edited by H. J. Wirz. V, 471 pages. 1975.

Vol. 42: H II Regions and Related Topics. Proceedings 1975. Edited by D. Downes and T. L. Wilson. XII, 488 pages. 1975.

Vol. 43: Laser Spectroscopy. Proceedings 1975. Edited by S. Haroche, J. C. Pebay-Peyroula, T. W. Hänsch, and S. E. Harris. X, 466 pages. 1975.

Vol. 44: R. A. Breuer, Gravitational Perturbation Theory and Synchrotron Radiation. VI, 196 pages. 1975.

Vol. 45: Dynamical Concepts on Scaling Violation and the New Resonances in e^+e^- Annihilation. Edited by B. Humpert. VII, 248 pages. 1976.

Vol. 46: E. J. Flaherty, Hermitian and Kählerian Geometry in Relativity. VIII, 365 pages. 1976.

Vol. 47: Padé Approximants Method and Its Applications to Mechanics. Edited by H. Cabannes. XV, 267 pages. 1976.

Vol. 48: Interplanetary Dust and Zodiacal Light. Proceedings 1975. Edited by H. Elsässer and H. Fechtig. XII, 496 pages. 1976.

Vol. 49: W. G. Harter and C. W. Patterson, A Unitary Calculus for Electronic Orbitals. XII, 144 pages. 1976.

Vol. 50: Group Theoretical Methods in Physics. 4th International Colloquium. Nijmegen 1975. Edited by A. Janner, T. Janssen, and M. Boon. XIII. 629 pages. 1976.

Vol. 51: W. Nörenberg und H. A. Weidenmüller. Introduction to the Theory of Heavy-Ion Collisions. IX, 273 pages. 1976.

Vol. 52: M. Mladjenović, Development of Magnetic β-Ray Spectroscopy. X, 282 pages. 1976.

Vol. 53: D. J. Simms and N. M. J. Woodhouse, Lectures on Geometric Quantization. V, 166 pages. 1976.

Vol. 54: Critical Phenomena. Sitges International School on Statistical Mechanics, June 1976. Edited by J. Brey and R. B. Jones. XI, 383 pages. 1976.